Basic HPLC and CE of Biomolecules

Robert L. Cunico, Karen M. Gooding and Tim Wehr

Basic HPLC and CE
of Biomolecules

Robert L. Cunico, Karen M. Gooding and Tim Wehr

Bay Bioanalytical Laboratory
Richmond, CA 1998

Library of Congress Preassigned Card Number Data

Basic HPLC and CE of Biomolecules by Robert L. Cunico, Karen M. Gooding and Tim Wehr.
Includes bibliographic references and index.

ISBN 0-9663229-0-8
Library of Congress Catalog Card Number: 98-70600

Copyright © 1998 by Bay Bioanalytical Laboratory, Inc.
Richmond, California

ISBN Number 0-9663229-0-8

Printed in the United States of America

ACKNOWLEDGEMENT

Karen Gooding would like to thank Bob and Tim for giving her this challenging and enjoyable project. She gratefully acknowledges Fred Regnier, with whom she studied HPLC of proteins. Karen would also like to thank her family and friends for their encouragement in this endeavor, in particular, Dave, Michael and Brian Gooding and Carol Epplin, for their editorial efforts and advice.

Bob and Tim would like to acknowledge their colleagues at Varian, Cetus (now Chiron) and Bio-Rad who contributed to our education in the separation sciences, including Ron Majors, Seth Abbott, Tom Alfredson, Tim Schlabach, Bob Stevenson, Ed Johnson, Terry Sheehan, Mike Kunitani, Mingde Zhu, and Roberto Rodriguez. Finally we would like to thank Dave Burke, Fred Klink, Debbie Johnson, Bill Usinger, and Ann Morrill for their excellent contributions to this work.

DEDICATION

This book is dedicated to all the separation scientists who had the foresight in the 1970's to develop techniques for HPLC of proteins, peptides and polynucleotides despite the lack of appropriate columns and equipment. It is their dedication and perseverance which now allows the routine use of HPLC in biomolecule analysis.

ABOUT THE AUTHORS

Robert L. Cunico holds degrees in chemistry from Colorado State University and San Francisco State University, where he also did postgraduate work. At Varian Associates, he developed HPLC methods, columns and instrumentation for biomolecules and served as the HPLC training manager. Bob then joined the analytical group at Cetus Corporation (now Chiron) where he developed and validated methods for protein therapeutics. In 1991, he founded Bay Bioanalytical Laboratory, Inc., a consulting and contract laboratory serving the biotechnology and biopharmaceutical industry. As president and principal scientist of BBL, Bob has been a contributor to numerous INDs.

Karen M. Gooding began her career in the laboratory of Fred Regnier at Purdue University, participating in the original development of columns and techniques for HPLC of proteins. In 1977, Karen and David Gooding founded SynChrom, Inc. specifically for the purpose of developing and manufacturing HPLC columns for protein analysis. As Analytical Director and President of SynChrom, Karen guided efforts in methods development with the express goal of expediting protein analysis. She has published extensively in the chromatography field and has served as an editor of *Journal of Chromatography, Trends in Analytical Chemistry*, and, with Fred Regnier, of the book, *HPLC of Biological Macromolecules: Methods and Applications*.

Tim Wehr received his Ph.D. in microbial physiology at Oregon State University and did postdoctoral research in molecular biology at UC Berkeley. He managed the HPLC applications lab at Varian Associates for eight years and worked on development of LC columns and HPLC-based analyzers. For the last eight years he directed the CE chemistry R&D group at Bio-Rad Laboratories, developing CE instrumentation, methods, and application kits. He has published extensively in the separation sciences, and served for nine years on the organizing committee of the *International Symposium on HPLC of Proteins, Peptides and Polynucleotides*.

PREFACE

The impetus to write this book grew from a course on separation of biomolecules taught by two of us (Bob Cunico and Tim Wehr) over the last ten years at University of California Berkeley Extension and other sites. After waiting several years for the appearance of a basic text on this subject, we decided to undertake the project ourselves. We wanted a complete yet inexpensive text for students or researchers new to chromatography. As a model, we picked "Basic Liquid Chromatography," written by Ed Johnson and Bob Stevenson in 1978 which grew out of HPLC courses taught at Varian Instruments and sold for $16. Our initial attempts, which included transcription of video tapes from our course lectures, proved woefully inadequate. We looked for editorial help and encountered Karen Gooding, who was beginning to focus on editing and writing in her career. Karen gave the book a single voice and added substantially to its content. She shouldered a multitude of tasks and helped carry the project to fruition.

This book represents knowledge accumulated from many sources. We were fortunate to learn chromatography from some of the HPLC pioneers. Bob and Tim honed their LC skills with Ron Majors and Seth Abbott at Varian, while Karen worked in the laboratory of Fred Regnier at Purdue. Bob left Varian to join Mike Kunitani's analytical development lab at Cetus (now Chiron) for a five-year immersion in chromatography of proteins, peptides and (to a lesser extent) nucleic acids. His perspective of biopharmaceutical separations has been broadened by managing a contract laboratory doing LC methods development in a GMP environment. Tim moved to Bio-Rad Laboratories and spent eight years developing CE instrumentation, separation techniques, and applications kits. Karen left Regnier's laboratory to co-found SynChrom Inc., a company which specialized in HPLC columns for biomolecule separations.

It has been about 20 years since we each made our first injections into an HPLC; it was fun and magical for us then. This book is an attempt to condense and distill that experience to help bring the fun and magic to a new generation of separation scientists.

Bob Cunico
Karen Gooding
Tim Wehr

TABLE OF CONTENTS

CHAPTER 1
INTRODUCTION

High performance liquid chromatography (HPLC) has become the workhorse of the biotechnology and pharmaceutical industries where it is used to identify, characterize and purify molecules at all stages of a process from research and development to quality assurance and validation. As a technique, it is unsurpassed in its potential to differentiate and purify molecules which are soluble in liquids. Consequently, the progress in HPLC since the 1970's has paved the way for many advances in biochemistry and biotechnology where separation of biomolecules is the heart of fruitful investigations.

Roles of HPLC in Biotechnology	
R & D	separation and characterization
Peptide Mapping	separation and characterization
Manufacturing	purification
Process Development	purification
Quality Assurance	qualitative and quantitative analysis
Quality Control	separation and identification

HPLC offers an array of versatile separation techniques which are ideal for the analysis and purification of soluble molecules ranging from drugs to proteins. For large biomolecules (10^3 - 10^6 Daltons) like proteins and nucleic acids, which are composed of units containing charged functional groups, separation must occur in a liquid environment. The earliest liquid chromatograph was developed to separate amino acids (1) and later models were specifically designed to separate proteins or nucleic acids. Currently, HPLC is a primary technique for analyzing most classes of biomolecules, including amino acids, peptides, proteins, carbohydrates, nucleic acids and lipids. Although HPLC is now more than 20 years old, there are few competing technologies that will soon displace it.

The elevation of HPLC as a key technique in biomolecule characterization and purification in virtually all biotechnology and pharmaceutical companies has led to a proliferation of information. It is the goal of this book to discuss the fundamental aspects of HPLC as they pertain to biomolecules so that the reader can choose and modify chromatographic methods using appropriate instrumentation. The book is divided into chapters which discuss instrumentation, columns, modes of chromatography, sample preparation, data analysis, and troubleshooting. Although the bulk of the book deals with HPLC, a chapter on capillary electrophoresis (CE), a complementary and related technique for high resolution of biomolecules, is also included.

I. HPLC AS A TECHNIQUE

The meaning of the acronym "HPLC" has evolved over the years. When first conceived, "HPLC" stood for "High Pressure Liquid Chromatography," which was bestowed to contrast with the low pressures produced by other liquid chromatography methods which were gravity-fed or driven with peristaltic pumps. In the early 1970's, liquid chromatographic procedures on columns containing small diameter particles were found to give better separations than had ever been achieved previously; however, solvents had to be forced through them because gravity-feed was not possible. Specialized pumps were developed to force the solvents (mobile phases) through the columns. The combination of the column, solvent viscosity, and flowrate usually generated 400 - 6000 pounds per square inch (psi) as read from a pressure gauge placed between the pump and the top (or head) of the column - yielding the designation of "High Pressure Liquid Chromatography". Gradually it was recognized that "high pressure" was not always present or even a central issue and, as a consequence, "High Performance Liquid Chromatography" was substituted as a more universal term for the technique. Today, when one specifies an HPLC separation, it usually denotes a separation using a column containing small diameter particles (≤ 10μm), a pump capable of pushing solvent through such a column, and a detector to measure the presence of the compound(s) of interest. "FPLC," or "Fast Protein Liquid Chromatography," is a subset of HPLC which refers to one specific operating system. Auto samplers (robotic or automatic devices used to inject samples) and data systems, which compile detector signals, are optional devices that often complete a contemporary HPLC.

II. HIGH PERFORMANCE SEPARATION OF BIOMOLECULES

A. Liquid Chromatography

The discovery of liquid chromatography is generally credited to Mikhail Twsett, a Russian scientist (2-3). The term, which comes from the roots, "chromato" (color) and "graphy" (writing), was coined by Twsett after he observed that colored plant pigments could be separated on a chalk (calcium carbonate) column when ether was run through as the mobile phase. When he described the process, he predicted that this method would have broad applications. He was so right! Today, HPLC is a vigorous industry accounting for about $1 billion dollars in worldwide sales per year. All major biotechnology and pharmaceutical companies use HPLC in research and development, manufacturing and quality control. It has become a key technique used in the discovery, characterization, quantification and manufacture of biomolecules.

Although the initial adaptation of chromatography for biomolecules was relatively slow, the progress since the 1970's has been astronomical. The general timeline of relevant discoveries includes:

1903-6: Mikhail Tswett separated plant pigments on a chalk column using ether with a technique he called "color writing" or "chromatography" (2-3).
1951: Martin and Synge received the Nobel prize for their theoretical description of chromatography (4).
1958: Stein and Moore developed an automated ion exchange separation of amino acids (1), leading to their reception of the Nobel prize in 1972 for characterization of proteins using this technique.
1959: Porath and Flodin discovered that cross-linked polydextrans could be used to separate biomolecules (5).
1965: Stahl published the *Thin Layer Chromatography Handbook* (6).
1969: *Journal of Gas Chromatography* became the *Journal of Chromatography*.
1972: Majors (7) and Kirkland (8) introduced small particles (≤ 10μm) as column packings.
1976: Protein separations by size exclusion and ion-exchange HPLC reported by Regnier (9-11). Moving belt interface made LC-MS possible (12).
1977: Peptide separations reported by Krummen and Frei (13) and Molnar and Horvath (14). Microprocessor controlled HPLC pumps introduced.
Use of trifluoroacetic acid (TFA) for reversed phase peptide separations reported by Bennett (15).
1978: Toya Soda introduced TSK columns for size exclusion chromatography (16). Rivier used triethylammonium phosphate (TEAP) in mobile phases for peptides (17).
1979: Protein separations on polymer-based high performance ion-exchangers reported by Mikes (18).
1980: Prediction of peptide retention times by Meek (19).
1981: Initial meeting of the *International Symposium on HPLC of Proteins and Peptides,* later to become the *International Symposium on HPLC of Proteins, Peptides and Polynucleotides*

(ISPPP), was held in Washington D.C.

1983: High performance hydrophobic interaction chromatography described by Kato (20).

1984: Microbore columns with 1mm ID used for protein separations (21).

1984-86: Mechanistic models of protein interactions in HPLC developed by Snyder, Hearn, Horvath, and Karger, among others. Many of these models and other developments have been described in symposium issues of the *Journal of Chromatography* associated with either the *International Symposium of HPLC of Proteins, Peptides and Polynucleotides (ISPPP)* or the *International Symposium on Column Liquid Chromatography*, both of which meet annually.

1985: Electrospray interface makes liquid chromatography-mass spectrometry (LC-MS) a viable technique (22).

These milestones in the history of HPLC, and all of the other developments of the past two decades, have resulted in its wide acceptance due to multiple advantages.

> **Advantages Of HPLC Over Traditional Techniques**
> - Speed (minutes vs. hours)
> - Resolution
> - Sensitivity (femtograms - nanograms)
> - Reproducibility ($\pm$ 1%)
> - Recovery
> - Accuracy
> - Automation

HPLC still has limitations for certain solutes or situations, as listed below; however, substitution of alternative methods can frequently overcome the problems and produce satisfactory results.

> **Possible Limitations Of HPLC**
> - Cost
> - Complexity
> - Sensitivity (for some compounds)
> - Serial analysis sometimes difficult
> - Irreversibly adsorbed compounds not detected
> - Coelution difficult to detect

B. Capillary Electrophoresis

Capillary electrophoresis (CE) was first introduced in the late 1960's. The basis of the technique is differential migration in an electrical field using concentration-sensitive detectors instead of the staining techniques used in conventional gel electrophoresis methods. CE procedures can often attain high resolution with minimal diffusion. Some selected historical contributions follow:

1967: Capillary electrophoresis using 2mm ID tubes first demonstrated by Hjerten (23).

1979-1981: Separations achieved in small bore capillaries (24-25).

1984: Micellar Electrokinetic Chromatography (MEKC) was developed (26).

1988: First commercial CE instrumentation introduced by Brownlee of Microphoretics.

1989: CE polymer sieving systems used to separate by molecular weight (27).

First International Symposium on High-Performance Capillary Electrophoresis convened by Karger in Boston.

III. SOURCES OF INFORMATION

Fig. 1.1 shows a plot of a keyword search of "HPLC" in *Chemical Abstracts* for the years 1972 - 1996.

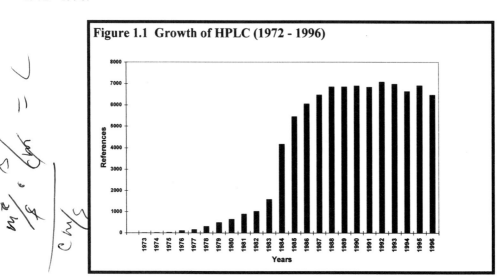

Figure 1.1 Growth of HPLC (1972 - 1996)

This sigmoidal growth curve is typical of fast evolving technology. There were 8 abstracts that contained the keyword "HPLC" in 1973 and a peak of 7071 in 1992. The exponential growth of activity during the 1970's and early 1980's reached a plateau about 1988. The slight decline in citations recently reflects the fact that HPLC is not necessarily specifically cited in abstracts because it is now a mature, well-established technology. There is no indication at this time that HPLC is being displaced by a competing technology.

In a similar manner, CE has grown as a technique since the early 1980's, as seen in Fig. 1.2. Now that the principles and operation are better understood, and commercial instrumentation is readily available, more scientists use it as a standard method.

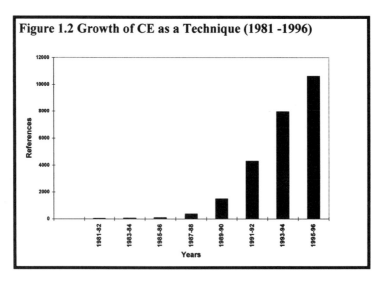

Figure 1.2 Growth of CE as a Technique (1981 -1996)

As with all science, the primary sources of information on HPLC or CE are books, journals, scientific meetings, and the Internet. Some selected books which currently serve as general references for HPLC of biomolecules are listed in Appendix A; this list continues to evolve.

The primary meeting for HPLC as a technique is the *International Symposium on Column Liquid Chromatography*, otherwise known as *HPLC '96, '97*, etc., which is held annually in early summer. It is a forum for the newest developments in instrumentation and methods. Additionally, there are topical symposia which examine specific classes of solutes or individual techniques. *International Symposium on HPLC of Proteins, Peptides and Polynucleotides (ISPPP)* is held every autumn and *International Symposium on HPCE* meets in the winter. All of these meetings provide both an educational experience and an opportunity to meet other scientists with similar interests.

At one time, *Journal of Chromatography* was the primary periodical for HPLC citations, but the universality of the technique has resulted in its frequent mention in most biological and pharmaceutical journals. Each year, *Journal of Chromatography* publishes symposium volumes about eight months after meetings such as *ISPPP, International Symposium on HPCE,* and *International Symposium on Column Liquid Chromatography*. *LC-GC*, a monthly magazine which is free for qualified subscribers, is an excellent source for articles on original research, in-depth discussion of methods and instrumentation, and troubleshooting.

Advances in computer technology have provided yet another information source encompassing three general categories of information: commercial, database, and educational. Many suppliers of instrumentation and services have sites on the Internet which may contain product information, technical assistance, application notes, and even catalogs and order forms. A central directory for many of these commercial sites can be found on the International Scientific Communications (ISC) web site which provides hyperlinks to thousands of products and manufacturers. The information on the *LC-GC* web site includes book reviews on recent publications.

Another well-defined source of information is the type of database service provided by both STN International (formerly CASONLINE) and Knight Ridder Information Services (formerly DIALOG). These two services are fee-based and provide access to most scientific information including copy delivery service. Mastery of search techniques is the key to success in using these databases (28); each supplier provides specific training materials and classes.

Accessing educational information about HPLC from the Internet is difficult due to the large (and growing) numbers of sites which include those of industries, universities and individuals. If information on methods for a specific compound is desired, searches can be easily narrowed. If the topic is more general, search engines like Lycos or Alta Vista, which allow limitation with wild card modifiers, permitting variants of a keyword (like "bio$") to be included, are most useful. Each search engine has its own unique formats and guidelines which should be consulted initially to obtain efficient and effective results. Appendix A lists some 1997 sites of interest for HPLC users.

IV. REFERENCES

1. D.H. Spackman, W.N. Stein, and S. Moore, *Anal. Chem.* 30 (1958) 1190.
2. M. Tswett, *Travl. Soc. Naturalistes Varisovic* 14 (1903).
3. M. Tswett, *Ber. Deut. Botan. Geo.* 24 (1906) 385.
4. A.J.P. Martin and R.L.M. Synge, *Biochem. J.* 35 (1941) 91.
5. J. Porath and P. Flodin, *Nature (London)* 183 (1959) 1657.
6. P.J. Schorn and E. Stahl, *Thin-Layer Chromatography. A Laboratory Handbook*, Academic Press, New York, 1965.
7. R.E. Majors, *Anal. Chem.* 44 (1972) 1722.
8. J.J. Kirkland, *J. Chromatogr. Sci.* 10 (1972) 593.
9. S.H. Chang, K.M. Gooding, and F.E. Regnier, *J. Chromatogr.* 120 (1976) 321.
10. S.H. Chang, K.M. Gooding, and F.E. Regnier, *J. Chromatogr.* 125 (1976) 103.

11. S.H. Chang, R. Noel, and F.E. Regnier, *Anal. Chem.* 48 (1976) 1839.

12. W.H. McFadden, H.L. Schwarz, and S. Evans, *J. Chromatogr.* 122 (1976) 389.

13. K. Krummen and R.W. Frei, *J. Chromatogr.* 132 (1977) 27.

14. I. Molnar and Cs. Horvath, *J. Chromatogr.* 142 (1977) 623.

15. H.P.J. Bennett, A.M. Hudson, C. McMartin, and G.E. Purdon, *Biochem. J.* 168 (1977) 9.

16. K. Fukano, K. Komiya, H. Sasaki, and T. Hashimoto, *J. Chromatogr.* 166 (1978) 47.

17. J.E. Rivier, *J. Liq. Chromatogr.* 1 (1978) 343.

18. O. Mikes, *Intl. J. Peptide Protein Res.* 14 (1979) 393.

19. J.L. Meek, *Proc. Natl. Acad. Sci. (USA)* 77 (1980) 1632.

20. Y. Kato, T. Kitamura, and T. Hashimoto, *J. Chromatogr.* 266 (1983) 49.

21. E.C. Nice, C.J. Lloyd, and A.W. Burgess, *J. Chromatogr.* 296 (1984) 153.

22. C.M. Whitehouse, R.N. Dreyer, M. Yamashita, and J.B. Fenn, *Anal. Chem.* 57 (1985) 675.

23. S. Hjerten, *Chromatogr. Rev.* 9 (1967) 122.

24. F.E.P. Mikkers, F.M. Everaerts, and T.P.E.M. Verheggen, *J. Chromatogr.* 169 (1979) 11.

25. J.W. Jorgenson and K.D. Lukacs, *Anal. Chem.* 53 (1981) 1298.

26. S. Terabe, K. Otsuka, K. Ichikawa, A. Tsuchiya, and T. Ando, *Anal. Chem.* 56 (1984) 111.

27. M. Zhu, D.L. Hansen, S. Burd, and F. Gannon, *J. Chromatogr.* 480 (1989) 311.

28. D.D. Ridley, *On-Line Searching: A Scientist's Perspective,* John Wiley and Sons, New York, 1996.

CHAPTER 2
THEORY OF HPLC

I. INTRODUCTION

Working familiarity with the fundamental theory of chromatography is a major tool for designing new schemes of separation or thoughtfully modifying current ones, even though the theoretical aspects of this or any scientific discipline often include a dizzying array of equations and assumptions which may seem irrelevant or even contradictory to practical experiences. The purpose of this chapter is to introduce the theoretical aspects of chromatography with a quantitative foundation as they relate to the qualitative and practical ramifications of the techniques. The chromatography process and components of resolution must be understood to allow effective decision-making. The discussion of band broadening in the chapter is more theoretical and superficially less important, but its mastery will result in intelligent choices in columns and operating conditions.

II. THE CHROMATOGRAPHY

In chromatography, a liquid sample is applied to a narrow cross section or zone of a column which is filled with a particular packing or support material. A solvent is pumped through the column, sweeping the sample from the inlet to the outlet and through a detector. It is the differential interaction of the various components in the sample with the stationary phase of the packing material which effects the separation. This is shown in Fig. 2.1 where a mixture of the molecules, A and C, have been injected into a narrow zone at the inlet of the column.

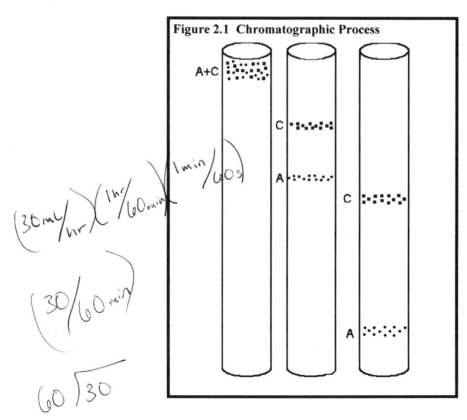

Figure 2.1 Chromatographic Process

While moving with the solvent (mobile phase), A and C become slightly separated because C shows a greater affinity for the stationary phase and moves more slowly than A. Eventually A elutes, distinct from and followed by C - this separation is the goal of the chromatography.

A chromatogram is a two dimensional plot of molecules eluting from a column; measurement of absorbance or some other defining property is plotted against elution time or volume, as seen in Fig. 2.2 for the molecules shown in the preceding figure.

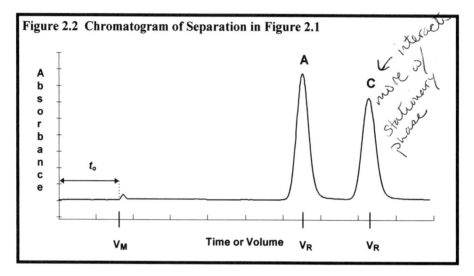

Figure 2.2 Chromatogram of Separation in Figure 2.1

Most organic molecules exhibit some absorption in the low ultraviolet area of the spectrum (210nm) and elution can often be monitored spectrophotometrically. The permeation volume or mobile phase volume (V_M) is the volume of the mobile phase in the column. This volume is the sum of the liquid outside the matrix (V_o) and the liquid inside the matrix (V_i). Importantly, V_M is the volume of fluid required for a completely unretained peak to pass through the column and is a major reference point for the chromatographic separation. This volume (V_M) is also called deadvolume; it is sometimes called "void volume", but this term has a different connotation in the context of size exclusion chromatography and will not be equated with deadvolume in this book. The deadtime (t_o) is the time required for the deadvolume to pass through the column. The V_M or t_o of a column can be determined in any particular run by several means (1). If the composition of the sample buffer differs slightly from the mobile phase eluent, a transient deflection due to changes in the refractive index may occur in the baseline; t_o should be measured where the signal begins to rise, as seen in Fig. 2.2. Modern detectors and flow cells often reduce this deflection to nearly zero; therefore, small UV-active compounds, which are predicted and confirmed to have no affinity for the stationary phase, are used to determine t_o. Specific compounds will be described in greater detail in Chapters 6 - 10, which discuss individual modes of chromatography. Deadtime and deadvolume can also be estimated to an accuracy of about 10% using the formula (1):

$$V_M = 0.5Ld_c^2 \qquad\qquad \text{Eq. 2.1}$$

where L is the column length and d_c is its diameter, both in cm. For a 4.6mm ID column, V_M is about one tenth of the length. To calculate the deadtime, the deadvolume in ml must be divided by the flowrate (F) in ml/min:

$$t_o = \frac{V_M}{F} \qquad\qquad \text{Eq. 2.2}$$

For a 250 x 4.6mm ID column at a flowrate of 1ml/min, the deadtime is about 2.5min.
  The total column volume (V_T) is the sum of the deadvolume and the volume occupied by the matrix or packing material; it is, therefore, the volume of the empty cylinder:

$$V_T = \pi r^2 L \qquad = Dead\ Volume\ +\ vol.\ stationary\ phase$$
$$\text{Eq. 2.3}$$

where r and L are the radius and the length of the column, respectively. For a 250 x 4.6mm ID column, V_T is 4.15cm^3 or 4.15ml. The matrix may occupy up to 50% of this volume; the exact proportion is dependent on the porosity and the packing density of the support.
  The point at which a sample is injected into a column is sometimes indicated by a characteristic spike which occurs in the baseline (usually downward). The spike, not to be confused with t_o which occurs later in the run, may be marked by the detector or the result of a transient pressure surge that occurs during the injection of the sample. This surge is transmitted immediately through the system, like a wave moving with the solvent front, and finally detected as a change in refractive index. Many data systems automatically begin the run at the injection, as was seen in Fig. 2.2.
  Retention volume, V_R, is the volume of solvent that passes through the column from the time of sample injection to the detection of the peak measured at its apex, as shown in Fig. 2.2. Flowrates are generally kept constant during a run, permitting a straightforward interconversion between time and volume of the eluent, as plotted on the x axis of the chromatogram. V_R for a particular component is thus associated with a distinct time, called the retention time, t_R, which is related through the flowrate:

$$V_R = F\, t_R \qquad\qquad \text{Eq. 2.4}$$

where F is the volumetric flow rate, usually in ml/min.
  Molecules which are effectively separated by any particular chromatographic system may actually have somewhat different retention times from run to run which can be caused by operational variance of the flowrate, mobile phase, or other parameter. Retention times can be normalized to remove this variance by defining the capacity factor or retention factor (k) which is the number of dead volumes or mobile phase volumes required to elute the molecule or peak of interest. Until recently, capacity factor was symbolized by k'. Most frequently, k is expressed in terms of retention time (t_R) because time is the usual unit of measurement:

$$k = \frac{V_R - V_M}{V_M} = \frac{t_R - t_o}{t_o} \qquad\qquad \text{Eq. 2.5}$$

Some simple cases will help explain this concept. If a molecule elutes with the nonretained peaks (which is commonly called the "void", in any mode except size exclusion chromatography), then $t_R = t_o$, indicating (correctly) that the molecule was not retained and $k = 0$. In all other cases, $t_R > t_o$, $V_R > V_M$ and $k > 0$, implying that the peak is retained and that it resides in the column for a positive number of deadvolumes. Rearranging Eq. 2.5:

$$t_R = t_o\,(1 + k) \qquad\qquad \text{Eq. 2.6}$$

# of dead volumes before the peak comes out

III. RESOLUTION

A. Concept

The goal of chromatography is separation of multiple peaks. The resolution (R_S) of two peaks, as illustrated in Fig. 2.3, can be defined quantitatively as:

$$R_S = \frac{2(t_2 - t_1)}{w_2 + w_1}$$

Eq. 2.7

where w_2 and w_1 are the peak widths measured at their respective bases in the same time units as the retention times. This is accomplished manually by drawing the tangents of the peaks to the baseline, as shown in Fig. 2.3, which also illustrates other basic parameters.

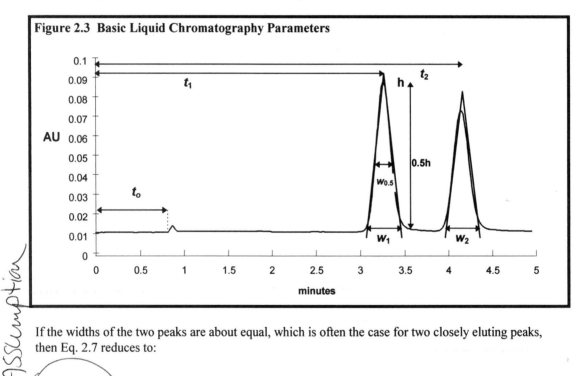

Figure 2.3 Basic Liquid Chromatography Parameters

If the widths of the two peaks are about equal, which is often the case for two closely eluting peaks, then Eq. 2.7 reduces to:

$$R_S = \frac{\Delta t_R}{w_2}$$

Eq. 2.8

The separation of two peaks can be visually assessed without any consideration or calculation of resolution, as illustrated in the sets of idealized chromatograms in Fig. 2.4.

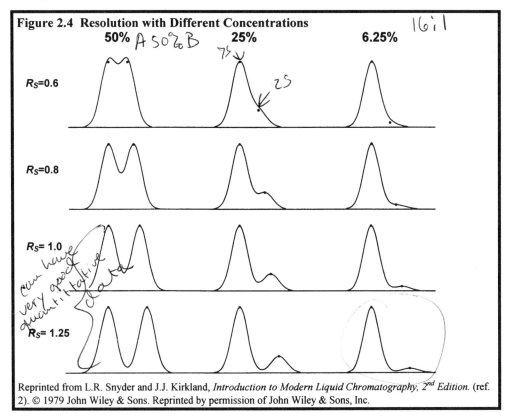

Figure 2.4 Resolution with Different Concentrations

Reprinted from L.R. Snyder and J.J. Kirkland, Introduction to Modern Liquid Chromatography, 2nd Edition. (ref. 2). © 1979 John Wiley & Sons. Reprinted by permission of John Wiley & Sons, Inc.

It is obvious that the peaks in the bottom row are totally separated from each other and those in the top row are barely resolved. At a resolution of one, the presence of two (or more) peaks is unmistakable - the peaks can be quantitated and are 98% separated, nearly to baseline. When the resolution is less than one, minor components may be masked by the major ones, as seen in the upper right example. The left group shows two compounds of equal band width and equal detector response with differential resolution, ranging from 0.6 to 1.25. If the resolution is less than 0.6, there is little, if any, indication of more than one peak.

Resolution is more difficult to establish if two peaks differ widely in their concentration or ability to be detected, as shown in the right examples where the difference in concentration factor is 16. The second peak is barely indicated at resolution values of 0.6 and 0.8 by the asymmetry of the first peak with a "growing tail." Resolution of 1 is required to totally discern the second peak, even though it is nearly completely separated from the first peak. In cases of low resolution, telltale signs of coincident elution, besides the presence of tailing, might be band widths which are greater than others in the chromatogram or slight signs of peak flattening at the crest. These would be suggestions, but not proof, of the presence of two compounds and alternative means of separation would be necessary for confirmation.

Resolution between two peaks will improve if the peaks are sharp with narrow base widths (high efficiency), if the peaks are far apart (good selectivity), or if they are highly retained, as seen in Fig. 2.5. Resolution is actually a function of the combined contributions of these parameters of a separation, which will be detailed in the following sections.

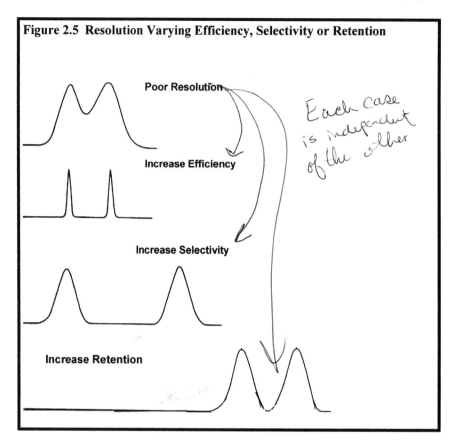

Figure 2.5 Resolution Varying Efficiency, Selectivity or Retention

Poor Resolution

Increase Efficiency

Each case is independent of the other

Increase Selectivity

Increase Retention

B. Retention

In a chromatographic process, molecules travel in a mobile phase except when they are interacting with a stationary phase. The differential interactions of molecules result in their separation. In HPLC, the mobile phase is a liquid and the stationary phase is usually a coating bonded to a small solid particle, called a support, which is contained in a metal cylinder or column. A molecule is retained if it partitions into the stationary phase rather than merely traveling with the mobile phase. The capacity factor or retention factor, k, is thus related to these phases:

$$k = K \frac{V_S}{V_M} \qquad\qquad\qquad \text{Eq. 2.9}$$

where V_S and V_M are the volumes of the stationary and mobile phases, respectively, and K is the partition coefficient for the solute in the two phases. In practice, k is calculated as in Eq. 2.5, by simply normalizing the retention time of the peak of interest to the deadtime. Retention, as indicated by k, gives a measure of the ability of the stationary phase to bind a particular molecule. If the capacity factor (k) of a molecule is zero, then it is not retained and it will not separate from the solvent front. As retention increases, the potential for separation is improved, as illustrated in Fig. 2.5. The kinetic aspects of the separation are shown in Fig. 2.6, emphasizing the key concept of selective "partitioning" of the molecules.

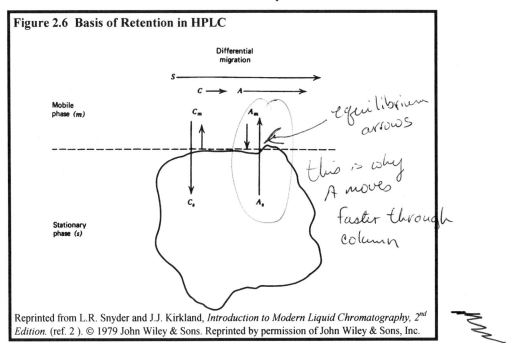

Figure 2.6 Basis of Retention in HPLC

Reprinted from L.R. Snyder and J.J. Kirkland, *Introduction to Modern Liquid Chromatography, 2nd Edition.* (ref. 2). © 1979 John Wiley & Sons. Reprinted by permission of John Wiley & Sons, Inc.

The length of the arrow labeled "S" indicates the rate of the mobile phase through the column, which is substantially faster than the rates indicated for C or A. Note that a molecule with no affinity for the stationary phase would have a velocity equal to the rate of the mobile phase or an arrow equal in length to S. The vertical arrows are classical indications of equilibrium. The "off rate" of A for the stationary phase, as indicated by a longer arrow showing A leaving the particle, exceeds the "on rate". And conversely, a longer arrow for C towards the particle indicates a greater interaction in this direction than with the mobile phase. The end result of the differential interaction of A and C with the stationary phase is their resolution into two separate bands as they leave the column, with A eluting before C, as was seen in Fig. 2.1 and 2.2.

C. Selectivity

The ability of the stationary phase to discriminate between two molecules is an essential aspect of chromatography. In Fig. 2.5, it can be seen that two peaks which were not separated under the initial conditions were resolved when the selectivity was increased. The peak widths were the same in both examples. Selectivity (α) is defined as the ratio of the capacity factor of the most highly retained compound to that of the less retained:

$$\alpha = \frac{k_2}{k_1}$$

Eq. 2.10

Selectivity is dependent on, and thus modified by, the physical and chemical structures of the analytes, the stationary phase, and the mobile phase. Selectivity **cannot** be changed by adjusting the flowrate or the column dimensions. In many cases, it is possible to substantially change selectivity by replacing the mobile phase with one with different properties. In other situations, it may be necessary to change the stationary phase (column) to achieve enough selectivity to resolve two peaks. Strategies for these modifications are mode-specific and will be discussed in detail in Chapters 6 - 10 which deal with the kinds of chromatography. Selectivity is the most important tool of successful chromatography.

D. Efficiency

From the chromatograms presented in Fig. 2.5, it is obvious that peak width is another integral component of resolution whose basis must be understood in order to achieve an acceptable separation. The peaks in the top two chromatograms have identical retention and selectivity but vastly different peak widths - resulting in the absence or presence of total resolution. The concept of "Theoretical Plates" was developed to explain such variations in peak width and the factors responsible (3).

Efficiency is a measure of the ease with which a compound moves through a column and, to the extent that it measures transition in and out of the mobile and stationary phases, it is a kinetic parameter. Efficiency is a statistical calculation of the standard deviation of the peak in unit time and is measured in theoretical plates (N) (4):

$$N = \left(\frac{t_R}{\sigma}\right)^2 \qquad\qquad \text{Eq. 2.11}$$

where σ is the standard deviation of the band in time units which can be calculated by using the peak width, generally at the baseline (w) or at half height ($w_{0.5}$), as seen in Fig. 2.3. Both the retention time and peak width must be measured in the same units, usually in length.

$$N = 16\left(\frac{t_R}{w}\right)^2 = 5.54\left(\frac{t_R}{w_{0.5}}\right)^2 \qquad\qquad \text{Eq. 2.12}$$

The different coefficients in these two equations reflect the fact that the statistical formula for the standard deviation varies according to where the peak width is measured. The second expression of efficiency is most often used to determine plates because measurement of the peak width at half its height ($w_{0.5}$) is more accurate than at the baseline (w) where ambiguities caused by noise and tailing are present.

Given that the basis of the theory is statistical, these equations are derived with the assumption that the peaks are Gaussian. This is often not the case in practice; therefore, in cases where substantial peak asymmetry is observed, calculation of N may be fraught with substantial error. Asymmetry is usually assessed independently and will be discussed later in this chapter. A large value for theoretical plates (N) indicates an efficient column, potentially capable of yielding sharp peaks; therefore, the performance of two similar columns can be meaningfully compared using this measurement with identical samples and conditions. Efficiency should not be the sole means of column selection, however. Simply choosing a column on the basis of plates (N), without careful consideration of other factors, such as packing material and bonding chemistry, could lead to the purchase of an efficient but useless column, due to inappropriate selectivity or unwanted secondary interactions.

The height equivalent of a theoretical plate (H) relates the plate number (N) to the column length (L):

$$H = \frac{L}{N} \qquad\qquad \text{Eq. 2.13}$$

where L is usually measured in mm. This is the preferred description of column efficiency because columns of any geometry can be compared as to their efficiencies per unit length. Typical plate height (H) values for HPLC are usually 0.01 to 0.1mm. It is obvious from Eq. 2.13 and the desire for a large number of plates, that small H values are preferable. An efficient process allows sample molecules to migrate in a narrow band in a short period of time with minimal dispersion. Plate count (N), therefore, gives an indication of the sharpness of a peak. It is used as a "figure of merit" for a column - high efficiency columns have plate counts that can exceed 80,000 plates per meter, while poor columns can be 10 - 20% of that value. The values of plates and/or plate height for standard compounds in a mixture are usually provided by column manufacturers as a measure of column performance and will vary for different column modes and dimensions.

E. Interrelationship of Factors

The three components of a separation - speed, resolution and retention or capacity - are interrelated; therefore, improving one will sometimes compromise the others. Optimization of a separation, therefore, lies in compromise and is usually achieved most efficiently by a stepwise examination of each parameter, without alteration of the others. The ultimate optimization of any method may actually vary dramatically between facilities. For example, the primary requirement of an analytical services laboratory may be speed with quantitation, whereas that of a commercial production group might be maximum yield of only one component.

Resolution is a function of the three factors, which can individually improve or destroy a separation, as was seen in Fig. 2.5:

1. retention - the ability for molecules to interact with the stationary phase or the affinity of the molecules for the matrix;

2. selectivity - the discriminating power of the matrix which produces differential interaction for at least two compounds.

3. efficiency - the ease of movement of the solute through the column.

Resolution is calculated by the formula in Eq. 2.7, but it can also be expressed in terms of efficiency, selectivity and retention:

$$R_s = \frac{\sqrt{N_1}}{4}(\alpha - 1)\left(\frac{k_1}{1 + k_1}\right) \qquad\qquad \text{Eq. 2.14}$$

where $(\alpha-1)$ is used as the selectivity term if the plates and capacity factor of the first peak are used. If the plates and capacity factor are associated with the second peak, then $(\alpha-1)/\alpha$ is used as the selectivity term: *eff*   *select*   *ret.*

$$R_s = \frac{\sqrt{N_2}}{4}\left(\frac{(\alpha - 1)}{\alpha}\right)\left(\frac{k_2}{1 + k_2}\right) \qquad\qquad \text{Eq. 2.15}$$

The relative effects of each of these parameters on resolution are clearly seen in the plots of Eq. 2.14 shown in Fig. 2.7 - 2.9. In these graphs, intermediate values are used for the constants: the efficiency (N) is 10,000 plates, the selectivity (α) is 1.5 and the capacity factor (k) is 3.

The graph of resolution vs. capacity factor (k) in Fig. 2.7 demonstrates that without retention there can be no resolution; however, there is negligible gain in resolution by increasing k above about 5, despite the cost of significantly increased time for the analysis. Generally, k values of 2 - 7 give the best compromise of resolution and retention time; these can be achieved by either adjusting the mobile phase or changing to a more (or less) retentive stationary phase (column).

In order to change K - play w/ mobile phase

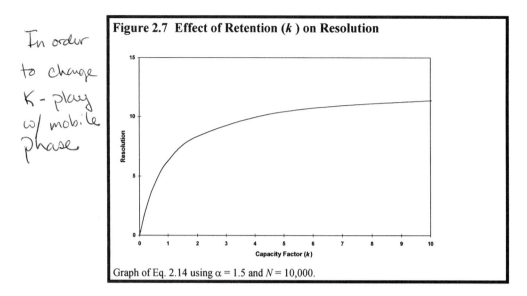

Figure 2.7 Effect of Retention (k) on Resolution

Graph of Eq. 2.14 using $\alpha = 1.5$ and $N = 10,000$.

It was obvious from Fig. 2.5 that increased plates (narrow peaks) is a factor in resolution, but Eq. 2.14 points out that this relationship is a function of square root. When the effect of plates on resolution is analyzed graphically, as shown in Fig. 2.8, it can be seen that increasing plates is not a very effective means of increasing resolution.

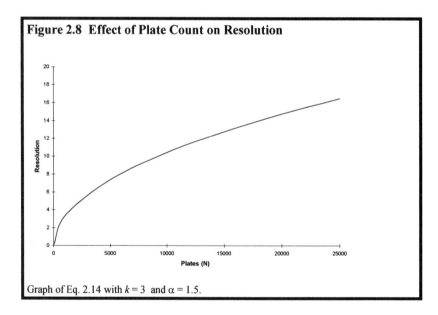

Figure 2.8 Effect of Plate Count on Resolution

Graph of Eq. 2.14 with $k = 3$ and $\alpha = 1.5$.

Because the relationship is to the square root of plates, increasing the plate number, by either purchasing a longer or more efficient column or joining two columns in series, is an inefficient means to achieve resolution. Doubling the plate number results in only a 40% increase in resolution, but it may also result in doubling the analysis time, pressure, etc. If other optimization techniques are not feasible or have been tried without success, this is one means of improvement; enhancement of plates by coupling columns is most frequently utilized in size exclusion chromatography where any other options for increasing resolution are fairly limited.

The most efficient avenue for increasing resolution is improving selectivity, as seen in Fig. 2.9.

Play w/ mobile phase to increase selectivity

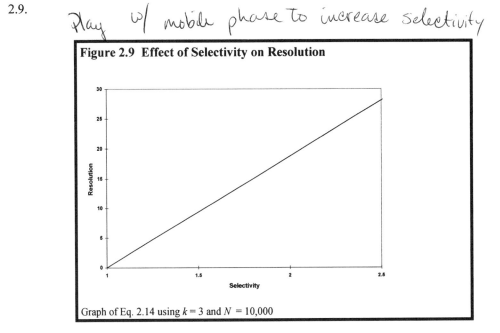

Figure 2.9 Effect of Selectivity on Resolution

Graph of Eq. 2.14 using $k = 3$ and $N = 10,000$

This is the parameter on which chromatographers should focus by employing rational and stepwise alterations of mobile phase composition on a column containing a carefully chosen stationary phase. Such mobile phase optimization will be discussed in the chapters focused on the particular modes of chromatography. If mobile phase adjustment is unsuccessful, substitution of a more selective column is probably necessary.

IV. BANDSPREADING

Column efficiency, as given in theoretical plates, is actually a measure of the broadening of a peak during passage through the chromatography column. The concept of theoretical plates (N) had its origins in distillation and its extrapolation into chromatography earned a Nobel Prize for Martin and Synge in 1941 (3). At the time, the height of a theoretical plate (H) was considered to be equivalent to the distance an analyte molecule traveled between successive points of equilibrium. This simplistic plate theory did not totally address contributions to peak broadening in chromatography and it has thus been expanded and replaced by more detailed and complex treatments. Unlike distillation, where molecules in the gas phase condense and reach equilibrium in a discreet step before continuing further, in chromatography there are flow variations, resistance to mass transfer, and continuous non-equilibrium along the column. Many scientists, including Giddings (5), Knox (6) and Snyder (4), have detailed these components and the band broadening that they induce as separate variables that can be individually analyzed and controlled.

The original concept of the theoretical plate has now been expanded to include key aspects of modern HPLC methodology, including particle size, diffusion, flow dynamics, and resultant effects on the mass transfer of a solute between the mobile and stationary phases of the column. The total column plate height, H, defined in Eq. 2.13, was described by van Deemter as the sum of various bandspreading contributions (7):

longitudinal diffusion

mass transfer into and out of stationary/mobile phase

$$H = A + \frac{B}{u} + Cu \qquad \text{Eq. 2.16}$$

Eddy diffusion *linear velocity*

where u is the linear flow velocity and A, B and C are constants, each associated with one mechanism of band broadening. This relationship of plate height to flow and diffusion parameters has been refined over the years and one form, presented by Snyder and Kirkland (2) is:

$$H = Au^{0.33} + \frac{B}{u} + Cu + Du \qquad\qquad \text{Eq. 2.17}$$

The "A" term represents band broadening due to multiple flow paths and mobile phase effects; the "B" term quantifies band broadening due to the axial or longitudinal diffusion of the solute; the "C" term encompasses the broadening caused by kinetic effects with the mobile phase within the particles; and the "D" term expresses the broadening effects attributable to interaction with the stationary phase. Each of these components will be discussed in detail in the following sections.

Why should one care about such a complicated equation? Why not just "do the chromatography"? The simple answer is that the quality of the data can improve substantially if the methods are optimized with these bandspreading factors in mind. Fig. 2.10 shows a simple example of this, using graphs of Eq. 2.17 for good columns containing porous and nonporous particles of 5μm diameter.

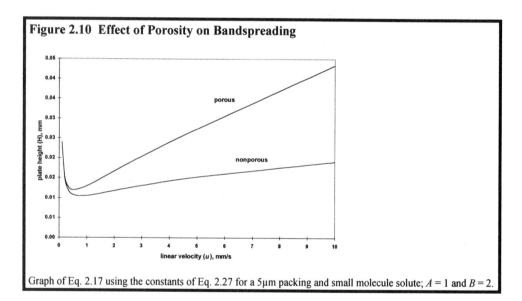

Figure 2.10 Effect of Porosity on Bandspreading

Graph of Eq. 2.17 using the constants of Eq. 2.27 for a 5μm packing and small molecule solute; $A = 1$ and $B = 2$.

It is obvious that the peaks will be twice as broad on the porous particles as on the nonporous at fast flowrates; at slow ones, the differences are much less. Additionally, there is an optimum flowrate which produces the narrowest peak ($H = 0.015$mm) for the porous column. A column with $H = 0.015$mm would have about 67,000 plates/m. In this simplistic example, the best efficiency on the porous column would be obtained at 0.5mm/s.

Giddings detailed the components of each of the terms of Eq. 2.16 and 2.17 in his treatise, *Dynamics of Chromatography* (5). Knox also investigated many of the parameters of bandspreading and developed the most widely used variation of the van Deemter equation (6). This particular equation uses reduced terms to normalize the parameters for particle diameter and diffusion; these will be discussed, with the equation, in Section IV.E. of this chapter.

A. Flow Variation

Anyone familiar with the preparation of low pressure chromatography columns knows that the production of a uniform bed is as much an art as a science. This is even more the case for microparticulate supports which must be packed under high pressure, explaining why most HPLC columns are packed and prepackaged at the factory. If all the support particles are not perfectly arranged and completely homogeneous, the sample and solvent will not move at a uniform rate through the column. In some cases, areas of loose packing will create channels which result in locally increased flow. Other zones of the column may be partially plugged due to aggregation, creating eddies which locally retard flow. These differences in flow patterns induce band broadening because different paths are available to identical molecules, as seen in Fig. 2.11 (2).

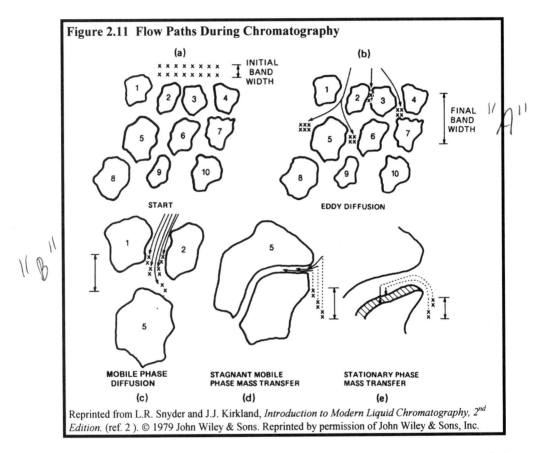

Figure 2.11 Flow Paths During Chromatography

Reprinted from L.R. Snyder and J.J. Kirkland, *Introduction to Modern Liquid Chromatography, 2nd Edition.* (ref. 2). © 1979 John Wiley & Sons. Reprinted by permission of John Wiley & Sons, Inc.

There are actually two kinds of flow variation which occur in the mobile phase paths through the column: eddy diffusion and laminar flow.

1. Eddy Diffusion

Eddy diffusion is caused by unequal flow patterns in the channels between the support particles packed in the column, as seen in Fig. 2.11b. The speed of the solute depends on the geometry and the diameter of these channels. The contribution of eddy diffusion to band broadening (H_{eddy}) is:

$$H_{eddy} = A_{eddy} d_p \qquad \text{Eq. 2.18}$$

where A_{eddy} is a measure of flow inequality in the column and d_p is the particle diameter. This factor is minimized when the support particles are homogeneous and packed perfectly into the column so that all flow paths are similar. Such bed homogeneity is best achieved using spherical particles with narrow particle diameter distribution.

2. Laminar Flow

Homogeneous flow is further disrupted by a phenomenon known to occur when fluid passes through a cylinder, a model for "the perfect pore". Laminar flow develops when a liquid travels down a cylinder, as shown in Fig. 2.12. The molecules in the center of the tube become the leading edge of the flow profile because their speed exceeds that of the average; molecules near the wall of the cylinder are retarded and their velocity approaches zero. Laminar flow primarily occurs in channels between the particles. Its contribution to bandspreading ($H_{laminar}$) is described by the following equation:

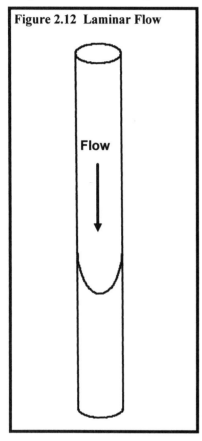

Figure 2.12   Laminar Flow

Flow

$$H_{laminar} = \frac{A_{laminar} d_p^2 u}{D_M}$$

Eq. 2.19

where d_p is the particle diameter, $A_{laminar}$ is a constant related to the column packing and D_M is the diffusion coefficient of the analyte in the mobile phase. This type of bandspreading is minimized by using small diameter support materials which produce narrow interparticle channels.

Although it would seem that band broadening due to the flow variation would be the sum of eddy diffusion and laminar flow contributions:

$$H_{flow} = A_{eddy} d_p + \frac{A_{laminar} d_p^2 u}{D_M}$$

Eq. 2.20

Giddings observed that the flow variation terms are, in fact, not independent but rather coupled (5). Consequently, :

$$\frac{1}{H_{flow}} = \frac{1}{H_{eddy}} + \frac{1}{H_{laminar}}$$

Eq. 2.21

Knox (6) proposed that the resultant complex equation be approximated by:

$$H_{flow} = A d_p \left(\frac{u d_p}{D_M} \right)^{0.33}$$

Eq. 2.22

where "A" is a measure of column packing and is on the order of unity for a good column. A very good column of uniform particles may have an "A" term as low as 0.5; whereas an "A" term greater than 2 or 3 indicates a poorly packed column. This "A" term can contribute significantly to the total band broadening, as seen in Fig. 2.13.

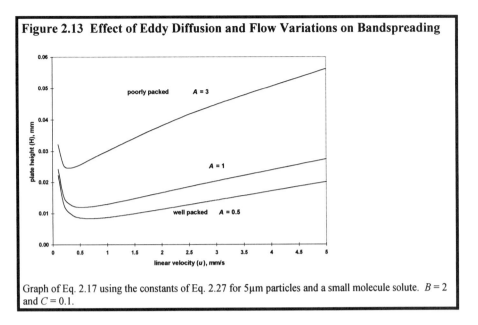

Figure 2.13 Effect of Eddy Diffusion and Flow Variations on Bandspreading

Graph of Eq. 2.17 using the constants of Eq. 2.27 for 5μm particles and a small molecule solute. $B = 2$ and $C = 0.1$.

It can be seen from these plots of 5μm supports, that the minimum achievable plate height is radically affected by the quality of the packed bed.

B. Axial or Longitudinal Diffusion

Axial or longitudinal diffusion is the ordinary molecular diffusion of the solutes in the mobile phase, as seen in Fig. 2.11c, and is irrespective of whether there is flow present. Because this type of diffusion does increase with time, it contributes most significantly at low flowrates. The contribution of axial diffusion to band broadening (H_{axial}) is defined as:

obstruction factor

$$H_{axial} = \frac{B_{mp}D_M}{u}$$

smaller than in GC (10^{-5} cm^2/s)

Eq. 2.23

where B_{mp} is a constant or obstruction factor related to axial diffusion in the mobile phase. Some diffusion may also occur in certain thick stationary phases, particularly for reversed phase materials (5). This may be a cause of bandspreading for solutes with large k values. An approximation of this term is (5):

$$H_{axial-sp} = \frac{2kB_{sp}D_S}{u}$$

Eq. 2.24

where B_{sp} is a constant related to the stationary phase and D_S is diffusion in that phase (quantitatively similar to D_M). This term is negligible in many modes of HPLC but could have some effect in thick stationary phases. It is more significant in gas chromatography, where diffusion coefficients are much larger than they are in liquids. In HPLC, the diffusion coefficient (D_M) for small molecules is on the order of 10^{-5} cm^2/s. For this discussion, axial diffusion will be approximated by Eq. 2.23. Fig. 2.14 shows the effect of the axial diffusion term on bandspreading for a good column of 5μm support.

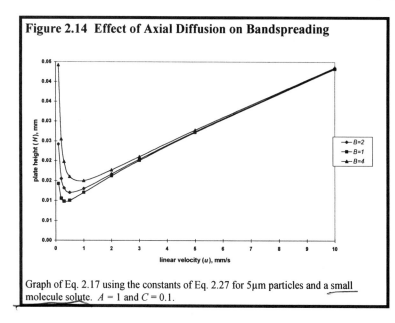

Figure 2.14 Effect of Axial Diffusion on Bandspreading

Graph of Eq. 2.17 using the constants of Eq. 2.27 for 5μm particles and a small molecule solute. $A = 1$ and $C = 0.1$.

Axial diffusion is obviously not a major source of bandspreading and only produces serious effects at velocities below the optimum velocity at the minimum plate height. Low flowrates (where the effect is large) are rarely used in HPLC except in size exclusion chromatography. If axial diffusion at low flowrates is suspected as a problem, increasing the flow velocity would be an effective solution.

C. Mass Transfer Kinetics

1. General Considerations

Molecules that interact with a stationary phase undergo a series of transitions - from being temporarily immobilized by the stationary phase to being transported with the solvent in the mobile phase. If this transfer between mobile and stationary phases were instantaneous, there would be no contribution to band broadening. In actuality, random fluctuations do occur in these transitions and cause otherwise identical molecules, moving in a single band, to proceed at different speeds. This random shifting creates dispersion among the molecules and is a major cause of band broadening in HPLC. Both the mobile and stationary phases may contribute to the dispersion.

2. Stagnant Mobile Phase

Stagnant mobile phase mass transfer is the kinetic hindrance due to penetration of the solutes into the pores of the support packings. The mobile phase in these crevices can be considered to be slowed or stagnated, as shown in Fig. 2.11d; therefore, solutes can fall behind the general flow in the column, leading to bandspreading. The effects on bandspreading caused by mass transfer into and out of the stagnant mobile phase (H_{pores}) can be described as:

$$H_{\text{pores}} = \frac{C_m d_p^2 u}{D_M} \qquad\qquad \text{Eq. 2.25}$$

where C_m is a mass transfer constant related to k and the porosity of the support (4, 8). Inspection of Eq. 2.25 shows that the use of small particle diameters will significantly reduce the bandspreading from mass transfer. Additionally, high flowrates will lead to bandspreading, especially with large biopolymers, which have very small diffusion coefficients. This is particularly critical in size exclusion chromatography. The effect of this mass transfer term on bandspreading can be very important, as seen in Fig. 2.15 for a column of 5µm particles.

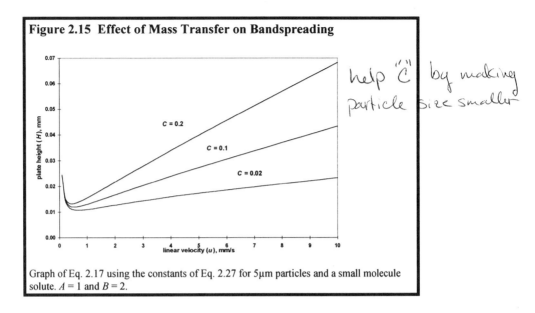

Figure 2.15 Effect of Mass Transfer on Bandspreading

help "C" by making particle size smaller

Graph of Eq. 2.17 using the constants of Eq. 2.27 for 5µm particles and a small molecule solute. $A = 1$ and $B = 2$.

The low "C" value is representative of a support with minimal stagnant mobile phase, such as, nonporous, perfusive or low porosity supports. The "C" value of 0.1 represents a support with average porosity. It can be seen that peaks can become very broad when high velocities are used for supports with stagnant mobile phase problems.

Several strategies have emerged to minimize the band broadening effects caused by stagnant mobile phase. At one time, thin crusts of derivatized porous supports were bound to inert nonporous cores to form 30 - 50µm particles - so called, "pellicular" packings (3). This approach was inadequate for most HPLC applications, however, because of the inefficiency of large particles and the inherently low surface area for interaction. Pellicular supports have seen some utility in guard columns because of minimum fouling and low pressure generation. Another, more recent solution, has been to use nonporous particles of very small particle diameters (1 - 3µm) which have higher surface areas than the large particles and the combined advantages for efficiency related to small particle diameter and elimination of pores (9, 10). These columns can yield very high efficiencies, but they have the drawback of generating extremely high pressure drops, frequently necessitating small columns and special equipment to minimize extra-column band broadening. One more solution which has been advanced for mass transfer problems in pores is the use of perfusive particles which contain flow-through pores with only shallow diffusive pores (11). These have primarily gained acceptance in preparative environments. The most popular solution to date for minimizing the problem of mobile phase stagnant pool formation is the use of small porous particles of 2 - 5µm so that the depth of the stagnant pools is not excessive.

3. Stationary Phase

The final term in the plate height equation of Eq. 2.17 accounts for the resistance to mass transfer which molecules experience during interaction with the stationary phase, as seen in Fig. 2.11e. Called the stationary phase mass transfer term (H_{sp}), it is defined as:

$$H_{sp} = \frac{C_s d_f^2 u}{D_S}$$
                                                          Eq. 2.26

where C_s is a constant related to stationary phase transfer, D_S is the diffusion of the solute in the stationary phase, and d_f is the thickness of the stationary phase or film on the surface of the particle, where molecules can interact. In modern packings for HPLC, this term is negligible because d_f is very small and D_S is similar in value to D_M (4). Stationary phase mass transfer can be a significant source of bandspreading in gas chromatography where the diffusion in the stationary phase is orders of magnitude smaller than that in the mobile phase, adversely effecting bandspreading. Although the effects of stationary phase on band broadening described in Eq. 2.26 and Eq. 2.24 are usually negligible in HPLC, they can be relevant if thick bonded phases are used on otherwise high efficiency columns.

D. Combination of Terms

        If all of the significant terms of bandspreading in HPLC are combined, the following version of Eq. 2.17 results from the sum of Eq. 2.22, Eq. 2.23 and Eq. 2.25:

$$H = A d_p \left(\frac{u d_p}{D_M} \right)^{0.33} + B \left(\frac{D_M}{u} \right) + C \left(\frac{d_p^2 u}{D_M} \right)$$
                                                          Eq. 2.27

It is obvious that the particle diameter, diffusion coefficient and linear velocity are all critical parameters of band broadening or efficiency. The importance of the particle diameter is graphically affirmed in Fig. 2.16 where columns packed with porous supports of different diameters are compared.

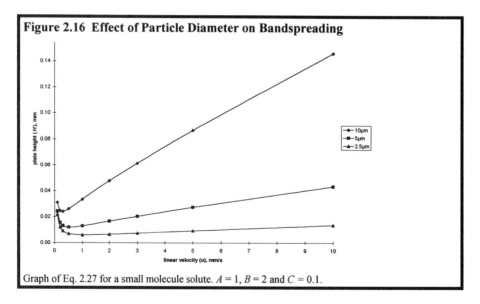

Figure 2.16 Effect of Particle Diameter on Bandspreading

Graph of Eq. 2.27 for a small molecule solute. $A = 1$, $B = 2$ and $C = 0.1$.

As expected from Eq. 2.27, particle diameter does indeed have a tremendous effect on bandspreading; the advantages of particle diameters of 5μm or less, especially at high flowrates, are readily apparent. The constants used for this graph and Fig. 2.13 - 2.15 were those which are attributed to a "good" column ($A = 1$, $B = 2$, $C = 0.1$).

        The other important constant in Eq. 2.27 is the diffusion coefficient. Most small molecules have diffusion coefficients of approximately 10^{-5}cm^2/s in the solvents used for HPLC; this value was used in plotting Fig. 2.13 - 2.16. Proteins and other biopolymers, however, have much lower diffusion coefficients, on the order of $10^{-7} cm^2/s$ or less, which makes this a major source of bandspreading. Fig. 2.17 illustrates how significant the effect of the diffusion coefficient on plate height can be for the column shown in Fig. 2.16 containing 5μm particles.

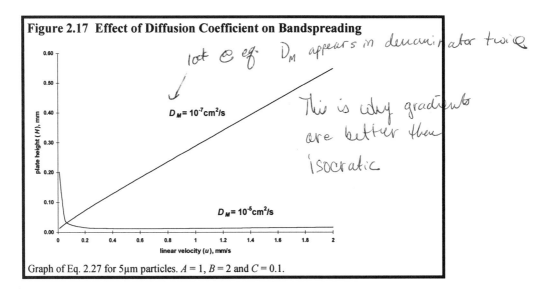

Figure 2.17 Effect of Diffusion Coefficient on Bandspreading

Graph of Eq. 2.27 for 5μm particles. $A = 1$, $B = 2$ and $C = 0.1$.

Note that the maximum velocity in this plot is only 2mm/s, yet the plate height for the macromolecule is 0.56mm, in contrast to the maximum plate height values of 0.15mm seen in Fig. 2.16 for small molecules at a linear velocity of 10mm/s. There is no question that bandspreading is tremendously increased for macromolecules because of their low diffusion coefficients. One reason that gradient elution is used for proteins is that it reduces some of the bandspreading which is seen isocratically; this will be discussed further in Section V.B. In size exclusion chromatography, where macromolecules must be analyzed isocratically, reduced flowrates greatly enhance resolution by minimizing peak widths.

E. Reduced Parameters

In the literature, plate height is often presented as a normalized term (h) rather than the actual value (H). The reduced plate height (h) is a consequence of the simplification by Knox of Eq. 2.27 where he defined reduced plate height and reduced velocity in terms of the particle diameter and diffusion in the mobile phase (6):

$$h = \frac{H}{d_p} \qquad\qquad \text{Eq. 2.28}$$

$$v = \frac{u\,d_p}{D_M} \qquad\qquad \text{Eq. 2.29}$$

In these equations, h and v are the reduced plate height and reduced velocity, respectively. They have no units. Substitution into the plate height equation (Eq. 2.27) thus yields (6):

$$h = \frac{B}{v} + Av^{0.33} + Cv \qquad\qquad \text{Eq. 2.30}$$

A plot of this equation, linearly or on a log scale, is most frequently used to describe bandspreading because columns of **all** particle diameters with **all** solutes and mobile phases can be directly compared for bandspreading characteristics. To illustrate this, Fig. 2.18 shows the plot of the eddy diffusion data seen in Fig. 2.13 using reduced plate heights. If reduced terms were used, the plots shown in Fig. 2.16 and 2.17 would all be identical because the differences were only in the particle diameters or the diffusion coefficients.

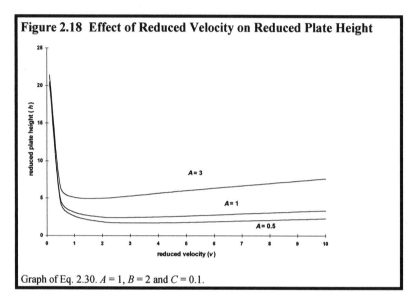

Figure 2.18 Effect of Reduced Velocity on Reduced Plate Height

Graph of Eq. 2.30. $A = 1$, $B = 2$ and $C = 0.1$.

A column with a reduced plate height of 2 - 5 is considered to be well packed; it can be seen in this figure that "A" values of one or less produce such columns.

V. ADDITIONAL INFLUENCES ON PEAK SHAPE

A. Peak Symmetry

As mentioned previously, although chromatographic peaks should ideally be symmetrical, they are frequently asymmetrical about an axis dropped from the apex to the baseline. These peaks possess tailing, as illustrated in Fig. 2.19, which can be caused by many factors, including column bed irregularities, nonspecific adsorption to the stationary phase, or problems with the mobile phase or sample.

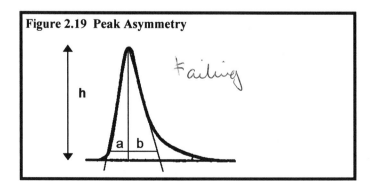

Figure 2.19 Peak Asymmetry

The height of the peak (h) is measured by the perpendicular axis of the peak to the baseline; the front and back distances to the peak from the axis (a and b) are determined at a height of 10% from the baseline to eliminate irregularities arising from baseline noise or drift. The peak asymmetry factor (A_s) is defined as:

$$A_s = \frac{b}{a} \quad \approx \quad 0.9 - 1.2 \qquad\qquad \text{Eq. 2.31}$$

For symmetrical peaks, a = b and the asymmetry is one. An acceptable range for asymmetry is usually 0.9 - 1.2. The peak in Fig. 2.19 has asymmetry greater than 1.5 and is generally unacceptable. Irregular column packing or secondary interactions caused by mixed modes or nonspecific adsorption are common sources of asymmetry or tailing. Values of asymmetry (A_s) that are less than one are sometimes due to column overloading, but they can also be caused by column packing problems. This phenomenon is termed "fronting" because the resulting asymmetric peak has a "front" instead of a "tail". Nearly coincident elution of a minor contaminating peak can resemble tailing, as was seen in Fig. 2.4, but this is a problem with selectivity rather than bandspreading.

Another definition of peak asymmetry is provided by the U.S.P. 1995 guidelines (12), as seen in Fig. 2.20.

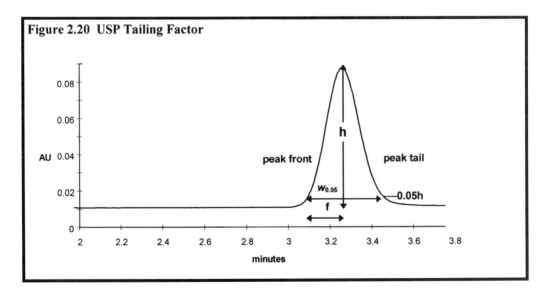

Figure 2.20 USP Tailing Factor

In this case, peak width is measured at 5% of the height, and the "tailing factor", T, is defined as:

$$T = \frac{w_{0.05}}{2f} \qquad\qquad \text{Eq. 2.32}$$

where f is the distance from the ascending peak to the axis at 5% of the peak height.

B. Peak Sharpening with Gradients

The discussion thus far has assumed all chromatography to be performed isocratically, which means that the mobile phase composition remains constant throughout the separation ("iso" = "constant" and "cratic" = strength). Isocratic separations are impractical, however, if several analytes with widely different affinities for the stationary phase must be separated in the same run. It is not unusual for a mixture to contain one compound which exhibits a useful k for a given column and mobile phase system, while another component exhibits a $k > 100$. Such a mixture can either be analyzed with two or more different chromatographic systems in series, or preferably, by use of a gradient to solvents of increasing strength. Gradient elution is a process of varying the elution strength of the mobile phase during the separation by continuous or stepwise blending of two or more solvents of widely different strengths. This approach usually results in all analytes eluting within a reasonable time. During the gradient, most compounds elute with earlier retention times than they would have isocratically; therefore, peak widths are decreased and peak heights are increased. The "band compression" shown in Fig. 2.21 results from the gradient and can never be achieved isocratically.

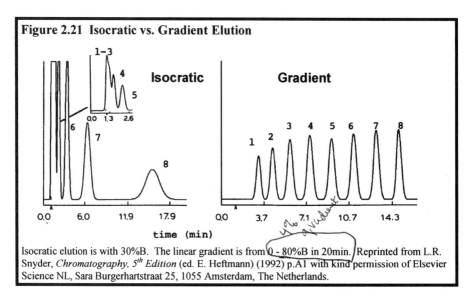

Figure 2.21 Isocratic vs. Gradient Elution

Isocratic elution is with 30%B. The linear gradient is from 0 - 80%B in 20min. Reprinted from L.R. Snyder, *Chromatography, 5ᵗʰ Edition* (ed. E. Heftmann) (1992) p.A1 with kind permission of Elsevier Science NL, Sara Burgerhartstraat 25, 1055 Amsterdam, The Netherlands.

The following equation describes the relationship between the retention obtained with a gradient (k) and the retention isocratically with water as the mobile phase (k_w) in reversed phase chromatography (13, 14):

$$\log k = \log k_w - \Phi S \qquad \text{Eq. 2.33}$$

In this equation, Φ is the volume fraction of organic solvent and S is a constant for a given solute/solvent combination.

The average value of k during elution, k^*, is actually (14):

$$k^* = 0.87\, t_G \left(\frac{F}{V_M \Delta \Phi S} \right) \qquad \text{Eq. 2.34}$$

where t_G is the time of the gradient, F is the flowrate, $\Delta \Phi$ is the change in solvent strength, and V_M is the total volume of mobile phase in the column. From this equation, it can be seen that elution time can be decreased by a faster gradient (shorter t_G) or larger change in mobile phase ($\Delta \Phi$). The relationships described in Eq. 2.34 can be confusing because they are contrary to those of isocratic separations where flowrate and mobile phase volume do not effect *k*. The interrelationship of all of these terms is the reason that changes in gradients have varied effects on the resolution of peak pairs in the chromatogram. Similar relationships to Eq. 2.34 have been derived for other modes of HPLC (14). Computer programs like *DryLab* have been developed to aid in gradient optimization; it is available from LC Resources, one of the first companies to compile such programs.

Developing initial gradient conditions with no previous experience is not difficult. If isocratic data are available, the gradient can go from conditions of no elution to those of no retention using the components of the isocratic mobile phase. For example, if 30% acetonitrile is used isocratically, the gradient could be 5 - 50% in 45min as a first trial. Table 2.1 lists some conditions for an initial gradient run when no information from literature is available. For a 4.6mm ID column, a flowrate of 1ml/min usually gives a good separation.

Table 2.1 Initial Gradient Conditions

Mode	t_G (min)	Composition	Weak or "A" solvent	Strong or "B" solvent
Reversed Phase	60	5 - 65% "B"	0.1% TFA in water	0.1% TFA in acetonitrile
Ion-Exchange	30	0 - 100% "B"	0.02 - 0.05M buffer	0.5 - 1M NaCl or sodium acetate in "A"
Hydrophobic Interaction	30	2M - 0M	2M ammonium sulfate in "B"	0.1M phosphate, pH 7

C. Peak Area and Flowrate

As discussed previously, the flowrate can affect the width of a peak. For some detectors, the flowrate (*F*) may also affect the peak area (*A*) (15):

$$S_i = \frac{AF}{Q_{inj}}$$

Eq. 2.35

where S_i is the sensitivity or calibration factor and Q_{inj} is the amount injected. In these cases, larger peaks will be obtained at lower flowrates. This effect is seen in flow sensitive detectors, like absorbance.

VI. SCALING A PROCESS

When scaling a particular process by changing to larger or smaller columns, it is often desirable to retain the same resolution and retention. To achieve the same retention times and performance on columns with different geometries, flow velocities must remain constant by adjusting flowrates accordingly. It is important to distinguish flowrate from flow velocity. Flow is almost universally described using volumetric flowrates measured in ml/min, rather than linear velocity in mm/s and the relative importance of velocity is often overlooked. To maintain identical separation conditions when scaling up or down, the volumetric flowrate must be adjusted to achieve the same linear flow velocity. The linear flow velocity (*u*) is defined by:

$$u = \frac{L}{t_o}$$

Eq. 2.36

When Eq. 2.36 is combined with Eq. 2.2:

$$u = \frac{LF}{V_M}$$ Eq. 2.37

Because the volume of mobile phase is proportional to the radius (or diameter) squared from Eq. 2.3,

$$\frac{F_2}{d_2^2} = \frac{F_1}{d_1^2}$$ Eq. 2.38

where F_1 and F_2 are the flowrates and d_1 and d_2 are the diameters of the standard and the new columns, respectively. To achieve the same velocity on a column with a different diameter, the flowrate used in the standard analysis (F_1) should be multiplied by the ratio of the squares of the diameters of the two columns:

$$F_2 = \frac{F_1 d_2^2}{d_1^2}$$ Eq. 2.39

Table 2.2 lists examples of flowrates which must be used on various standard column diameters to maintain specific linear velocities.

Table 2.2 Factors for Scale-up or Scale-down			
flow velocity (u) (mm/s)	0.7	1.4	2.8
column diameter (d_c) (mm)	flowrate (ml/min)		
1	0.024	0.05	0.1
2.1	0.1	0.2	0.4
4.6	0.5	1	2
10	2.4	4.8	9.6
deadtime (t_o) (min)	6	3	1.5

VII. CONCLUSIONS

The theoretical concepts presented in this chapter are the basis for adjusting operational parameters to improve chromatographic separations. The following tables offer guidelines for improving resolution which are based on these ideas. Chapter 6 - 10, which discuss modes of chromatography, will detail specific suggestions, especially regarding selectivity issues.

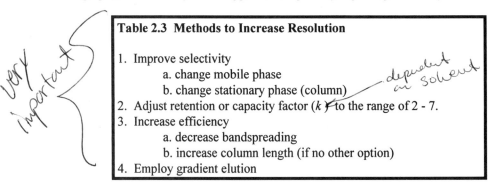

Table 2.3 Methods to Increase Resolution

1. Improve selectivity
    a. change mobile phase
    b. change stationary phase (column) _dependent on solvent_
2. Adjust retention or capacity factor (k) to the range of 2 - 7.
3. Increase efficiency
    a. decrease bandspreading
    b. increase column length (if no other option)
4. Employ gradient elution

very important

also pretty important

Table 2.4 Methods to Decrease Band Broadening	
Problem	**Solution**
eddy diffusion	use column that has been packed better
	reduce particle diameter
	use support with narrower particle size distribution
laminar flow diffusion	reduce particle diameter
axial diffusion	increase flow velocity above that for minimum plate height
mass transfer	reduce particle diameter
	decrease flowrate, especially for isocratic elution of macromolecules
	use supports with less porosity
	employ gradient elution

VIII. REFERENCES

1. J.W. Dolan, *LC-GC* 13 (1995) 24.
2. L.R. Snyder and J.J. Kirkland, *Introduction to Modern Liquid Chromatography, 2nd Edition*, John Wiley & Sons, New York, 1979, Chapter 5.
3. A.J.P. Martin and R.L.M. Synge, *Biochem. J.* 35 (1941) 91.
4. L.R. Snyder, "Theory of Chromatography" in *Chromatography, 5th Edition* (ed. E. Heftmann), Elsevier, Amsterdam, 1992, A1.
5. J.C. Giddings, *Dynamics of Chromatography*, Marcel Dekker, New York, 1965.
6. J.H. Knox, *J. Chromatogr. Sci.* 15 (1977) 352.
7. J.J. van Deemter, F.J. Zuiderweg, and A. Klinkenberg, *Chem. Eng. Sci.* 5 (1956) 271.
8. J.H. Knox and H.P. Scott, *J. Chromatogr.* 282 (1983) 297.
9. K. Kalghati and Cs. Horvath, *J. Chromatogr.* 398 (1987) 335.
10. K.K. Unger, G. Jilgs, J.N. Kinkel, and M.T.W. Hearn, *J. Chromatogr.* 359 (1986) 61.
11. N.B. Afeyan, N.F. Gordon, I. Mazsaroff, L. Varady, S.P. Fulton, Y.B. Yang, and F.E. Regnier, *J. Chromatogr.* 519 (1990) 1.
12. *United States Pharmocopeial,* U.S.P. 23 (1995), Rockville, MD, p. 1776.
13. L.R. Snyder, "Gradient Elution Separation of Large Biomolecules" in *HPLC of Biological Macromolecules: Methods and Applications* (ed. K.M. Gooding and F.E. Regnier), Marcel Dekker, New York, 1990.
14. L.R. Snyder and J.W. Dolan, "Gradient Elution Separation of Large Biomolecules" in *HPLC of Biological Macromolecules: Methods and Applications, 2nd Edition* (ed. K.M. Gooding and F.E. Regnier), Marcel Dekker, New York, in press.
15. H. Poppe, "Column Liquid Chromatography" in *Chromatography, 5th Edition* (ed. E. Heftmann), Elsevier, Amsterdam, 1992.

CHAPTER 3
HPLC HARDWARE

I. INTRODUCTION

The focus of discussions about HPLC is often on the column, which provides the mechanism of separation; however, the design and operation of the instrumentation are also critical to the success of an analysis. High performance liquid chromatographs, as diagrammed in Fig. 3.1A and B, include solvent delivery, sample introduction, separation, and detection components which together determine the quality of the chromatography in terms of efficiency, accuracy and reproducibility. Understanding the functions and operation of each of these components aids in their selection, effective usage, and troubleshooting, when necessary.

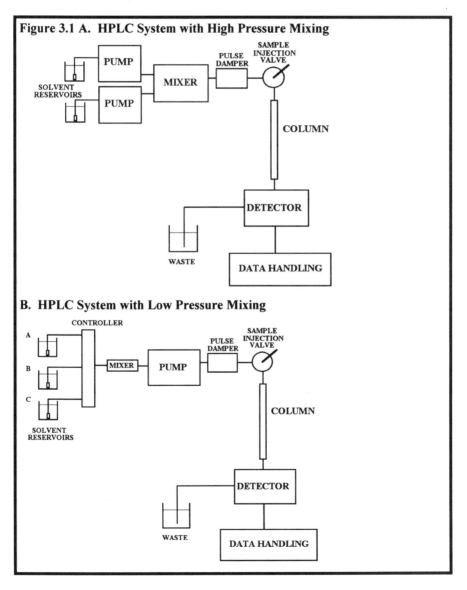

Figure 3.1 A. HPLC System with High Pressure Mixing

B. HPLC System with Low Pressure Mixing

This chapter will discuss the solvent delivery and sample introduction components of an HPLC; detectors will be discussed in Chapter 4. The primary component of the solvent delivery system is the pump; however, it often includes gradient formers, mixers, pulse dampers and degassers. Sample introduction basically comprises the injector, but fittings and filters are also relevant. Temperature control devices are adjunct, but can be critical to the system. Characterization of the whole system, especially aspects relevant to protein analysis or to the use of microbore or capillary columns, will conclude this discussion.

II. PUMPS

A. General

The pump in an HPLC is expected to yield invariable performance for a wide range of flowrates (from μl/min to ml/min) at pressures up to 5000psi. Critical comparisons of relative retention times (RRT) of multiple sample peaks between different runs demand that pumps deliver very precise flowrates. Ideally, an HPLC pump should produce infinitely accurate flowrates with little or no pulsation and it should also be compatible with all solvents, with fast and convenient changeovers (1). After more than twenty years of development and refinement, principle HPLC manufacturers offer pumps which embody many of these requirements.

Pumping systems that have been used in HPLC either deliver constant volume or constant pressure. It is easy to measure pressure at almost any point in the system; however, flowrate can only be assessed at specific limited locations. Nonetheless, constant flowrate pumps are almost exclusively used for HPLC. Constant pressure pumps are rarely used except to pack columns and will therefore not be discussed further.

The compressibility of liquids is an important physical parameter which can affect the accuracy of pumping systems, especially when operating at high pressures. The compressibility, which is the change in volume per unit pressure, can vary by a factor of three for the solvents commonly used in HPLC. Water, which has low compression, differs substantially from hexane and cyclohexane which compress more than threefold higher. In fact, at HPLC pressures, cyclohexane turns into a solid - obviously not a good choice for "liquid" chromatography. Compressibility is primarily a concern if a variety of solvents are run under high pressure conditions. In these cases, to achieve accurate flowrates, pumping systems must be adjusted manually or automatically to compensate for compressibility.

B. Syringe Pumps

Syringe pumps, initially developed in the 1960's, have the simplest design of commonly used HPLC pumps, as diagrammed in Fig. 3.2.

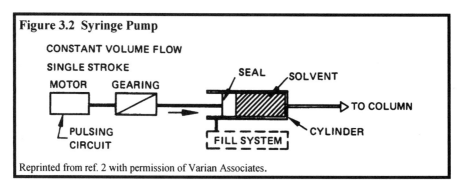

Figure 3.2 Syringe Pump

CONSTANT VOLUME FLOW

SINGLE STROKE

MOTOR GEARING SEAL SOLVENT

TO COLUMN

PULSING CIRCUIT

FILL SYSTEM CYLINDER

Reprinted from ref. 2 with permission of Varian Associates.

A motor-driven piston (plunger) delivers fluid contained in the reservoir (the syringe). The major advantage of this system is an accurate delivery of solvent, even at very low flow rates, which is unsurpassed by other pumping methods. This has been a boon for microbore HPLC (1 - 2mm ID columns) and, more recently, capillary HPLC (< 1mm ID), where small sample size, very low flowrates and concomitant ultrafine sensitivity are the norm. The flow from syringe pumps is also pulse-free. A major drawback of these pumps is that the syringe is a finite size; if it empties during a run, the run is ruined and must be abruptly terminated. Although automatic refill systems have been developed to expedite repeated runs, avoidance of intrarun emptying must always be a priority. The Eldex MicroPro® incorporates a reciprocating multi-pump system which allows one pump to refill while another is pumping. Pressure variation during changeover is minimal, as seen in Fig. 3.3.

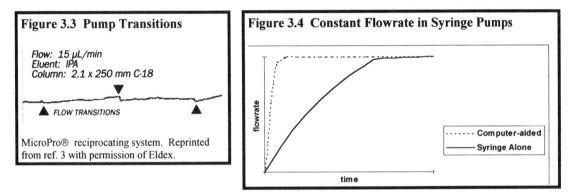

Figure 3.3 Pump Transitions

Flow: 15 μL/min
Eluent: IPA
Column: 2.1 x 250 mm C-18

▲ FLOW TRANSITIONS

MicroPro® reciprocating system. Reprinted from ref. 3 with permission of Eldex.

Figure 3.4 Constant Flowrate in Syringe Pumps

flowrate

time

- - - - - Computer-aided
—— Syringe Alone

Solvent compressibility is a particular problem for syringe pumps, especially during gradient operation. It takes a specific amount of time, dependent on the volume of the pump, to attain constant flow vs. pressure for the syringe (4). To precisely generate gradients, prepressurization of the solvent within each syringe is required before high pressure mixing. Rapid and accurate attainment of constant operating pressure and flowrate can be achieved using microprocessor-controlled pressurization to compensate for compressibility, as seen in Fig. 3.4.

C. Reciprocating Pumps

1. Single Piston

A reciprocating pump simply utilizes a piston driven back and forth by a rotating cam, linked to it by a shaft, as illustrated in Fig. 3.5 (5, 6).

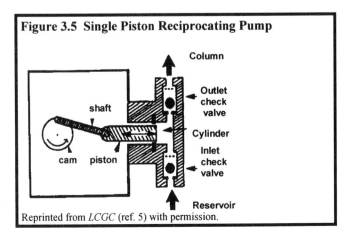

Figure 3.5 Single Piston Reciprocating Pump

Column

Outlet check valve

Cylinder

Inlet check valve

shaft

cam   piston

Reservoir

Reprinted from *LCGC* (ref. 5) with permission.

The "reciprocating" movement of the piston creates alternating negative and positive pressures which draw solvent from a reservoir into a collection cylinder during the intake and expel it towards the column during the delivery or compression cycle, as diagrammed in Fig. 3.6.

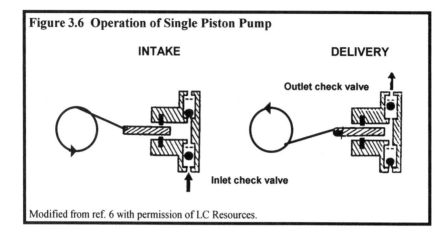

Figure 3.6 Operation of Single Piston Pump

INTAKE   DELIVERY

Outlet check valve

Inlet check valve

Modified from ref. 6 with permission of LC Resources.

Unlike syringe pumps, which have fixed, and usually small, solvent reservoirs, those of reciprocating pumps can be virtually any size, permitting unattended weekend runs or recycling of mobile phase, if necessary.

Using Fig. 3.6, the action of the check valves can be discussed. Standard check valves create unidirectional seals to regulate flow from the inlet to the outlet side of the pump. During the intake, the action of the piston causes the ball in the inlet check valve to rise from its seat, allowing liquid to flow, and the ball in the outlet to sit in its seat, stopping flow. During delivery, the action of the piston pushes the inlet check valve into its seat, stopping flow from the reservoir, and causes the outlet to rise, allowing flow towards the column. Obviously, proper operation of the check valves is critical to the function of the pumps. If they stick or cannot seat, poor pump performance results. Most check valves will fail to operate in the presence of air bubbles. Prior to operation, it is important to remove any air in the intake lines and to thoroughly degas the solvent. Placing the solvent reservoir above the pump will assist the valves in passing small air bubbles without loss of prime and result in more reliable operation. Clogged solvent inlet filters can cause cavitation and loss of prime; to avoid this problem, they should be cleaned or replaced periodically. Check valves which eliminate most of these difficulties are available for many pumps from Analytical Scientific Instruments. These self-priming check valves will not lose prime, even with large air bubbles. They are also not susceptible to damage by overtightening.

The reciprocating motion of single piston pumps produces cycles of regularly varying delivery volume. This discontinuous flow can create substantial baseline fluctuation in both UV and refractive index (RI) detectors due to changes in flowrates. The light levels which a UV detector measures can vary with the refractive index of a solvent, which changes as it is pressurized. Tapering and other improvements in flow cell design have minimized this fluctuation in the baseline, but they have not completely eliminated it. Pulse dampers, which will be discussed in a subsequent section, can also be used to reduce the effects. Because of this fluctuation problem, the use of simple single piston pumps is limited to less demanding applications in which pulsed flow is acceptable. Fast refill modifications, which will be discussed, overcome some of the pulsation.

2. Dual Piston

A major improvement to the pressure fluctuations of a single piston reciprocating pump is the addition of a second piston. Discontinuities in flow are reduced substantially by having the pistons operate in opposite directions, 180° out of phase. The pistons can be of equal size, creating equal duty cycles, or of unequal size. If the pistons are of equal size, the cyclical wave pattern of flow caused by the first piston cancels much of the wave pattern of the second, as seen in Fig. 3.7 (5, 6).

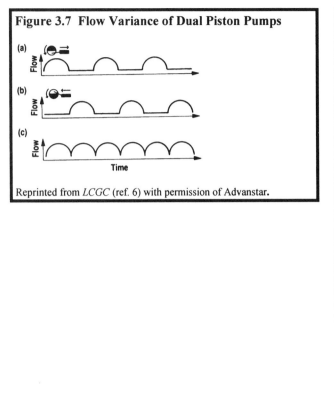

Figure 3.7 Flow Variance of Dual Piston Pumps

(a)

(b)

(c)

Time

Reprinted from *LCGC* (ref. 6) with permission of Advanstar.

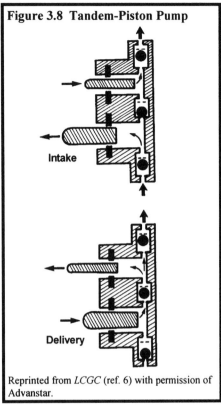

Figure 3.8 Tandem-Piston Pump

Intake

Delivery

Reprinted from *LCGC* (ref. 6) with permission of Advanstar.

The maximum flow from the first piston coincides with the minimum flow introduced by the other. This "wave cancellation" dampens the flow fluctuations and makes the flow more constant to the column.

In another design, called a tandem-piston pump, the large piston is twice the size of the smaller one, displacing twice the volume. As shown in Fig. 3.8, the outstroke of the larger piston both forces fluid to the column and fills the cylinder of the smaller piston (5, 6). In this design, there is fairly constant flow to the column. When precision gradients under low flowrates ($\leq 100\mu l/min$) are required, the flow characteristics of the dual piston designs render them generally less acceptable than syringe pumps.

3. "Fast Fill" Single Piston

In the 1970's, a new design of pumps was developed, exemplified by the Beckman Model 110 series and the Varian 5000/9000 systems. The problem of flow fluctuations when pumping a constant volume of fluid at high pressure was resolved with a single piston, "fast fill" pump design. In this design of pump, the single piston, made of sapphire or ruby like those found in all reciprocating pumps, is returned by spring action, after being driven through a guide and housing into a Teflon seal via a motor driven cam (7, 8). A recent model of this design is shown in Fig. 3.9.

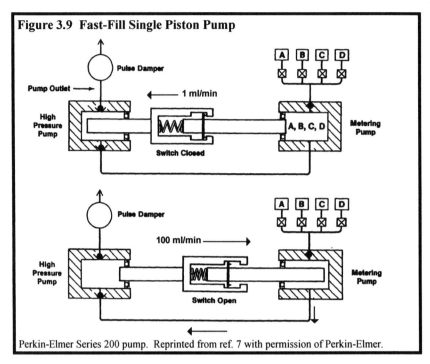

Figure 3.9 Fast-Fill Single Piston Pump

Perkin-Elmer Series 200 pump. Reprinted from ref. 7 with permission of Perkin-Elmer.

The fill cycle (bottom) is substantially shorter than the delivery cycle (top), inducing a dampening effect. A microprocessor monitors the pressures to ensure that the pressure during the intake cycle matches the pressure that was in the pump head during delivery. The pump is automatically modulated during the fill/delivery cycles to compensate for solvent compressibility and achieve accurate volumetric flow. For example, the Varian LC Star® employs an electronic flow feedback to ensure constant repeatable flows (9). A fast-fill single piston system requires only one pump, and is therefore less expensive than dual pump systems. It creates reasonable control of flowrate, but may require a pulse dampening system downstream.

D. Diaphragm Pumps

In a diaphragm pump, a single piston pumps oil to activate a diaphragm, which moves to deliver the solvent to the column, as shown in Fig. 3.10.

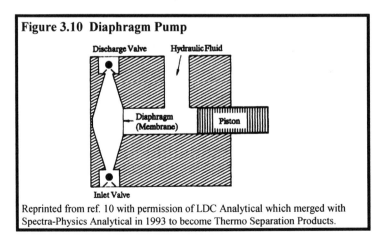

Figure 3.10 Diaphragm Pump

Reprinted from ref. 10 with permission of LDC Analytical which merged with Spectra-Physics Analytical in 1993 to become Thermo Separation Products.

This type of pump cycles quickly (~10 hertz) and delivers pulse-free flow with the aid of pulse dampers before and after the pump. The very low delay volume (as little as 50-100µl between the pump and the column) is well suited to the small volumes and gradient delivery used in microbore work. The Hewlett Packard Model 1090 pump is an example of this design. One disadvantage of this system is that if any oil leaks from the diaphragm pump into a column, the column is ruined and the system must be shut down until a complete cleanup is performed.

E. Commercial HPLC Pumps

Table 3.1 lists current pump designs of some commercially available HPLC systems. New systems are constantly being developed; therefore, the most recent information should be evaluated before making purchasing decisions.

Table 3.1 Selected HPLC Systems		Syringe	Single Piston Diaphragm	Fast-Fill	Dual Piston
Beckman	System Gold®			X	
Eldex	MicroPro®	X			
Gilson	Model 305			X	
Hewlett Packard	Model 1100 binary				X
ISCO	µLC-500®	X			
Hewlett Packard	Model 1090		X		
ISCO	Model 2350			X	
Perkin-Elmer	series 200			X	
ThermoSeparations	Spectra Vision®				X
Varian	LC Star®, 9000			X	
Waters	Model 515				X
Waters	Model 626			X	

All pumps will have the best and most reliable performance if they are maintained properly (11). This will be discussed in section VII of this chapter and in Chapter 15.

F. Pulse Dampers

The goal of the solvent delivery system in an HPLC is to deliver constant flow to the column at moderate to high pressures without pulsations. The unwanted cyclical surges caused by many pumps, particularly single piston pumps, can be reduced with pulse dampers. The simplest design employs a flattened coil of thin-wall stainless steel, which dampens pressure fluctuations by expanding with pressure to a more circular configuration, and returning to its original configuration upon release of the pressure (2). The Handy-Harmon pulse damper, named after its inventors, is a common version of this design. The flexible coil works well at pressures below 100atm but deforms irreversibly at high pressures, similarly to an overstretched spring. The design is suitable for isocratic applications but its dead volume of 2 - 15ml introduces unacceptable gradient delay.

Another approach to reducing pulsations when delivering constant volume utilizes the "liquid spring" design. In this pulse damper, the solvent is routed via flexible tubing through a chamber containing a compressible liquid like isopropanol. The inner tube flexes and relaxes with each pressure cycle of the pump, while the liquid acts as a capacitor to "store" and "release" pulsation energy during pump delivery and fill strokes, respectively. This damper works efficiently at high pressures, but any leaks which develop between the flexible tube and the chamber will add small amounts of the compressible fluid to the chromatography column, altering the separation dramatically.

A popular pulse damper is that patented by SSI. This design, shown in Fig. 3.11, utilizes an inert Teflon diaphragm surrounded by a compressible fluid, which flexes with each pump pulse. It has only a 0.9ml sweep-out volume at low pressure but about 3ml at high pressure (> 3500psi).

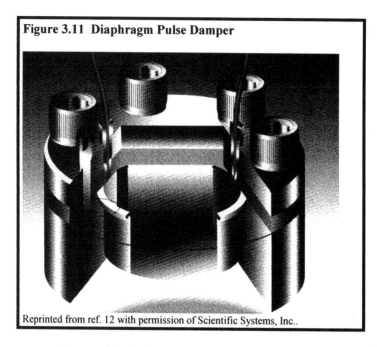

Figure 3.11 Diaphragm Pulse Damper

Reprinted from ref. 12 with permission of Scientific Systems, Inc..

Some pumps are dampened by their second piston or by feedback during the "fast fill" cycle of the pump, as in the Waters Model 626 pumps. When these dampening measures are successful, they eliminate the need for pulse dampers and the additional dead volume they add to a system.

III. GRADIENT FORMATION

A. Concept

Isocratic elution utilizes a single mobile phase and thus requires only one pump. Solvents are often completely premixed mechanically and delivered without need of further "blending". For more complex applications, especially those involving peptides or proteins, where solutes have substantially different affinities for the stationary phase and cannot be adequately analyzed with a single mobile phase, a change in solvent is necessary. Sometimes a "step" to a second solvent is adequate. More often, two or more solvents must be systematically blended to achieve retention, elution and ultimately, resolution. This is generally accomplished with a linear gradient which uses a mixing device, blending control, and possibly, two (or more) pumps. For most HPLC systems, gradient rate and shape are controlled by a microprocessor or computer program so that virtually any variation is possible.

B. Multipump, High Pressure Mixing

In high pressure gradients, flows from multiple pumps are blended at high pressure, with each pump delivering a single solvent into a mixing chamber, as illustrated in Fig. 3.1A. The features of the gradients are controlled by varying the flowrates of the individual pumps; therefore, the accuracy and reproducibility of the gradient is dependent upon the precision of these flows. For most of the range of compositions, this method is exact; however, at very high concentrations of either solvent, it may not be. For example, at a mixing ratio of 1% Solvent A and 99% Solvent B, the A pump would have to pump at 10µl/min if the overall flow rate were 1ml/min. At concentrations below 1% or above 99%, the required flowrate is even lower. These low flowrates may be inherently inaccurate for some pumps, particularly reciprocating pumps whose check valves function suboptimally at low pressure; thus, they may introduce substantial error into the content of the mixture. A simple solution to this problem is to begin the gradient at high enough levels of B solvent to be in an accurate flowrate range and to end similarly before 100%, permitting the minimum required flowrates for each pump to be higher. Minimum accurate flowrates for a pump, given by the manufacturer's specifications, should be considered when assessing the problematic gradient percentages. Table 3.2 compares some HPLC systems which employ high pressure gradient formation.

Table 3.2 Selected High Pressure Gradient HPLC Systems		Pulse Damper	Mixer
Beckman	System Gold®	none	dynamic
Eldex	MicroPro®	none	static or dynamic
Gilson	Model 305	hydraulic	dynamic
Hewlett Packard	Model 1100 binary	yes	in-line
ISCO	µLC-500®	none	static

C. Single Pump, Low Pressure Mixing

1. Proportioning Valves

A gradient system using low pressure mixing employs only one high performance pump, as diagrammed in Fig. 3.1B. One variation of low pressure gradient formation involves the connection of a single piston reciprocating pump to one or a series of proportioning valves on the inlet side, as seen in Fig. 3.12 (1).

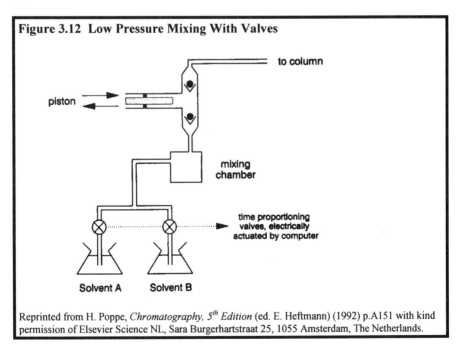

Figure 3.12 Low Pressure Mixing With Valves

to column

piston

mixing chamber

time proportioning valves, electrically actuated by computer

Solvent A   Solvent B

Reprinted from H. Poppe, *Chromatography, 5th Edition* (ed. E. Heftmann) (1992) p.A151 with kind permission of Elsevier Science NL, Sara Burgerhartstraat 25, 1055 Amsterdam, The Netherlands.

The solenoid valves are microprocessor-controlled to permit exact volumes of solvent to enter at a given time on the inlet stroke of the high pressure pump. For example, mixing 10% Solvent A and 90% Solvent B would be achieved by having the A valve open for 10% of the duration of the fill stroke, and the B valve open for 90% of the time. One problem with mixing two (or more) solvents under low pressure is creation of a different volume in the mixture than the sum of the original volumes of the individual solvents. This occurs with some common solvent blends such as 50% water/50% methanol. Low pressure blending can create significant cavitation, bubble formation and outgassing, but these can be prevented with the degassing of each solvent. Low pressure mixing with proportioning valves has been implemented by numerous instrument companies, as seen in Table 3.3.

2. Metering Pumps

Another approach to low pressure blending employs metering pumps to deliver solvents proportionally to the main pump. The metering pump approach has been used by ISCO and others. A variation of this design, used by Varian on the 9000 Series, employs an electronically or mechanically-actuated valve in place of a check valve between the low pressure metering system and the high pressure delivery pump. This minimizes cavitation problems and the requirement for degassing.

A complicated, but effective, way to achieve multi-solvent delivery is to use a low pressure, dual piston syringe pump to meter each solvent to a high pressure pump, as exemplified by the HP 1090 system. The pumping rate for each pump is dictated by the respective solvent and the desired blend. The metering pumps deliver their liquids via a valve into a single low pressure mixer, which also serves to dampen. From this device, the mixture enters a high pressure diaphragm pump, which ultimately delivers the final mixed solution to the column.

Table 3.3 Low Pressure Gradient HPLC Systems		Pulse Damper	Mixer	Gradient Means
Hewlett Packard	Model 1090	hydraulic	in-line	metering pumps
Perkin-Elmer	series 200	none	in-pump	valves
ThermoSeparations	Spectra Vision®	none	in-pump	valves
Varian	LC Star®, 9000	none	in-pump	valves
Waters	Model 626	none	static	valves

D. Mixers

To attain reproducibility in gradient elution, it is essential that the blended solvents be mixed thoroughly. Mixers, which can be static or dynamic, are located downstream from the solvents and gradient formers.

1. Static

Static mixers, as the name suggests, have no moving parts. Mixing of two or more fluids is achieved by a chamber with a carefully designed geometry where these fluids are combined. The "T" mixer produces such mixing - when solvent A goes into one arm and solvent B into the other, the mixture comes out the third. Some pumps with low pressure gradient formers accomplish the mixing within the pump cavities, as was seen in Fig. 3.9. The very low dead volumes of static mixers make them suitable for microbore work.

Another method of achieving fairly good mixing is to interface multiple sets of "tortuous" flow paths for the fluids. One example of this kind of mixer is a small column packed with large diameter non-reactive beads. It is critical that the packing be non-reactive with respect to the particular solvents being used, or undesirable demixing or hysteresis effects may occur. This kind of mixer has high volume and is usually used for postcolumn mixing for derivatization, rather than gradient formation.

2. Dynamic

If solvent pairs have substantially different viscosities, like water/isopropanol or water/THF, static mixing may be inadequate. In these cases, dynamic mixers, which utilize mechanical means such as magnetically driven stir bars or "propellers," are needed to achieve uniform blending. These mixers often have larger volumes than static mixers because they must allow for the stirring mechanism. For example, a small Eldex static mixer has a volume of 2µl, whereas the volume of the small dynamic mixer is 15µl. Dynamic mixers are usually more expensive than static, due to their more complex designs.

E. Degassers

All mobile phases contain dissolved gases which may potentially affect the operation of pumps or detectors; electrochemical and fluorescence detectors are particularly sensitive to such gases. Certain mobile phase systems like methanol/water promote bubble formation during mixing. Other systems, such as aqueous mobile phases, may not require degassing, especially with UV detection.

Helium sparging has been a popular method of deoxygenating a mobile phase. In this technique, helium is bubbled into the mobile phase, removing the oxygen. Because the mobile phase is in a helium atmosphere, oxygen is not reintroduced until the mobile phase enters the pumping system. This apparatus is fairly easy to set up using helium and appropriate containers with tubing; however, helium is expensive and a rare gas. Additionally, helium sparging may remove volatile components, such as trifluoroacetic acid, from the mobile phase. Its effectiveness is also related to the rate of flow of helium (13).

A second technique for removing gas from the mobile phase uses a vacuum. Solvent is pumped through membrane tubing contained in a vacuum chamber (13). This method achieves a consistently low concentration of oxygen and no solvent vapors are released into the laboratory. Fig. 3.13 illustrates the effect of various degassing methods on the background absorbance of methanol.

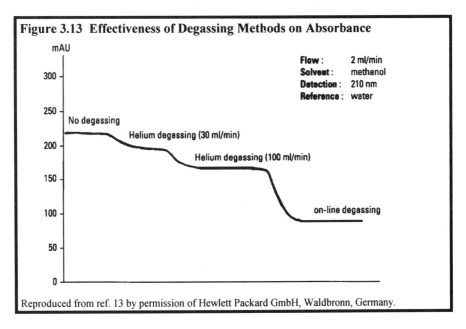

Figure 3.13 Effectiveness of Degassing Methods on Absorbance

Flow :	2 ml/min
Solvent :	methanol
Detection :	210 nm
Reference :	water

mAU

No degassing

Helium degassing (30 ml/min)

Helium degassing (100 ml/min)

on-line degassing

Reproduced from ref. 13 by permission of Hewlett Packard GmbH, Waldbronn, Germany.

IV. SYSTEM CHARACTERIZATION AND VALIDATION

It is very important to measure critical operational parameters of an HPLC system to verify that they are within the manufacturer's specifications or the limits of acceptable operation defined within one's laboratory. Such measurements are also major troubleshooting tools when the system is malfunctioning because the current performance can be compared to the original. Several simple tests can assess basic parameters.

A. Volumetric Flowrate

Volumetric flowrate is readily measured by collecting eluent on a time basis. Several different rates of flow in the range being used for analysis should be evaluated because flowrate variations may not be consistent across the range of the pump. In common practice, the time required to fill a volumetric flask is measured, and the flowrate calculated (flowrate = volume/time). Accurate measurement of volume can be confirmed by weighing the fractions and using the density of the solvent (volume = mass/density).

B. System Volumes

1. General

Each HPLC system has inherent volumes which cause a delay in the start or the finish of a gradient. These must be known if data between different systems are compared. Measurements are made without any column in-line. If air bubbles are a problem during the test, low volume restrictors can be placed in-line before the detector to generate back pressures that simulate those found in typical runs ($\geq$ 1000psi). Volumes are measured by using solvents spiked with a readily detectable material like acetone, a strong UV absorber. Reservoir A usually contains pure distilled water while Reservoir B holds the water containing acetone (1%, or an amount which exhibits full scale absorption at 100% B at a suitable wavelength like 254 or 280nm).

2. Hold Up or Delay Volume

Hold up (or delay) volume is defined as the breakthrough volume required for the mobile phase to reach the column from the reservoir. It is measured by determining the time (t_d) required for the detector to respond, after initiation of the acetone-spiked solvent B, and multiplying by the flowrate. To accomplish this, a system is equilibrated with pure solvent A (water) and the detector set at baseline. After confirming equilibration with a straight baseline, the solvent is switched to pure B and the time (volume) measured until the baseline shows equilibration with the second solvent. Hold up volume is measured at the time when the absorbance reaches half scale, as illustrated for two different systems in Fig. 3.14.

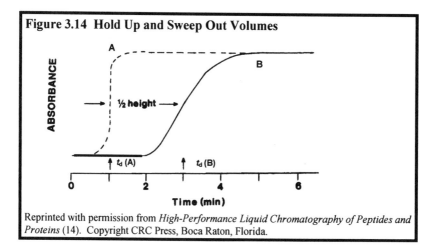

Figure 3.14 Hold Up and Sweep Out Volumes

Reprinted with permission from *High-Performance Liquid Chromatography of Peptides and Proteins* (14). Copyright CRC Press, Boca Raton, Florida.

In this example, system A has very low dead volume, resulting in the second solvent reaching the detector in one minute. In system B, the dead volume is higher, with the mobile phase not reaching the detector until three minutes have elapsed.

3. Sweep Out Volume

Sweep out volume is the volume required to totally change from one solvent to another. It is measured by the volume required for detector response to rise from baseline to 100% full scale, after delivery of solvent B is initiated. In Fig. 3.14, the sweep out volume for system A is only slightly larger than its delay volume; this system has very low dead volume components. System B has a large sweep out volume of about 4.5ml (if the flowrate is 1ml/min); a high volume mixer will often give this type of response.

C. Proportioning Accuracy

When gradients are utilized, many factors, including malfunctioning check valves, mixers, and inlet filters, can cause suboptimal performance. The accuracy of a linear gradient (0%B to 100% B) can be easily evaluated, as shown in Fig. 3.15 (14).

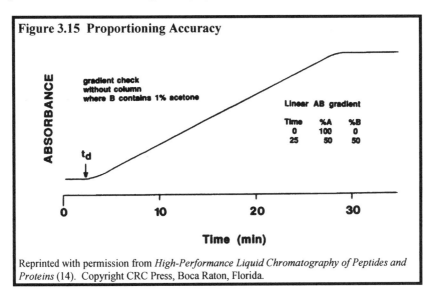

Figure 3.15 Proportioning Accuracy

Reprinted with permission from *High-Performance Liquid Chromatography of Peptides and Proteins* (14). Copyright CRC Press, Boca Raton, Florida.

The doped solvent system of water and acetone, described in the previous section, is used for this test. Proportioning inaccuracy is indicated by deviations from linearity for response vs. time in the linear gradient test at 270nm. Multiple runs can verify a system's consistency.

V. OTHER DEVICES

A. Injectors

1. Manual

Loop injectors, which permit the injection of a wide range of volumes with no change in pressure and flow, are now universally used in HPLC. As shown in Fig. 3.16, the loop acts as a bypass so that the sample can be introduced. It is then delivered onto the column in a smooth, nearly continuous fashion by a single switch of a valve, which diverts the fluid flow in the appropriate direction.

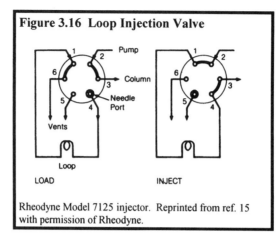

Figure 3.16 Loop Injection Valve

Rheodyne Model 7125 injector. Reprinted from ref. 15 with permission of Rheodyne.

In this design, the sample is not measurably diluted because it remains in the part of the loop near the column. Most loop injectors are manufactured by Rheodyne, Valco and SSI.

Contrary to common belief, the amount injected onto the column with a loop injector is not necessarily the amount delivered by the syringe into the loop, or even the fixed capacity of the loop (1, 15). Variation in delivery volume is caused by laminar flow within the loop, as shown in Fig. 3.17.

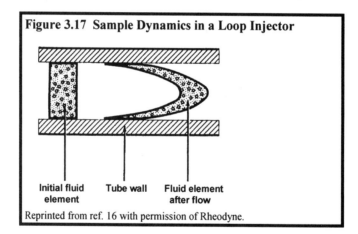

Figure 3.17 Sample Dynamics in a Loop Injector

Reprinted from ref. 16 with permission of Rheodyne.

An initial cylindrical "slug" of sample forms a parabola as it travels through the tubing, due to laminar flow, diffusion, and the wall effects of the tubing (16). There are several consequences of this sample spreading:

1. Volumes less than 50% of the loop volume enter the column quantitatively.
2. Volumes of 50% to 200% of the total loop volume may suffer losses to the overflow port and retention of sample along the walls of the tube. Portions of the sample may never be injected onto the column and, obviously, never recovered.
3. The best run-to-run reproducibility for the full loop is obtained when sample volumes are 2 - 5 times the loop volume. In this range, all the sample is swept from the loop and 99[+]% of the total loop volume reaches the column.

The linearity of injection for a 20μl loop, illustrated in Fig. 3.18, shows that consistent total peak area for a full loop was only obtained at injections over 60μl.

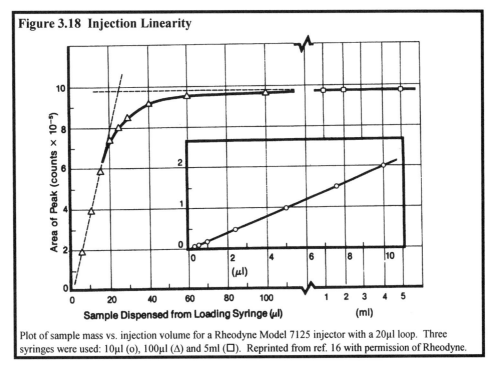

Figure 3.18 Injection Linearity

Plot of sample mass vs. injection volume for a Rheodyne Model 7125 injector with a 20µl loop. Three syringes were used: 10µl (o), 100µl (Δ) and 5ml (□). Reprinted from ref. 16 with permission of Rheodyne.

In this example, samples up to about 15µl exhibit a linear relationship between the volume of sample loaded and the observed peak area, indicating that all the sample that was loaded reached the column. Between about 15 and 60µl, part of the sample was lost to overflow, and therefore, linearity was no longer obtained. From 60 to 100µl (or 3 to 5 times the loop volume), a plateau of peak area was observed. Therefore, for good linearity with this 20µl injector, sample volumes would have to be less than 15µl; for good consistency of full loop or 20µl injections, sample volumes must exceed 60µl. The excess required for reproducible full loop injection limits its suitability to samples which are large. Less abundant samples are best analyzed with partial fill techniques of no more than 50% of the loop volume. For this injector, even the very low injection volumes are delivered accurately, as seen in the inset. This valve has no filler port, but those that do have an offset volume equal to that of the port, at which linearity begins (16).

There is a special variation of loop injector for very small volumes, such as those used for microbore columns. In this design, there is an internal loop so that submicroliter injections can be made with a fixed volume. A common injection volume for 1mm ID microbore columns is 0.5µl. For accurate injection with this injector, volumes of 4µl should be used (16).

Manual injectors should generally be washed out with about 1ml of mobile phase between injections using the needle port cleaner, not a needle (17). For gradient elution, the beginning solvent should be used. Periodically, injectors and syringes should be cleaned with a stronger agent, such as 0.1% trifluoroacetic acid (TFA) or organic solvent, to remove contaminants. Troubleshooting measures for specific injector problems are detailed in a guide by Rheodyne (18).

2. Autosamplers

Automated sample injection can be achieved with an autosampler. This instrument allows loading of many samples containing volumes from submicroliters to milliliters. Samples can be injected as frequently as several per minute. Multiple injections from a single vial, washing between samples, variation in procedures between samples, and other customizations are simply performed with programmers. In many cases, the vials can be cooled to enhance sample stability. Autosamplers generally use the same loop injectors discussed in the previous section; therefore, precautions about quantitative delivery volumes are equally valid. Fig. 3.19 illustrates the operation of this type of injection device.

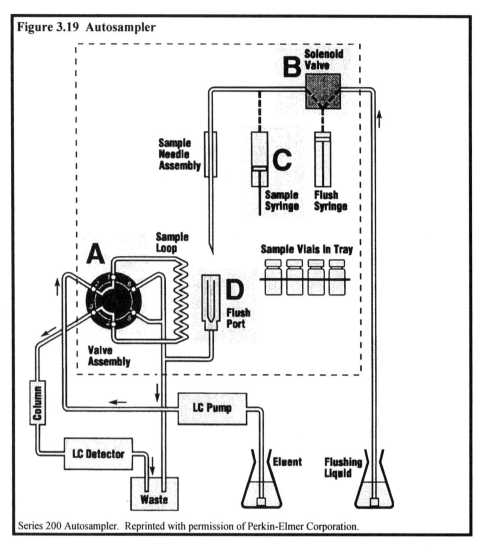

Figure 3.19 Autosampler

Series 200 Autosampler. Reprinted with permission of Perkin-Elmer Corporation.

B. Fittings

The flow of liquid along the path from the solvent reservoir to the inlet of the column encounters numerous connections with low and high pressure fittings, all of which need to be free of leaks and, preferably, of zero dead volume. Such connections are accomplished with combinations of nuts and ferrules which, unfortunately, are of several distinct varieties. Because fittings and ferrules look quite similar but are usually not interchangeable, it is important to know their subtle differences to identify them. Certain fittings, like Parker and Waters, may seem compatible because they do not leak when used together, but they usually have extra dead volume. Other fittings leak and are not interchangeable at all (most notably, Rheodyne and any other fitting). To insure elimination of dead volume, as well as leaks, it is best to use the proper nuts and ferrules designed for any specific fitting. Upchurch Scientific manufactures fittings which conform to each individual manufacturer's specifications. Fig. 3.20 illustrates these specifications for many of the nuts and ferrules commonly used in HPLC.

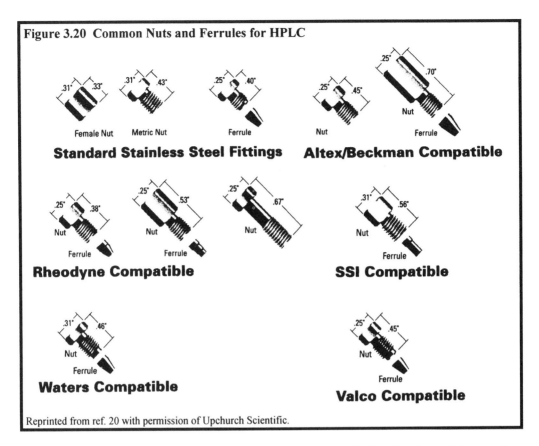

Figure 3.20 Common Nuts and Ferrules for HPLC

Reprinted from ref. 20 with permission of Upchurch Scientific.

It can be seen that although some of these nuts or ferrules are very similar, they have slight to major differences in configuration and dimensions.

Proper connection of nuts and ferrules to tubing and fittings is vital to achieving a low dead volume and leak free system. A connection using nuts and ferrules is shown in Fig. 3.21 (21).

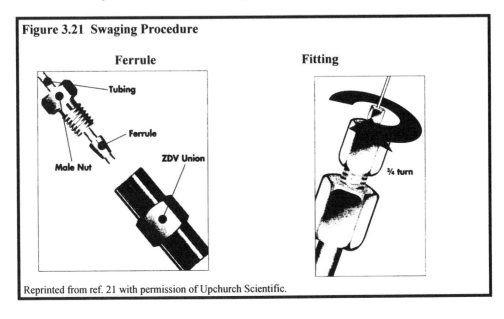

Figure 3.21 Swaging Procedure

Reprinted from ref. 21 with permission of Upchurch Scientific.

All parts are loosely assembled and then the tubing is seated onto the fitting, pushing firmly by hand. The nut is subsequently tightened to the fitting completely by hand, and then the ferrule is swaged by applying an additional 1/4 to 1/2 turn using wrenches. Most problems in this process occur by overtightening. If the fitting leaks after applying pressure with the system pump, the nut should be retightened with an additional 1/8th turn. If leaking still persists, a change of ferrule or fitting is probably required. Further tightening is unlikely to fix the leak, and may even shear the nut and create a serious, time-consuming plumbing problem. Overtightening causes distortion of ferrules, making them only suitable to the particular fitting. It can also lead to shearing of tightening nuts. With proper seating, the amount of tubing which emerges from the ferrule is of fixed, but different, lengths, as seen in Fig. 3.22 for tubing with swaged ferrules of each of four manufacturers (20).

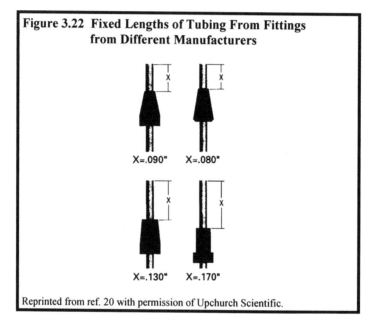

Figure 3.22 Fixed Lengths of Tubing From Fittings from Different Manufacturers

X=.090"   X=.080"

X=.130"   X=.170"

Reprinted from ref. 20 with permission of Upchurch Scientific.

The differences in these fixed lengths are responsible for the common problems of improper seating diagrammed in Fig. 3.23.

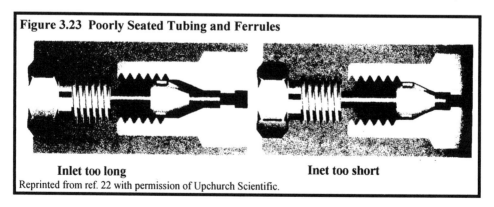

Figure 3.23 Poorly Seated Tubing and Ferrules

Inlet too long   Inet too short

Reprinted from ref. 22 with permission of Upchurch Scientific.

If the tube is too long to seat, the poorly seated ferrules will cause leaks. If the tube is too short, dead volume occurs in the fitting, resulting in retention time variation and band broadening.

It has been observed that slight variations in the geometry of injection valve seats may impart irreversible changes to the respective fitting, preventing interchange of loops or transfer lines between valves. It is a good idea to label loops with the loop size and specific injector to avoid any problems.

The fitting and ferrule compatibility problem has been overcome substantially by recent innovations in design, most notably, construction of nuts and ferrules from polymers which, when used properly, can deform enough to make a good seat in virtually any fitting. Once tightened, polymeric fittings are dedicated to the particular connection because they become molded to the exact shape and dimensions. The ferrule must still be seated properly so that no dead volume is produced. It is important to use only polymeric fittings designated for HPLC so that they can withstand high pressures. Polyetheretherketone (PEEK) is an inert polymer which is frequently used in chromatography applications. Designs range from solid PEEK to fittings with stainless steel collars over PEEK ferrules.

A well-maintained parts "organizer" or cabinet is worth the small investment to expedite quick and leak-free replacement of connections. An inventory of fitting and ferrule replacements should be stocked for each HPLC. Attention to these small but important details can prevent numerous, time-consuming delays and problems.

C. Connecting Tubing

All of the various components of an HPLC system must be connected by tubing through which the mobile phase, and possibly the sample, must pass. Usually, this tubing is made from polymer on the low pressure side and stainless steel on the high pressure. Tubing from the injector to the column and the column to the detector should be of small diameter (≤ 0.01" ID) and of short lengths so that the sample band has minimal spreading. Tubing before the injector can be somewhat larger in diameter, but in gradient systems, it is best to keep it small (0.02 - 0.03" ID) to help minimize the hold up and sweep out volumes.

Connecting tubing is generally 1/16" OD; however, smaller tubing, which is available from Upchurch, can be used to connect capillary columns to mass spectrometers. The smaller tubing greatly enhances performance by minimizing bandspreading; the low volume also produces decreased elution times. The tubing is attached by threading it through a short 1/16" tubing sleeve which is connected with appropriate nuts and ferrules; this type of connection is adequate because the system is not at high pressure after the column.

It is very important that all connecting tubing be of good quality, usually 316 stainless steel, and that it be clean and cut perfectly so that the fittings will seat properly. Pieces of 1/16" OD tubing can be cut by scoring the circumference with a knife (2) or using tubing cutters designed for this purpose. PEEK tubing can be cut with a razor blade. It is often easiest to buy such tubing ready-cut from a parts supplier like Upchurch, Alltech or Supelco.

D. In-Line Filters

The discriminant use of filters in an HPLC system can result in better and more consistent performance overall. Any particulate matter which invades the system can stop check valves from seating, plug columns, and damage pistons or injection valves, any of which can be costly in terms of time and materials. The first filter in the system is usually on the solvent inlet line, submerged in the mobile phase. Such filters prevent particulate matter from entering the pumps; they are absolutely necessary when using biological buffers which are prone to bacterial growth and may contain insoluble impurities. When inlet filters become plugged, they can be a source of air bubbles or inaccurate flow; therefore, they need to be cleaned or replaced periodically.

There are usually one or more in-line filters before the injector in an HPLC system. Filters placed in-line prevent insoluble particles larger than 0.5µm from entering and blocking valves or clogging the column. These serve to backup the inlet filters and also to protect against particulate matter that may be generated by the system, such as degeneration of seals. Salts can cause such deterioration by abrasion, especially when leaks are present.

Lastly, filters for the sample can be placed after the injector. Usually it is better to filter the sample before injection or to use guard columns rather than filters in this position in the system to prevent sample loss or bandspreading due to the filter. In-line filters used between the injector and guard or analytical column must be low-volume to minimize band spreading. The filter will protect the guard and analytical columns from particulates originating from the sample, as well as from the solvent delivery system. Large volume in-line filters increase system volumes and should only be used upstream from the injector.

E. Guard Columns

A guard column is a small column containing a support with the same bonded phase as the analytical column. It serves to protect the analytical column by binding molecules which might be irreversibly adsorbed and by filtering particulate matter which may be present in the sample. For substantially pure samples like pharmaceutical drugs, no filters or guards are necessary. For cruder samples, such as those which arise from early stages of natural product isolation, guard columns are highly recommended. Guard columns often contain the same column packing as the analytical column in terms of particle diameter, pore and bonded phase; this offers optimum protection but is also the most expensive option. Alternatively, for samples with minor amounts of impurities, nonporous supports with a bonded phase corresponding to that of the analytical column are used. These have less capacity than porous supports but are also less expensive and can be dry-packed.

Guard columns can introduce bandspreading into the systems; therefore, it is important that they be well packed and connected with minimal dead volume. Prevention of column problems caused by impurities is well worth any compromise guard columns or filters can induce in overall system efficiency. Guard columns should be replaced when they are plugged or defective to avoid resultant poor performance. If the guard columns or frits are even slightly clogged, higher back pressure and tailing of peaks can result.

F. Column Heaters

Temperature control of HPLC columns has gained popularity in the past few years for several reasons (23):
1.   Certain columns containing small particles have better performance and lower pressure drops with elevated temperatures.
2.   The stability of some biological molecules is improved with subambient temperatures.
3.   Some bonded phases have temperature dependent selectivity.
In all these cases, temperature control may be essential to reproducibility.

There are three primary kinds of column heaters for HPLC: block, forced-air and water jackets. Water jackets are effective, but messy and difficult to implement. Block heaters are very effective, but for the best operation, they must conform exactly to the column so that the whole surface is heated evenly. Forced air ovens are the simplest heaters to use, but they are also the least effective because of their poor heat transfer properties (23, 24). For good temperature control with all designs of heaters, the mobile phase must be preheated, usually by a coil that is in the oven or on a heating block.

The smaller the column diameter, the more efficiently a column and its contents can be heated. As the diameter increases, the temperature may vary radially. There may also be a longitudinal temperature gradient if heating is not uniform along the length of the column. Any temperature gradients will tend to yield irreproducible results.

VI. SPECIAL CONSIDERATIONS

A. Biocompatibility

One option that is available for HPLC systems is "biocompatibility". What this term actually denotes is that components are made from materials other than stainless steel (SST). Whether those components, be they glass, titanium, PEEK, or other polymers, are more compatible with biological activity than stainless steel is a factor of the specific molecules, mobile phases, etc. Because of this, the need for biocompatible HPLC systems, which can be more expensive and sometimes less robust than stainless steel systems, has been debated over the years (25). It is certainly well known that glass, for example, can adsorb proteins nonspecifically and irreversibly. Originally, all HPLC systems were composed of stainless steel components which could give high performance under pressures up to 5000psi. Using these systems, most proteins and other molecules gave quantitative yields; however, occasionally an iron-sensitive molecule lost activity. This loss of activity may have been due to the column frits or even metal impurities in the mobile phase rather than the components of the instrument. In cases where mobile phase constituents used for biomolecules, like lithium chloride, cause deterioration of only certain stainless steel parts of HPLC instruments, substitution of appropriate alternative materials for those components can eliminate the problems (26).

Pharmacia developed a version of liquid chromatograph with no stainless steel parts that come in contact with the mobile phase or the sample; this fast protein liquid chromatograph (FPLC) also has lower pressure limits than standard HPLC systems. Since that time, other manufacturers have developed "biocompatible" systems which replace all in-line stainless steel components with titanium or nonmetal materials. Similarly, biocompatible columns are composed of materials like PEEK rather than stainless steel. Table 3.4 lists the materials which compose wettable parts of standard HPLC systems compared to those in biocompatible versions (26).

Table 3.4 Standard vs. Biocompatible HPLC

	Standard HPLC System	**Biocompatible HPLC System**
Solvent Delivery		
Reservoir filters	316 SST	Polymer (Teflon®, PEEK)
Pump		
Liquid Ends	316 SST, Hastalloy C®	Titanium
Piston Seals	Polymers (frequently designated for specific solvent components like salt or organic)	
Pulse Dampers	316 SST, Fluorocarbon	Polymer (PEEK)
In-line Filters	316 SST	Polymer (PEEK)
Tubing	316 SST	Polymer (PEEK)
Sample Injection		
Injection Valve	316 SST	Polymer (PEEK, Tefzel® and Alumina-ceramic)
Sample Filter	316 SST	Polymer (PEEK, Teflon®)
Column		
Tubing	316 SST	PEEK
Frits	316 SST	Titanium, ceramic, Polymer (PEEK)

Portions from ref. 26.

Most HPLC systems and columns are constructed of stainless steel components and are very durable, giving excellent resolution and recoveries. Stainless steel columns have higher pressure ratings than biocompatible columns so that various diameters of columns and packings can be used effectively. The added expense, and sometimes, lower performance or pressure limitations of "biocompatible" HPLC systems need not be incurred unless the applications of interest have been seen to have problems with standard hardware. When corrosive mobile phases damage only specific parts of an HPLC, it is sometimes possible to modify only the relevant components (26).

B. Microbore and Capillary Columns

Reducing the diameter of an HPLC column has the significant advantages of greater sensitivity, compatibility with smaller samples, low solvent consumption, and usability with low flowrates. These factors make them ideal for very small samples and usable with specialized detectors like mass spectrometers (4). "Microbore" generally denotes columns which are 1mm ID but it may represent columns from 0.2 - 2.0mm ID. While the advantages of such columns are obvious, the systematic problems with running them can be equally great. To achieve the same linear velocities on two columns, the reduction in flowrate is related to the radius squared; therefore, if a 4.6mm ID column is run at 1ml/min, a corresponding 1mm ID column must be run at about 50μl/min. As discussed previously, such flowrates are not compatible with many standard HPLC systems, especially in a gradient mode. Generally, syringe pumps can handle these flowrates most successfully and the limitation of their running out of liquid during a run is not really pertinent. An additional problem with microbore columns is that the volume in the column is so low that extracolumn volumes can significantly add to bandspreading. Volumes in the injector, detector, and connecting tubing must be as nearly zero as possible. The effect of extracolumn volume on bandspreading is substantial for 1mm ID columns. These columns are also extremely sensitive to injection volume and injection techniques (17). One design for microbore HPLC actually eliminates the connectors and inserts the column directly into the injector and detector (4).

Capillary columns extend the limits of HPLC even further. These columns are commonly 300 - 500μm ID, although even lower diameters are available. The dimensions require that the operational flowrates must be further reduced from those of 1mm columns by at least a factor of four to nine. Special adaptations of equipment are needed to achieve adequate performance under these conditions. Eldex uses 2ml syringe pumps with 2 - 10μl static mixers or 15μl dynamic mixers to achieve high quality gradients at 1 - 20μl/min (27). Excellent resolution and reproducibility can be achieved on such systems, as seen in Fig. 3.24.

Figure 3.24 Resolution and Reproducibility of Capillary HPLC for an
α-chymotrypsinogen A tryptic digest

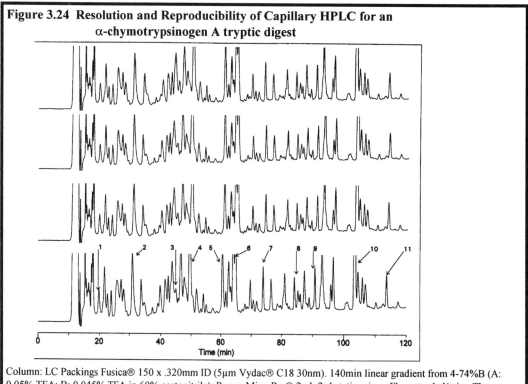

Column: LC Packings Fusica® 150 x .320mm ID (5μm Vydac® C18 30nm). 140min linear gradient from 4-74%B (A: 0.05% TFA; B: 0.045% TFA in 60% acetonitrile); Pump: MicroPro® 2ml; 2μl static mixer; Flowrate: 1μl/min. The numbered peaks were used to compare reproducibility of retention time. Reprinted from ref. 27 with permission of Eldex.

An alternative to using specially designed pumps is to implement a mixing and splitting device with standard HPLC pumps to deliver flowrates of 1 - 10μl/min. The design of a device from LC Packings and its flowrate characteristics are seen in Fig. 3.25 (28).

Figure 3.25 Microflow Processor

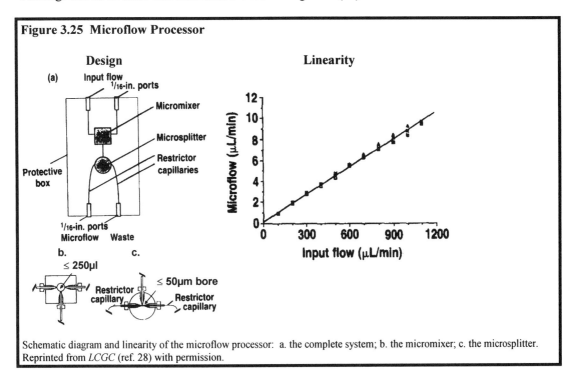

Schematic diagram and linearity of the microflow processor: a. the complete system; b. the micromixer; c. the microsplitter. Reprinted from *LCGC* (ref. 28) with permission.

Before using microbore or capillary applications, it is best to consult the HPLC manufacturer's specifications for flow and gradient accuracy, as well as precision, in the microliter range. For capillary LC, injection valves must have even smaller volumes than the 500nl used for microbore columns. Similarly, detectors must have very low volume cells, as discussed in the next chapter. Connecting tubing as small as 0.025" ID may be used for extremely sensitive connections, such as the inlet to a mass spectrometer.

VII. TROUBLESHOOTING

A detailed discussion of troubleshooting for HPLC systems will be presented in Chapter 15. The most important tools for troubleshooting HPLC problems are assessing the optimum performance characteristics, performing preventative maintenance, and keeping a log for each instrument which records this information. Instrument manufacturers usually recommend specific performance tests and preventative maintenance procedures. John Dolan is an excellent resource for information about the proper use, care, and maintenance of HPLC equipment through his monthly articles in *LC-GC* and his books (6, 8).

The most frequent cause of failure in any HPLC pump is the check valve. Indications of check valve problems are excessive pulsations, irreproducible retention times, or lower than normal operating pressure. Check valve problems can usually be attributed to contamination because most check valves cannot tolerate any particulate matter in the mobile phase. The primary source of such contamination is pump seal wear debris; therefore, it is very important to ensure that the sapphire piston is not worn or scratched, and that the seals are replaced periodically (consult manufacturer for seal service life and compatible mobile phases). Salt deposits are a significant source of system deterioration. Pumps must always be washed thoroughly with deionized water prior to shutting them off for an extended period of time (evening or weekend). A final source of contamination is the mobile phase itself. It is good practice to use a 10μm solvent filter on the inlet line; if using non-HPLC grade solvents, or if the solvent reservoir is open to air, then a 2μm filter should be used. Some common problems with HPLC systems and simple preventative measures are listed in Table 3.5.

Table 3.5 Common Problems with HPLC Systems

Problem	Preventative Measures
Dirty check valves	filter mobile phase
	flush salts from pumps
Plugged filters (inlet, in-line)	filter mobile phase
Deteriorated pump seal	flush salts from pumps
	flush behind seal
	use appropriate seal for mobile phase
	replace at timely intervals, e.g. every 6 months
Inadequate mixing	use dynamic mixer if solvents not readily miscible
Pump pulsing	use pulse damper
Air bubbles in check valve	degas mobile phase
Broken or scored piston	flush salts from pumps
Blockage in analytical column	use guard column
Leaks	use correct fittings/ferrules
Pump failure	lubricate and maintain as recommended
Injector leaking	flush salts (18)

VIII. EQUIPMENT VALIDATION

Meaningful and valid data can only be obtained if equipment is operating within its specifications. FDA regulated laboratories must have written standard operating procedures (SOPs) for maintenance and calibration procedures, as well as SOPs noting the steps required to check out equipment after maintenance or repair (29). These policies vary from laboratory to laboratory but will always include procedures relevant to installation, instrument qualification, performance qualification, and maintenance. Some HPLC systems have built-in software which assists in implementing procedures and produces documentation for performance qualification (30). System components should always be tested initially, evaluated periodically, and requalified after maintenance and repair. It is often easiest to carry out the performance tests which the equipment manufacturer recommends, but these can be replaced or supplemented by other procedures relevant to the day-to-day use. Some common tests for equipment evaluation are listed in Table 3.6.

Table 3.6 Equipment Validation Procedures

Solvent Delivery

flow accuracy	test flow rate volumetrically with back pressure
flow precision	reproducibity of retention time or peak area for 5 injections

Sampling Device

precision	reproducibility of peak areas for 5 consecutive injections
accuracy	linearity of peak area vs. concentration or volume

Absorbance Detector

linearity	absorbance vs. concentration
wavelength accuracy	measurement of wavelengths for maximum and minimum absorbance of a standard compound
noise	measure in mAU

Refractive Index Detector

linearity	RIU vs. concentration
accuracy	calculation from linearity data of dn/dv, a calibration constant relating the change in RIU to that in voltage
noise	measure in RIU
drift	measure change in RIU for at least one hour

Temperature Control

accuracy	measure temperature of eluent
reproducibility	measure temperature of eluent at 5 different times

Gradient Formation and Mixing

accuracy	evaluate linear gradient as outlined in Section IV.C

In the SOP, repair procedures to be implemented in the event of failure of any of the specifications should be outlined. It is often convenient to test the pump flow accuracy and sampling devices during system suitability studies. System suitability is usually specified for individual test methods, to include area or retention time reproducibility, tailing factor, capacity factor, resolution, selectivity, or signal-to-noise ratio. All equipment validation procedures and results should be reported in the appropriate equipment log book.

IX. REFERENCES

1. H. Poppe, "Column Liquid Chromatography" in *Chromatography, 5th Edition* (ed. E. Heftmann), Elsevier Science Publishers, Amsterdam (1992) A151.
2. E.L. Johnson and R. Stevenson, *Basic Liquid Chromatography*, Varian Associates, Palo Alto, 1978.
3. Eldex MicroPro® Pumping System, Eldex.
4. K. Potter and J. Tehrani, *LC-GC* 8 (1990) 862.
5. J.W. Dolan, *LC-GC* 9 (1991) 344.
6. L.R. Snyder and J.W. Dolan, *Getting Started in HPLC, User's Manual*, LC Resources, 1985.
7. *Series 200 LC Pumps*, Perkin-Elmer, 1994.
8. J.W. Dolan and L.R. Snyder, *Troubleshooting LC Systems*, Humana Press, Totowa, 1989.
9. *LC Star®*, Varian, 1995.
10. C. Gertz, *HPLC Tips and Tricks*, LDC Analytical, 1990.
11. J.W. Dolan, *LC-GC* 15 (1997) 110.
12. Form 1100A, Scientific Systems, Inc.
13. *The HP 1050 Series On-line Degasser*, Hewlett Packard, 1993.
14. C.T. Mant and R.S. Hodges, "HPLC Terminology: Practical and Theoretical" in *High-Performance Liquid Chromatography of Peptides and Proteins* (ed. C.T. Mant and R.S. Hodges), CRC Press, Boca Raton (1991) 69.
15. *Operating Instructions for Model 7125 Syringe Loading Sample Injector*, Rheodyne, 1984.
16. Technical Notes 5, Rheodyne, 1983.
17. S.R. Bakalyar, C. Phipps, B. Spruce, and K. Olsen, *J. Chromatogr. A*, 762 (1997) 167.
18. *Troubleshooting Guide for HPLC Injection Problems, 3rd Edition*, Rheodyne 1993.
19. *Series 200 Autosampler*, Perkin-Elmer, 1995.
20. *Catalog of Chromatography and Fluid Transfer Fittings*, Upchurch Scientific, 1996.
21. P. Upchurch, *HPLC Fittings*, Paul Upchurch, 1992.
22. *Catalog of Chromatography and Fluid Transfer Fittings*, Upchurch Scientific, 1995.
23. B. Ooms, *LC-GC* 14 (1996) 306.
24. P.-L. Zhu and J.W. Dolan, *LC-GC* 14 (1996) 944.
25. S. Hattangadi, *LC-GC* 7 (1989) 108.
26. M.V. Pickering, *LC-GC* 6 (1988) 800.
27. *Applications Notes*, Eldex, 1996.
28. J.P. Chervet, C.J. Meijvogel, M. Ursem, and J.P. Salzmann, *LC-GC* 10 (1992) 140.
29. J.P. Boehlert, *LC-GC* 15 (1997) 636.
30. *HP 1100 Series Modules and Systems for HPLC*, Hewlett-Packard, 1996.

CHAPTER 4

HPLC DETECTORS

I. INTRODUCTION

A. General

A key component in an HPLC system, as depicted in Fig. 3.1, is the detector. Any molecules which are separated by chromatography must be qualitatively and/or quantitatively assayed to yield meaningful results. The function of a detector is to produce an electrical signal proportional to the concentration of the sample; this is then routed to a recording device for display and storage. In addition to producing a response which is linear with concentration, the ideal detector should not contribute to band spreading nor be affected by temperature. An ideal detector should produce little or no signal for commonly used eluents, such as buffering salts, organic modifiers, detergents, or chaotropic agents, yet be responsive to all analytes across the full concentration range found in typical samples. In preparative applications, the detector must be non-destructive, permitting recovery of sample. Unfortunately, no detection system exhibits all of these ideal properties; therefore, the best mode of detection must be chosen in terms of selectivity and sensitivity within mobile phase and chromatographic limitations.

LC detectors can be distinguished by the physical or chemical principle(s) by which they operate. There may be more than one method of detection that permits adequate tracking of a particular chromatographic separation. Conversely, more than one detector can be connected in series to utilize the differences in selectivity each may provide. Some detectors are very selective and give a response which varies greatly according to the chemical nature of the sample - a response which is significantly greater than the response caused by the solvent, sample matrix, or other components in the mixture. Other detectors are more universal in their response - signaling the presence of nearly all compounds, including the solvent and sample matrix, but at the expense of a limited dynamic range of sensitivity. For example, ultraviolet-visible (UV-VIS), fluorescence, and radioactivity detectors are selective and refractive index (RI) detectors are universal.

This chapter will briefly describe the most commonly used detectors in HPLC, including the chemical or physical principles of their operation, and significant advantages and shortcomings of each. Table 4.1 lists these detectors. The next section presents an overview of the important general features of detectors.

Table 4.1 Detection Modes for HPLC

1. Ultraviolet/Visible (UV/VIS)
2. Refractive Index (RI)
3. Fluorescence
4. Electrochemical (EC)
    a. Amperometric
    b. Conductivity
5. Radioactivity
6. Light Scattering
    a. Multi-Angle Laser Light Scattering (MALLS)
    b. Evaporative Light Scattering (ELSD)
7. Mass Spectrometry (MS)

B. Performance

Detector performance must be carefully assessed because it directly affects the quantitation of analytes. Performance encompasses many factors including selectivity, sensitivity, noise, and baseline drift (1-3). The selectivity of the detector reflects its ability to detect a component over the background of eluent and matrix signal, including the noise; it is the basis for each mode of detection and will be discussed in detail in this chapter. In developing an HPLC method, it is prudent to evaluate the detector selectivity for the analyte against that for the sample matrix to make certain that the analyte can be detected and differentiated from any other components.

1. Sensitivity

The sensitivity of a detector is the response factor - the rate of signal to concentration in the detector (2). This is generally derived from the graph of detector response vs. sample concentration and is only a constant over the linear range of the detector. This relationship must be known before quantitative experiments are performed. One criterion used to compare detection methods is the limit of detection (LOD) (4). This is the lowest quantity which can be detected and can be in terms of mass (MLOD) or concentration (CLOD). The values are usually in moles or moles/liter.

The limits of detection are related to the signal to noise ratio (S/N). For qualitative work, the limit of detection or minimum detectable quantity (MDQ), is defined to be a signal of 2 - 3 times that of the noise. The minimum required for quantitative work may be higher, up to a tenfold signal to noise ratio, as shown in Fig. 4.1.

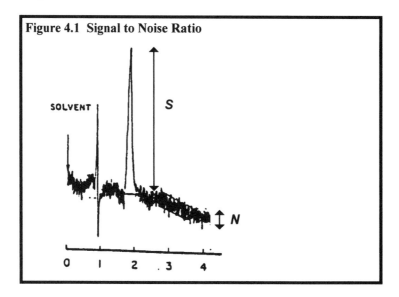

Figure 4.1 Signal to Noise Ratio

Quantitation for a sample is usually achieved by constructing a calibration curve as in Fig. 4.2.

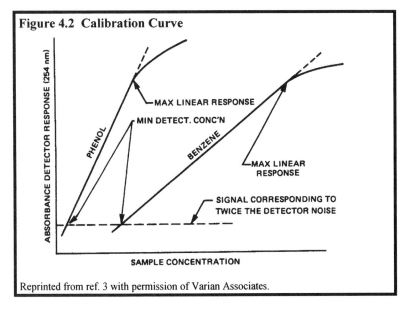

Figure 4.2 Calibration Curve

Reprinted from ref. 3 with permission of Varian Associates.

This example illustrates that each compound exhibits a unique linear detector response for a range of concentrations. The upper limit of linearity is not necessarily the same for each compound because it is related to molar absorptivity in this example. Linearity is critical for quantitative analysis where the concentration of an unknown is being interpolated. Ideally, the linearity should extend over a wide range that finally tapers off at high concentrations of analyte. Samples are often quantitated by comparing the response of the sample to internal or external standards, as discussed in Chapter 14.

2. Noise

Noise is the variation in output not directly attributed to the solute, but rather to factors such as electronics, temperature, pump pulsations, or air bubbles (1). High frequency noise, like that seen in Fig. 4.1, is often filtered out electronically by the detector or data reduction processor. A detector time constant controls the electronic filtering of high frequency noise. The time constant should be less than 10% of the peak width at half height to give an accurate depiction of the peak (3). The response time is the time required for the output signal to reach a new equilibrium value following a stepwise change in the composition in the flow cell. If the response time is slow, the signal being recorded will not be representative and will lag behind what is currently in the flow cell. If detector response is not faster than the rate of change of the solute concentration, distortion of the peaks occurs, as seen in Fig. 4.3.

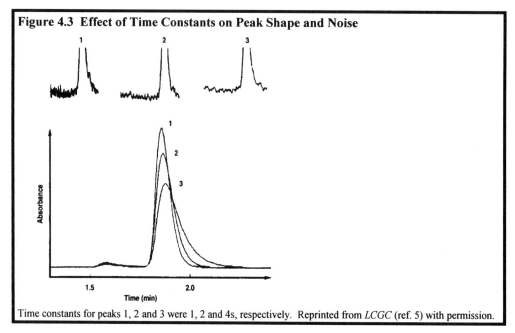

Figure 4.3 Effect of Time Constants on Peak Shape and Noise

Time constants for peaks 1, 2 and 3 were 1, 2 and 4s, respectively. Reprinted from *LCGC* (ref. 5) with permission.

As the time constant is increased, the noise in the baseline is reduced, but the peaks broaden - at the cost of reduced sensitivity. This may lead to an unacceptable flattening and tailing, particularly with very efficient columns - a very high price to pay for chromatograms with smoother baselines! High efficiency columns may require a time constant of 50ms or less to achieve optimum peak shape at high flowrates. If data reduction methods are used, the sampling rate must be similarly adjusted or peaks may be deformed.

Noise can sometimes be caused by contamination of the column or leaks in the system. Noise produced by air bubbles, pump pulsations, or mobile phase contaminants must be eliminated by specifically addressing the source of the problem. For example, a degasser or pulse dampener can be implemented.

3. Baseline Stability

Baseline drift often results from temperature variation or inadequate warming up of the detector. Certain detectors like the refractive index are very sensitive to changes in ambient temperature. Drift can also be produced by conditioning or removing contaminants from the column. These sources of potential baseline abnormalities can be minimized by giving proper attention to instrument configuration and maintenance. The manufacturer's recommendations for detector warm-up, sensitivity to temperature, maintenance, etc. should be heeded for optimum performance.

4. Flow Cell

The flow cell must possess sufficient volume to produce a measurable signal; however, if the volume of a flow cell and/or the connecting tubing is too large, peak broadening may result, as illustrated in Fig. 4.4.

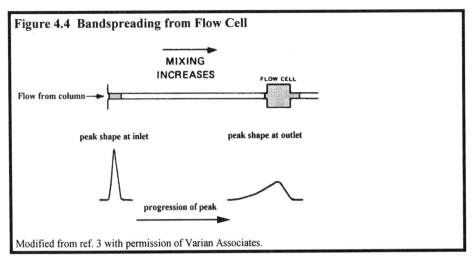

Figure 4.4 Bandspreading from Flow Cell

MIXING INCREASES

FLOW CELL

Flow from column

peak shape at inlet   peak shape at outlet

progression of peak

Modified from ref. 3 with permission of Varian Associates.

The flow cell and the connecting tubing can broaden or deform a peak which eluted from the column as a narrow band. Flow cells for each specific detection mode are designed to offer a compromise between maximum sensitivity and minimum dispersion. Fig. 4.5 illustrates a design used for ultraviolet absorption detectors which meets those requirements.

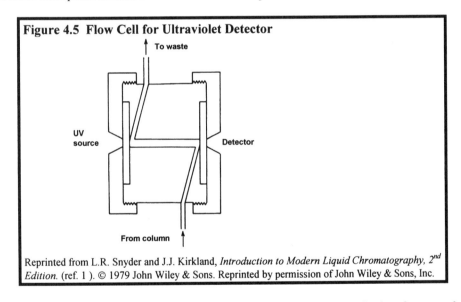

Figure 4.5 Flow Cell for Ultraviolet Detector

To waste

UV source   Detector

From column

Reprinted from L.R. Snyder and J.J. Kirkland, *Introduction to Modern Liquid Chromatography, 2nd Edition.* (ref. 1). © 1979 John Wiley & Sons. Reprinted by permission of John Wiley & Sons, Inc.

Frequently, flow cell volumes can be matched to column geometry and sample size; for example, small volume cells are usually used for microbore columns. Very small detector cell volumes significantly reduce bandspreading but also decrease sensitivity due to smaller path lengths (6). A compromise must often be made to adjust for the specific sample and column.

II. HPLC DETECTORS

A. Absorbance (UV/VIS) Detectors

1. General

The most commonly used detectors for HPLC are based on ultraviolet (UV) and visible (VIS) spectrophotometers. These are classified as selective detectors because only compounds with appropriate spectral characteristics are detected. UV/VIS detectors operate according to Beer's Law. The absorbance (*A*) is related to the fraction of light transmitted (1):

$$A = \varepsilon\, b\, c = \log\left(\frac{I_o}{I}\right) \qquad\qquad \text{Eq. 4.1}$$

In this equation, ε is the molar absorptivity or molar extinction coefficient, *b* is the cell path length in cm, *c* is the sample concentration in moles/liter, I_o is the reference intensity and *I* is the sample intensity. Because the path length and molar extinction coefficient for a particular compound in a given detector are constants, absorbance is only dependent on the concentration. The linear range of the absorption vs. concentration curve is 10^4 - 10^5 (2), falling off at absorbance values of 2 - 2.5. When the absorbance is higher than the linear range, the concentration cannot be accurately assessed, and the sample must be diluted.

Molar extinction coefficients are specific for each molecule, based on its structure and the wavelength used for detection. Published values for molar absorptivity (ε) are typically standardized to reflect the absorbance of a 1M solution (correcting for any solvent absorbance) at the wavelength which elicits the greatest absorbance (the absorbance maximum). Molar absorptivity varies greatly among molecules at any single wavelength and between wavelengths for any given molecule. This explains the large differences in sensitivity and selectivity for UV absorbance of different compounds. This last point is demonstrated in the chromatograms presented in Fig. 4.6 where a mixture of drugs is detected at both 250nm and 280nm. Only theophylline and carbamezapine have measurable absorbance at both wavelengths.

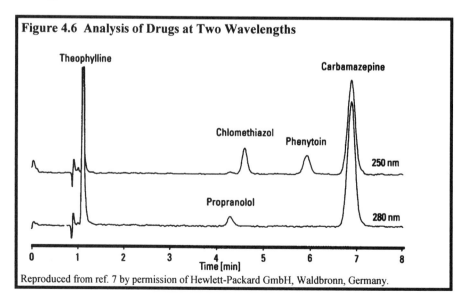

Figure 4.6 Analysis of Drugs at Two Wavelengths

Reproduced from ref. 7 by permission of Hewlett-Packard GmbH, Waldbronn, Germany.

The wavelengths generated by a UV/VIS detector are dependent on the specific detector lamps. Mercury, cadmium, and zinc discharge lamps have all been used for UV detection (4). Mercury lamps generate major spectral lines at 254 and 314nm, zinc lamps at 214nm, and cadmium at 229nm (8). Deuterium lamps, which have a continuous spectrum from 190 - 350nm, are usually employed for operation at low wavelengths (≤ 300nm), whereas for wavelengths in the visible range, a tungsten lamp is frequently used. Their spectral characteristics are shown in Fig. 4.7.

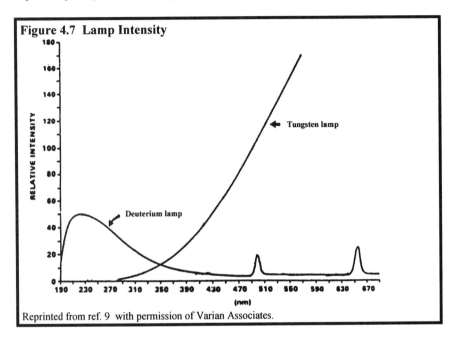

Figure 4.7 Lamp Intensity

Reprinted from ref. 9 with permission of Varian Associates.

Detector wavelengths can be calibrated with solutions of known spectral properties. The signal to noise ratio of the system and the linearity of response for individual components should be measured in each analysis being developed.

2. Mobile Phase

The minimum practical working wavelength of a UV detector is the wavelength at which the mobile phase absorbs. Below this limit, termed the UV cutoff of the solvent, background is too high to allow compensation by the detector electronics. The UV cutoff is the wavelength at which the absorbance of the solvent in a 1cm cell is equal to 1 AU, using water in the reference cell. Water is transparent to UV light, but the addition of salts and ion pairing reagents can raise the UV cutoff wavelength (5, 10). Table 4.2 lists the absorbance at 210nm, 254nm, and 280nm for selected solvents used in biological separations. The UV cutoff or the lowest useable wavelength, and its related absorbance (often less than 1 AU), are also specified.

Table 4.2 Absorbance of Mobile Phase Components

	UV Cutoff		Absorbance		
	Wavelength	Absorbance	210nm	250nm	280nm
Solvents					
Acetonitrile	190nm		0.02	none	
Methanol	205nm	1.00	0.53	0.02	none
degassed	205nm	0.76	0.35	< 0.01	none
Isopropanol	205nm	0.68	0.34	0.03	0.02
Acids and Bases					
Acetic acid, 1%	230nm	0.87	2.61	0.01	none
Trifluoroacetic acid					
0.1% in water	205nm	0.78	0.54	< 0.01	none
0.1% in acetonitrile	< 200nm	0.28	0.37	0.04	< 0.01
Triethylamine, 1%	240nm	0.50	2.50	0.12	< 0.01
Buffers and Salts					
Potassium phosphate					
monobasic, 10mM	< 200nm	0.03	none		
dibasic, 10mM	< 200nm	0.53	0.05	none	
Sodium acetate, 10mM	205nm	0.96	0.52	none	
Sodium chloride, 1M	210nm	0.40	0.40	none	
Sodium phosphate, 100mM, pH 6.8	205nm	0.75	0.19	0.01	< .01
Tris-hydrochloric acid, 20mM					
pH 7	205nm	0.77	0.28	none	
pH 8	215nm	0.43	1.11	none	
Detergents					
CHAPS, 0.1%	215nm	0.80	1.48	0.02	0.01
SDS, 0.1%	< 200nm	0.02	< 0.01	none	

Reprinted from *LCGC* (ref. 5) with permission.

As seen in Table 4.2, salts and additives can increase the absorbance of water or another solvent. The absorbance of salts can also vary with pH and may be significant at low wavelengths (5, 11), as seen in Fig. 4.8. The absorbance for Tris at pH of 8 or higher seriously limits its use at low wavelengths.

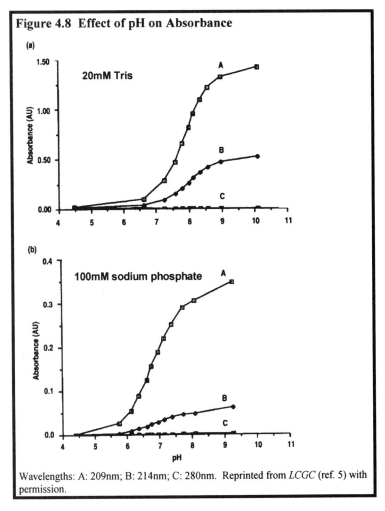

Figure 4.8 Effect of pH on Absorbance

(a)

20mM Tris

(b)

100mM sodium phosphate

Wavelengths: A: 209nm; B: 214nm; C: 280nm. Reprinted from *LCGC* (ref. 5) with permission.

A spectral shift for additives or buffers can occur at distinct organic modifier concentrations. For example, the absorbance of trifluoroacetic acid (TFA) shifts to a lower wavelength at high organic modifier concentrations, resulting in the characteristic hump in acetonitrile gradients on reversed phase columns. This is sometimes eliminated by using a lower concentration of TFA in the organic solvent than in the aqueous. It is extremely important to use very pure solvents for the mobile phase because impurities may produce background absorbance or baseline drift (10). Most solvents and salts normally used for HPLC can be obtained in HPLC grade.

3. Instrument Designs

a. Fixed Single Wavelength Detectors

Fixed wavelength detectors operate at a single wavelength determined by a filter and/or the light source, commonly the 254nm line from a mercury lamp (2). These detectors are inexpensive and sensitive, having minimal noise due to low scattering of light. If the operational wavelength of the detector is not optimal for a particular analyte, however, poor sensitivity will result. Fixed wavelength detectors sometimes employ filters and/or interchangeable lamps to produce different wavelengths from the source illumination for particular applications. Multiple wavelength detectors offer detection at two or three specific wavelengths through the use of different lamps or filters.

b. *Variable Wavelength Detectors*

In variable wavelength detectors, a specific wavelength is isolated by the use of a monochromator incorporating either a grating, a prism, or multiple filters. Light is directed through a series of filters, slits, focusing devices, and the monochromator. Frequently the beam is split so that it hits both a reference and a sample diode. Fig. 4.9 illustrates the design of a variable wavelength detector.

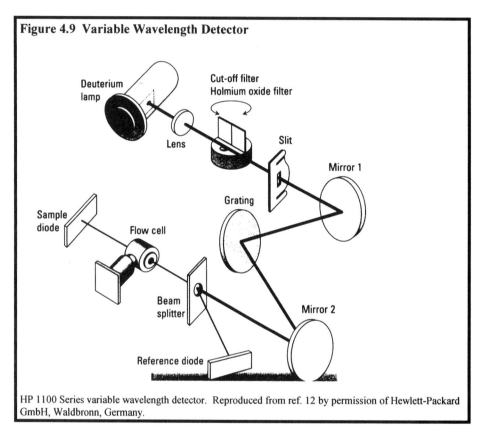

Figure 4.9 Variable Wavelength Detector

HP 1100 Series variable wavelength detector. Reproduced from ref. 12 by permission of Hewlett-Packard GmbH, Waldbronn, Germany.

The variable wavelength detector is the most versatile UV/VIS design, offering selectable wavelengths to maximize selectivity and response for different components in a sample. They use a deuterium lamp for the UV wavelengths and sometimes a tungsten lamp for the visible. These detectors can go from the far UV, where olefins, phospholipids, and carbohydrates can be detected, to the visible range. The isolation of wavelengths in the monochromator can be achieved by either a grating or a prism, but the former is currently most popular. As with all UV/VIS detectors, response is dependent on the wavelength and the sample. To increase sensitivity, some designs have automated wavelength selection which can be changed during a run, and auto zero to compensate for a changing baseline.

c. *Scanning Detectors*

Scanning detectors enable spectral information to be collected as the peak is migrating through the flow cell. In a fast-scanning "forward-optics" detector, the monochromator is rotated through an angle to vary the wavelength across a defined spectral region within a short time span. This type of detector allows acquisition of spectra up to 10 - 20 times per second, depending on the sampling rate and the spectral range covered, but it may also require stopped flow during scanning (2). Currently, photodiode array detectors are preferred for scanning in the UV.

d. *Photodiode Array Detectors*

The most popular type of scanning detector is the photodiode array (PDA) detector (13). In this design, which is deemed "reverse optics", full-spectrum light from the light source is used to illuminate the flow cell, and light transmitted from the flow cell is directed to a grating called a polychromator which disperses the light onto an array of photodiodes, each of which samples a narrow spectral range. An example of a PDA detector is shown in Fig. 4.10.

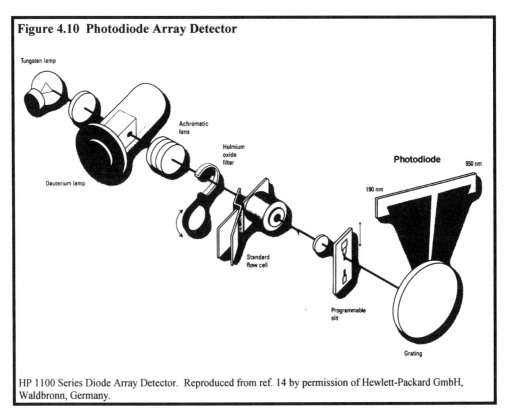

Figure 4.10 Photodiode Array Detector

HP 1100 Series Diode Array Detector. Reproduced from ref. 14 by permission of Hewlett-Packard GmbH, Waldbronn, Germany.

The advantages of this optical approach are high rates of spectral acquisition and lack of any moving parts, because both the grating and the photodiode array are fixed. Spectra acquired during the movement of the analyte through the detector flow cell provide information about the identity and purity across the peak. For example, tailing impurities will cause variations in the spectral profile. Fast-scanning and PDA detectors are usually equipped with sophisticated spectral analysis software, which enables comparison of an unknown analyte spectrum with a library of known spectra and allows quantitative assessment of peak purity using algorithms. For biopolymers, such as proteins, the spectral data may also provide information about protein folding (15).

4. Biochemical Applications

The utility of UV detectors can be illustrated with purine and pyrimidine bases. Five of these bases and their absorption spectra are shown in Fig. 4.11.

Figure 4.11 UV Absorption of Nucleosides

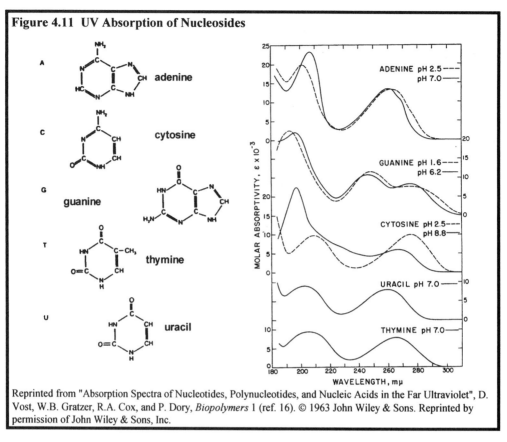

Reprinted from "Absorption Spectra of Nucleotides, Polynucleotides, and Nucleic Acids in the Far Ultraviolet", D. Vost, W.B. Gratzer, R.A. Cox, and P. Dory, *Biopolymers* 1 (ref. 16). © 1963 John Wiley & Sons. Reprinted by permission of John Wiley & Sons, Inc.

Not only do the individual absorption maxima of the bases differ, but so do the spectra in the lower UV. Because of these variances, one means of nucleoside or nucleotide identification is an absorbance ratio, usually at 280nm and 254nm (11). Nucleosides, nucleotides and their bases can also exhibit specific spectral changes with pH, which can aid in their identification (11).

      Proteins and peptides are usually detected at 280nm or 254nm based on the presence of the aromatic amino acids, tryptophan (Trp), tyrosine (Tyr), and phenylalanine (Phe) (17). As seen in Fig. 4.12A and B, a wavelength of 280nm will yield maximum absorption for tryptophan and tyrosine but will not detect phenylalanine. Proteins or peptides without these amino acids are transparent at wavelengths above about 240nm, as shown in Fig. 4.12C.

Figure 4.12 Absorption Spectra of Aromatic Amino Acids and Peptides
A. Aromatic Amino Acids   **B. Peptide with Tyrosine**   **C. No Aromatic Residues**

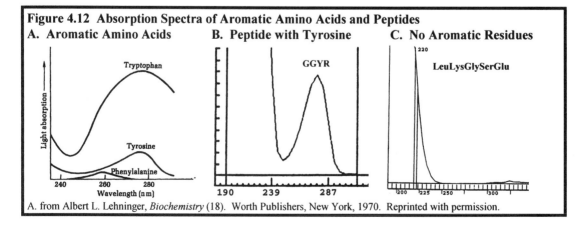

A. from Albert L. Lehninger, *Biochemistry* (18). Worth Publishers, New York, 1970. Reprinted with permission.

The chromatogram in Fig. 4.13 illustrates that detection at two wavelengths, 214nm (where the peptide bond can be detected) and 280nm, yields the total peptide profile and also identifies aromatic peptides (19).

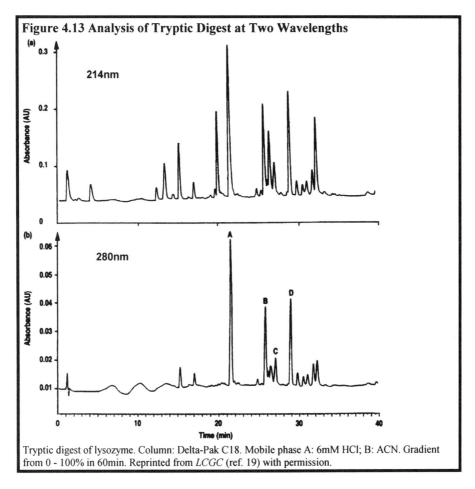

Figure 4.13 Analysis of Tryptic Digest at Two Wavelengths

Tryptic digest of lysozyme. Column: Delta-Pak C18. Mobile phase A: 6mM HCl; B: ACN. Gradient from 0 - 100% in 60min. Reprinted from *LCGC* (ref. 19) with permission.

The UV/VIS detector is particularly selective for metal-containing molecules which absorb in the visible range (17). One notable example is the routine quantitation of hemoglobin variants in hemolyzed whole blood (20). By using a wavelength of 415nm, the hemoglobins can be distinguished and accurately quantitated despite the presence of a plethora of other molecules in the sample.

5. Advantages and Disadvantages

In summary, some of the advantages and disadvantages of UV/VIS absorbance detection are listed in Table 4.3.

Table 4.3 UV/VIS Absorbance Detection

Advantages
- High sensitivity (10^{-10} to 10^{-11}g).
- High selectivity
- Nearly universal at low wavelengths (≤ 200 nm)
- Low background with many HPLC solvents, allowing gradient elution without excessive background drift.
- Nondestructive to samples.
- Easy operation.

Disadvantages
- Analyte must have absorbance in the UV or visible range.
- Cannot operate at wavelengths below the UV cutoff of the solvent.
- At a given wavelength, response varies between molecules based on their absorptivity.

B. Refractive Index Detectors

1. Operation

Refractive index (RI) detectors compare the refractive index (light bending) of the pure mobile phase with that containing the analyte. The angle of refraction is a function of each constituent of a solution and can be either positive or negative. The total refractive index is the sum of all of the indexes of the components in a solution, irrespective of whether they are monomers or in aggregates. Refractive index detectors cannot compensate for baseline changes caused by varying mobile phase composition, and thus they can only be used for isocratic separations.

There are three primary designs for RI detectors - deflection, Fresnel, and interference (1, 3, 4). The deflection type measures the diffraction of a light beam passing through a prism shaped cell (2), as seen in Fig. 4.14 (21).

Figure 4.14 Deflection RI Detector

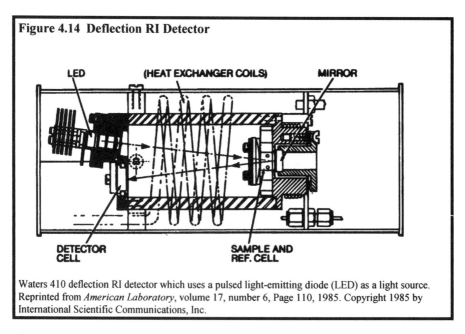

LED   (HEAT EXCHANGER COILS)   MIRROR

DETECTOR CELL   SAMPLE AND REF. CELL

Waters 410 deflection RI detector which uses a pulsed light-emitting diode (LED) as a light source. Reprinted from *American Laboratory*, volume 17, number 6, Page 110, 1985. Copyright 1985 by International Scientific Communications, Inc.

The operation of deflection instruments is based on Snell's Law of refraction (3). As diagrammed in Fig. 4.15, the angles of the incident and refracted beams are related by :

$$n \sin \theta = n' \sin \theta'$$ Eq. 4.2

where n is the refractive index of the reference, θ is the angle of incidence, n' is the refractive index of the sample and θ' is the angle of refraction.

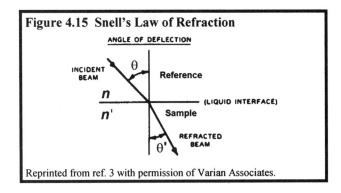

Figure 4.15 Snell's Law of Refraction

Reprinted from ref. 3 with permission of Varian Associates.

Because the refractive indexes of the components in the sample cell are additive, that due to the solvent in the sample cell, can be canceled by the use of a reference cell. Both the Waters Model 410 and the Hewlett Packard 1047A RI detectors employ the deflection design. Deflection RI detectors have a wide range of linearity of 10^5 (22, 23).

The second type of RI detector is the Fresnel reflection design, which measures the intensity of a reflected beam at the interface of a liquid and a glass block (2). Two beams of light pass through a glass prism and then into one of two photo cells, a reference cell or a sample cell. The detector measures the difference in their reflected light. This kind of detector comes equipped with two distinct prisms for different refractive index ranges; however, most analyses use the prism for the low refractive index range (1). The cells for this design can be very small, but the linear range is lower than the deflection type (1).

The third RI detector is the interferometric design, which uses beam splitters to divide the beam before the sample and reference cells and to recombine them before detection (1). If the contents of the sample cell are different than those of the reference, a change in optical path length is produced which can be either constructive or destructive. The interference design is the most sensitive RI detector, with a CLOD of about 10^{-6}M (4). Fig. 4.16 shows an example of an interference RI detector.

Figure 4.16 Interference RI Detector

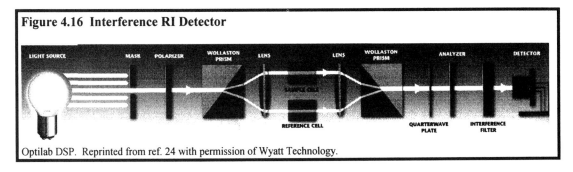

Optilab DSP. Reprinted from ref. 24 with permission of Wyatt Technology.

2. Applications

Refractive index detection is especially useful for size exclusion (SEC) methods because they are isocratic. With this type of detection, polymers and proteins can both be analyzed in a mixture, whereas with UV, generally only proteins are detected. An example of a relevant application is the analysis of proteins conjugated with polyethylene glycol (PEG). RI detection is frequently used for carbohydrates because they do not absorb well in the UV. Although RI detection is considered a low sensitivity method, it is still possible to detect small quantities, as seen for the quantitation of 20ng of saccharose in Fig. 4.17. In such sensitive assays, temperature control is compulsory.

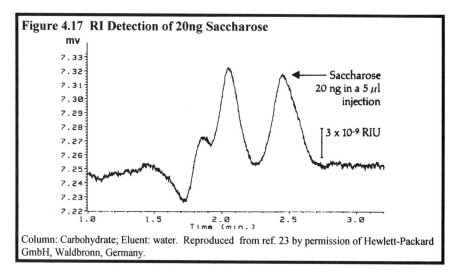

Figure 4.17 RI Detection of 20ng Saccharose

Column: Carbohydrate; Eluent: water. Reproduced from ref. 23 by permission of Hewlett-Packard GmbH, Waldbronn, Germany.

3. Advantages and Disadvantages

RI detectors are nearly universal in that they potentially respond to everything except compounds which possess a refractive index close to that of the solvent. Refractive indices are the same for similar biomolecules, such as proteins or nucleic acids; therefore, a method is not likely to miss sample components. Small volume flow cells, approximately 5 - 10μl, are used to minimize band spreading. Because RI detection is nondestructive and has a moderate detection limit of about 10^{-6} g, it is useful for preparative scale chromatography.

The refractive indexes of all compounds, including solvents, change with temperature because of its effect on density; this causes most of the drift and noise associated with RI detection. Most RI detectors are equipped with temperature control units, as shown in Fig. 4.14, which must be used to attain the lower detection limits. Back pressure changes the density and thus, also effects refractive index measurements. RI flow cells generally have low tolerance to pressure compared to cells from other kinds of detectors. These limitations, plus the lower sensitivity compared to UV detectors and incompatibility with gradients or any changing mobile phase composition, render RI detection less suitable for routine analytical HPLC than some other more robust methods.

Table 4.4 Operating Tips for RI Detectors
1. Never use a back pressure regulator because pressure changes affect RI detection.
a. Use a small piece of large ID Teflon tubing as a waste line to prevent pressure build up.
b. Place the RI detector last in a series of detectors to keep pressure constant.
2. Always pre-mix and then degas solvents to prevent changing composition.
3. Control the temperature of the cell to prevent expansion and contraction of solvent.
a. Let RI detectors warm up 2 - 8 hours before using.
b. Use a temperature control system to prevent changing detector response.
4. Keep RI cells clean, as specified in manual, because response is sensitive to impurities.
5. Use pulse-free pumps or pulse dampers.

C. Fluorescence Detectors

1. General

Some molecules naturally fluoresce. They are capable of absorbing energy at a specific wavelength called the excitation wavelength (λ_{ex}), rising to an excited state, and then returning to ground state with the concomitant emission of light at a longer wavelength termed the emission wavelength (λ_{em}). Fluorescence detectors operate by exciting the sample at λ_{ex} and then detecting the fluorescence emitted by the sample at λ_{em}.

Some of the best examples of naturally fluorescing compounds are benzene and its derivatives. Table 4.5 lists some commonly assayed molecules. Fluorescence can vary with ring substitution, as well as with pH and the composition of the solvent.

Table 4.5 Molecules With Native Fluorescence	
Aromatic amino acids (Trp, Tyr)	Polynuclear Aromatic Hydrocarbons
Aflatoxins	LSD
Coumarins	Phenols
Catecholamines	Porphyrins
Estrogens	Quinolines
Flavins	Riboflavins
Indoles	Vitamin A
Indole Alkaloids	

As with UV absorption, the presence of tryptophan and tyrosine in proteins permits their ready detection by fluorescence; Fig. 4.18 shows the excitation and emission spectra of tryptophan.

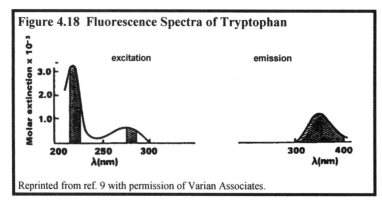

Figure 4.18 Fluorescence Spectra of Tryptophan

Reprinted from ref. 9 with permission of Varian Associates.

Excitation of tryptophan can occur at either 220nm or 280nm with emission about 350nm. Peptides without tryptophan or tyrosine require pre- or post-column derivatization, as discussed later in this chapter, before they can be detected by fluorescence. For example, primary amines, which normally do not fluoresce, can be derivatized with reagents like dansyl chloride, fluorescamine, and o-phthalaldehyde to make them amenable to fluorescence detection (17). Nucleic acids require the addition of a substance such as ethidium bromide, which fluoresces strongly when intercalated into the helix, before they can be detected by this method.

Fluorescence offers high selectivity due to the use of two wavelengths. There are few, if any, interfering peaks because the detection wavelength is different than the excitation wavelength and the combination is quite selective for any particular compound. Sensitivity is also high (10^{-12}g) because the change in light level is at a wavelength with negligible background light. This contrasts with absorption, for example, where the detector is forced to distinguish small differences in substantial levels of light. Fluorescence detection is not affected by changes in refractive index and is compatible with gradient elution and strongly UV-absorbing solvents.

2. Detector Design

Fluorescence detectors employ a series of filters and/or monochromators and lenses to focus the excitation light and, ultimately, collect the emission beam. They can either measure at distinct wavelengths or offer scanning capabilities. Fig. 4.19 illustrates the design of a fluorescence detector.

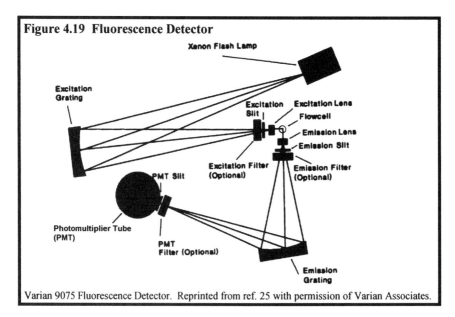

Figure 4.19 Fluorescence Detector

Varian 9075 Fluorescence Detector. Reprinted from ref. 25 with permission of Varian Associates.

Filters may be used in place of a monochromator to select the excitation and emission wavelengths. Glass band filters, which transmit across broad spectral ranges, are commonly used to select the excitation wavelength in the far UV. Glass cut-off filters, which filter out all radiation below a specific cut-off wavelength and pass up to 85% of the radiation above it, are used to collect fluorescence emission. Interference filters are excellent for both excitation and emission because they offer a relatively narrow transmission band with little leakage outside the bandpass, especially in the red end of the spectrum.

The lamp intensity of the excitation light source must be adequate at the required wavelength; a xenon lamp is usually used. The emission filter or monochromator setting is chosen to fall in a region relatively free of both scattering and background fluorescence. Although the fluorescence of both the solvent and the solute are usually measured to determine optimum emission conditions, performance is primarily based on the fluorescence of the analyte.

3. Applications

The very high selectivity and sensitivity of fluorescence detection makes it an ideal technique for determination of contaminants in food or environmental samples. Fig. 4.20 shows the analysis of aflatoxins at a level under 1 ppb after postcolumn addition of iodine.

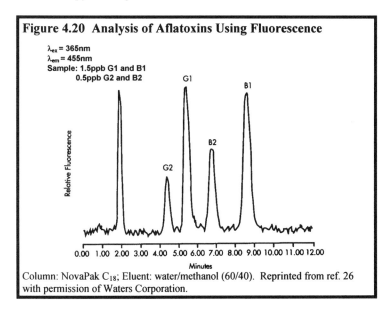

Figure 4.20 Analysis of Aflatoxins Using Fluorescence

λ_{ex} = 365nm
λ_{em} = 455nm
Sample: 1.5ppb G1 and B1
0.5ppb G2 and B2

Column: NovaPak C_{18}; Eluent: water/methanol (60/40). Reprinted from ref. 26 with permission of Waters Corporation.

Some fluorescence detectors offer the possibility of changing the excitation and emission wavelengths at different points in the analysis to optimize selectivity and sensitivity. Scanning detectors are used either to scan standards to determine optimum wavelengths or to identify peaks by their fluorescent properties (26).

4. Advantages and Disadvantages

Fluorescence detection has excellent selectivity and very low limits of detection (< 1pg) for appropriate molecules. There is virtually no interference from the mobile phase and most other sample components. The main disadvantage of this mode of detection is that the distinct wavelengths necessary for excitation and emission require optimization for each compound (27). Application of fluorescence to nonfluorescent compounds requires derivatization which can be both time-consuming and complicated. Additionally, it is often necessary to warm-up detectors overnight for optimum performance (27).

D. Electrochemical Detectors

Electrochemical detectors measure electrical data from solutes using two electrodes. In amperometric and coulometric detectors, the second electrode is downstream from the first and a current consisting of the electrons released from a redox reaction is measured. In conductivity detectors, the current conducted by the effluent, including mobile phase and sample, is measured as the flow goes through a cell containing two electrodes.

1. Instrument Designs

a. Amperometric Detectors

The term electrochemical detection is usually applied to amperometric detection, where the sample undergoes an electrolysis reaction. After a constant potential is applied, the resultant current is measured over time (1-3). Compounds that are either readily oxidized or reduced, such as aromatic amines and phenolic molecules, are often detected by this method; the signal varies proportionally to the concentration of analytes which were reacted. This is an extremely selective method because only molecules which can be oxidized or reduced within the working potential of the electrode can be detected. The negative potential limits are usually -0.2 to -1.0V and the positive are 0.5 to 2.0V (2). Reversed phase chromatography is most frequently used with amperometric detectors because they require a polar mobile phase which possesses electrical conductivity. The mobile phase must be electrochemically inert and capable of supporting a dissociated electrolyte (0.01 - 0.1M) (28). This technique offers high sensitivity (10^{-11}g), and selectivity which is controlled by adjusting the potential of the flow cell.

Amperometric detectors operate by detecting electrochemical reactions that take place in thin-layer cells, as illustrated in Fig. 4.21. A generator electrode causes an electrochemical reaction whose products are detected downstream. Under normal flow conditions, only a fraction (0.2 - 10%) of the molecules contact the electrode and undergo the redox reaction (2).

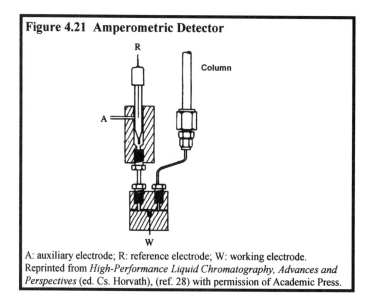

Figure 4.21 Amperometric Detector

A: auxiliary electrode; R: reference electrode; W: working electrode.
Reprinted from *High-Performance Liquid Chromatography, Advances and Perspectives* (ed. Cs. Horvath), (ref. 28) with permission of Academic Press.

A modification of this electrochemical technique is pulsed amperometric detection (PAD). In PAD, a three step potential wave form is applied to the sample in the flow cell to produce amperometric detection with alternating cathodic and anodic polarization. The current is only measured at a specific voltage (15). This approach minimizes fouling of the electrode surfaces. PAD is quite specific for the detection of sugars, their alcohol and amino derivatives, and alkali-stable polysaccharides; sensitivity of 1-10 pmol can be readily achieved (29).

b. *Coulometric Detectors*

Coulometric detection is a modification of amperometric detection in which total conversion of all molecules which can undergo reaction is achieved (2). This results from the use of a large porous graphite working electrode or low flowrates, and allows stoichiometric calculation of concentration (29). Detection at levels less than 5pg is possible, but the method produces more noise than amperometric detection; therefore, the limits of detection are actually similar (15).

c. *Conductivity Detectors*

Conductivity detectors are electrochemical detectors used to detect ions. The principle of operation is simple - solutions rich in ions are better conductors of current than solutions with few ions. This type of detector applies a constant potential at the flow cell and measures current over time. Conductivity detectors have seen their greatest application in the analysis of organic and inorganic ions eluting from ion-exchange columns. This somewhat paradoxical use stems from the fact that modern conductivity detectors can detect 1 part per million (ppm) changes in ionic strength in the presence of high salt. Sensitivity can be improved by the use of stripper columns or hollow fiber devices between the analytical column and the detector to reduce background signal from the eluent (30). These detectors are quite sensitive (10^{-11}g), but lack selectivity.

2. Applications

Electrochemical detection has been applied to certain difficult applications, such as metabolite analysis, with success unsurpassed by most other methods of detection. Catecholamines from very small biological samples like brain perfusates or atrial tissue can be quantitated by this means. Fig. 4.22 shows the analysis of standard catecholamines and components of rodent atrial tissue with reductive coulometric detection. In this study, 0.1 - 50mg of tissue were analyzed with detection of components present in amounts less than 5pg per injection (29).

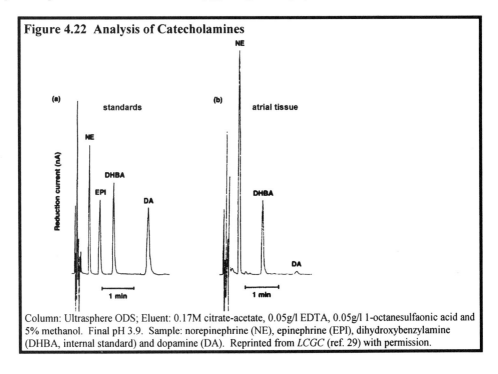

Figure 4.22 Analysis of Catecholamines

Column: Ultrasphere ODS; Eluent: 0.17M citrate-acetate, 0.05g/l EDTA, 0.05g/l 1-octanesulfaonic acid and 5% methanol. Final pH 3.9. Sample: norepinephrine (NE), epinephrine (EPI), dihydroxybenzylamine (DHBA, internal standard) and dopamine (DA). Reprinted from *LCGC* (ref. 29) with permission.

Photochemical derivatization is a method which increases the selectivity of electrochemical detectors by postcolumn bombardment of the column effluent with UV radiation, followed by detection of the changes in conductivity caused by the ionic products (31). In this procedure, the column effluent is split, with half serving as reference and the other subjected to the UV radiation. Five to 500pg of nitrosamines, sulfonamides, and halogenated organics have been detected by this technique - providing unique selectivity for these important classes of compounds.

3. Advantages and Disadvantages

Amperometric and coulometric detection are extremely selective with low limits of detection, and thus they can be used to detect trace components of complex biological mixtures. Conductometric detection is unsurpassed for detection of ions.

Generally, amperometric detection is not amenable to gradient elution because changes in ionic concentration cause drift. Polar or conducting solvents must be used and the analyte must be electroactive. In some cases, it is a destructive method. The working electrode must be polished daily because electrodes are easily fouled. Overnight warm-up is required to obtain maximum performance (32). Electrochemical detectors seem to require more skill and care than instruments based on absorbance or fluorescence.

E. Radioactivity Detectors

1. General

Radioactivity detectors count the radioactivity of the solution as it passes through a flow cell and are most commonly used to detect ^{14}C, ^{32}P and ^{3}H (2). This method is very selective because only appropriately labeled molecules are detected and all other substances, including the mobile phase, are transparent.

2. Instrument Design

Modern radioactivity detectors offer both the versatility of interchangeable flow cells for efficient detection of different kinds of radiation and the capability of minimizing radioactive waste. Fig. 4.23 illustrates such a design.

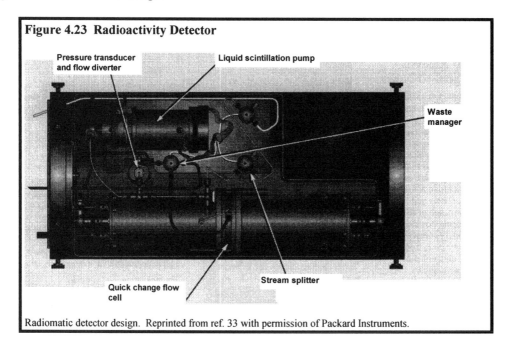

Figure 4.23 Radioactivity Detector

Pressure transducer and flow diverter

Liquid scintillation pump

Waste manager

Quick change flow cell

Stream splitter

Radiomatic detector design. Reprinted from ref. 33 with permission of Packard Instruments.

Because efficient detection is dependent on the specific radioisotopes and appropriate scintillating materials, this detector is available with a variety of interchangeable cells (34). In homogeneous detection methods, a scintillation cocktail is mixed with the effluent before passage through the flow cell. This provides high sensitivity and efficiency and the lowest background. Flow of the scintillation cocktail can be adjusted during the run to prevent waste. The column effluent stream can also be split, so that only a fraction is contaminated with the cocktail.

Alternatively, heterogeneous detection methods can be implemented. In these, a solid scintillator is packed into the flow cell and the eluate passes through it (34). To work effectively, the scintillator must be inert to the analytes and efficiently packed to minimize bandspreading. Various scintillation media are available. The relative efficiency of homogeneous and heterogeneous detection can be seen in Table 4.6. It is obvious that tritium requires the liquid homogeneous method for efficient detection, whereas the distinction for ^{14}C is not as great. Another heterogeneous option is used for the detection of gamma emitters. In this case, the flow cell contains scintillating windows to count the radiation (34).

Table 4.6 Radioactivity Detection Efficiency

	Homogeneous Liquid		Heterogeneous Solid	
	^{3}H	^{14}C	^{3}H	^{14}C
Background	4 CPM	6 CPM	8 CPM	15 CPM
Efficiency	52%	93%	2.90%	70%
MDA	120 DPM	60 DPM	2000 DPM	200 DPM

Data from ref. 34 with permission of Packard Instruments.

3. Applications

Radioactivity is frequently used to monitor tagged metabolites in pharmacokinetic studies. Proteins can either be chemically modified with radioactive labels or physically associated with tagged molecules. Estrogen and progestin receptor proteins have been quantitated in cancer tissue by the latter method. After incubation with radioactive substrates (steroids), the protein isoform/steroid complexes are separated by ion-exchange (35) or hydrophobic interaction chromatography (36). Fig. 4.24 illustrates the assay of estrogen receptors using a radioactivity detector.

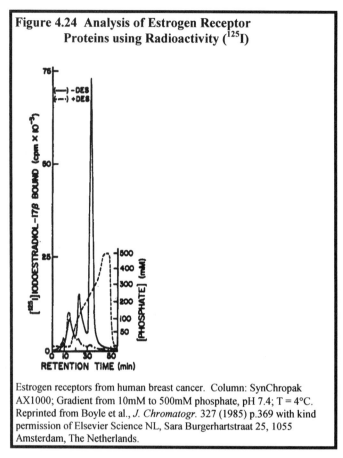

Figure 4.24 Analysis of Estrogen Receptor Proteins using Radioactivity (^{125}I)

Estrogen receptors from human breast cancer. Column: SynChropak AX1000; Gradient from 10mM to 500mM phosphate, pH 7.4; T = 4°C. Reprinted from Boyle et al., *J. Chromatogr.* 327 (1985) p.369 with kind permission of Elsevier Science NL, Sara Burgerhartstraat 25, 1055 Amsterdam, The Netherlands.

4. Advantages and Disadvantages

Radioactivity is a very selective technique without normal background interference; therefore, it can detect trace components, if they are properly labeled. Because detection of radioactivity is time based, detection limits depend on residence time in the flow cell which may compromise peak efficiency or analysis time. A serious disadvantage of radioactivity detection is that every component in the separation system which contacts the sample may have inherent radioactive contamination. Generally, columns used for radioactive samples are dedicated members of the system. Waste disposal of the mobile phase, column, etc. may be a problem if they possess radioactivity. Detectors which isolate radioactive waste can greatly reduce the disposal problems. If separated components are collected, they may pose interference in downstream testing, especially if they are used in bioassays using radiomarkers. For samples with high specific activity, autodegradation by internal radiation can be an additional problem.

F. Light Scattering Detectors

1. General

Light scattering detectors are based on the interaction of light with matter. When light hits matter, it induces a temporary dipole in the molecule which oscillates at the frequency of the incident light. When the molecule returns to its ground state, it emits the light in different directions (38, 39). If the frequencies of the emitted and incident light are the same, this is called elastic scattering - the primary basis of light scattering detectors. The Rayleigh factor is the measure of scattering based on the relative light intensities. The amount of Rayleigh factor in excess of that of the mobile phase can be related to the molecular weight, given the concentration, refractive index, and wavelength of the incident light. For this reason, light scattering detectors must be run in series with a refractive index or UV detector to determine the concentration, after which compilation of the data must be undertaken with appropriate software.

The relationship of molecular weight to light scattering is complex (38, 39). The Rayleigh factor (R_θ) is the intensity of scattered light (I_θ) of a given material per unit volume at defined angle, θ, and total light intensity (I_o) using vertically polarized light:

$$R_\theta = \frac{I_\theta r^2}{I_o V} \qquad\qquad \text{Eq. 4.3}$$

V is the volume and r is the radius of gyration (39) or root mean square radius of the solute (40). The Rayleigh constant is the specific turbidity for the light scattered at angle θ (41) and is related to the molecular weight through:

$$\frac{Kc}{R_\theta} = \frac{1}{\overline{M}P(\theta)} + 2A_2 c \qquad\qquad \text{Eq. 4.4}$$

where

$$K = \frac{2\pi^2 n^2}{\lambda^4 N_A}\left(\frac{dn}{dc}\right)^2 \qquad\qquad \text{Eq. 4.5}$$

In these equations, c is the concentration, n is the refractive index, N_A is Avogadro's number, and dn/dc is the value of n extrapolated for a concentration of zero. $P(\theta)$ is a particle scattering function dependent on the geometry and size of a macromolecule with respect to the incident wavelength (λ) and A_2 is the second virial coefficient which measures ideality (39).

These equations can be simplified if compact proteins are detected under identical conditions and if the following assumptions are made (42): First, $P(\theta) \approx 1$ for compact biopolymers smaller than 10^6 Da; second, under normal HPLC conditions, the virial coefficient (A_2) is negligible; and third, dn/dc is the same for all proteins in the same system. With these simplifications, the relationship of concentration to molecular weight becomes:

$$M = \frac{(I_\theta)k}{c} \qquad\qquad \text{Eq. 4.6}$$

where M is the molecular weight, I_θ is the intensity of the scattered light, k is a constant of the system, and c is the concentration (42). Fig. 4.25 shows this relationship for a series of proteins.

Figure 4.25 Molecular Weight vs. Intensity of Scattered Light

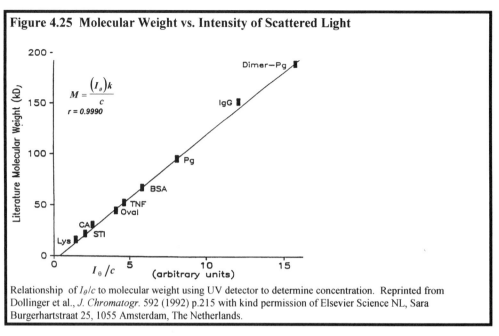

$$M = \frac{(I_\theta)k}{c}$$
$r = 0.9990$

Relationship of I_θ/c to molecular weight using UV detector to determine concentration. Reprinted from Dollinger et al., *J. Chromatogr.* 592 (1992) p.215 with kind permission of Elsevier Science NL, Sara Burgerhartstraat 25, 1055 Amsterdam, The Netherlands.

2. Instrument Design

Lasers are usually used as the light source in light scattering detectors so that a high intensity beam can be produced without stray light. The low-angle laser light scattering detector (LALLS) was the first commercial design. This detector measured light scattering at one angle. In newer designs, light scattering is measured at multiple angles; the variation of the excess Rayleigh ratio with angle is a measure of molecular size (40). The total scattered light is directly proportional to the product of molar mass and concentration. Using appropriate software, these values can be readily calculated. Fig. 4.26 shows the design of a multi-angle light scattering detector (MALLS).

Figure 4.26 Multi-Angle Laser Light Scattering Detector

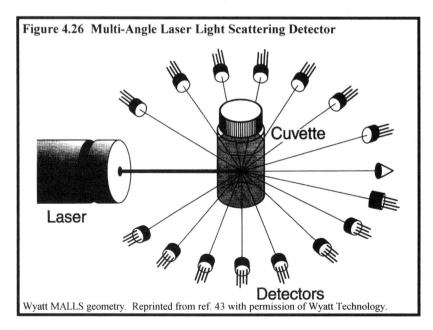

Wyatt MALLS geometry. Reprinted from ref. 43 with permission of Wyatt Technology.

Using the MALLS detector, the plot of Kc/R_θ vs. $\sin^2(\theta/2)$ yields both the molecular weight, which is given by the y-intercept, and the radius of gyration, which can be derived from the slope (40). The minimum detectable concentration using a light scattering detector is inversely proportional to the molecular weight of the solute. A solute with a molecular weight of 10^6 Da can be detected at a level of about 2.5μg/ml while one which is 10^4 Da can be detected at about 250μg/ml (44).

In some cases, a right angle fluorescence detector can be modified to measure light scattering at a 90° angle (42); in this case, wavelengths are limited to about 400nm and the excitation and emission wavelengths are set to be equal.

3. Applications

By comparing the relative signals for a molecule by light scattering with those of UV or RI, the amount of aggregation can be determined. The high sensitivity of light scattering to aggregation is demonstrated in Fig. 4.27 where human albumin and its aggregates are separated by SEC (41). The molecular weight distribution of the aggregates, which elute in the void volume, is indicated by the signal from light scattering.

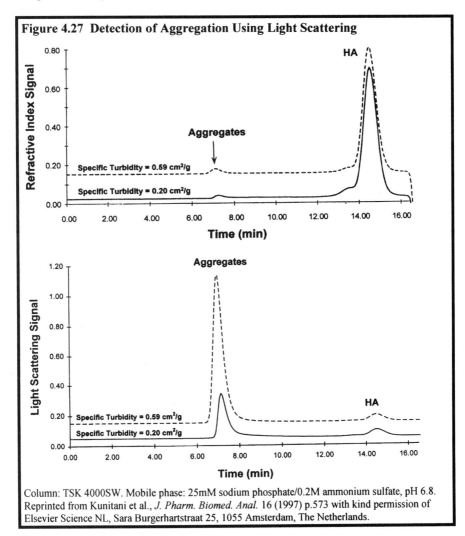

Figure 4.27 Detection of Aggregation Using Light Scattering

Column: TSK 4000SW. Mobile phase: 25mM sodium phosphate/0.2M ammonium sulfate, pH 6.8. Reprinted from Kunitani et al., *J. Pharm. Biomed. Anal.* 16 (1997) p.573 with kind permission of Elsevier Science NL, Sara Burgerhartstraat 25, 1055 Amsterdam, The Netherlands.

Light scattering measures the state of aggregation, assembly, or denaturation of the analyte. Quasi-elastic light scattering can also measure physical parameters, such as hydrodynamic radius and diffusion coefficients (38).

4. Advantages and Disadvantages

Light scattering detection is accurate for molecular weights greater than 10,000 Da, but is not as precise for those less than 5,000 Da (38). Sensitivity offered by low or multi-angle light scattering is variable because it is based on molecular weight; however, this is not generally a problem because the detectors are designed for large polymers which have good detection limits (10^{-4} - 10^{-6}g). Laser light scattering detectors are fairly expensive due to both the lasers and the sophisticated software.

One critical aspect of light scattering detectors is the accurate measurement of delay volume between the two detectors (45). Any error will result in a concomitant error in molecular weight due to an incorrect correlation with concentration. It is also important to eliminate peak broadening between the two detectors (45). Both samples and solvents must be prefiltered and clarified to prevent extraneous light scattering. Filters may be placed in-line, but if adsorption occurs on them, data will be skewed (38).

G. Evaporative Light Scattering Detector

1. Instrument Design

Evaporative light scattering detectors (ELSD) are universal detectors which can detect all compounds less volatile than the mobile phase. In these detectors, the column effluent is nebulized to form a uniform mist, which is heated to remove volatile solvents. The non-volatile analytes flow through a cell where they scatter an incident beam from a light source (2, 46, 47). Fig. 4.28 illustrates the design of this detector.

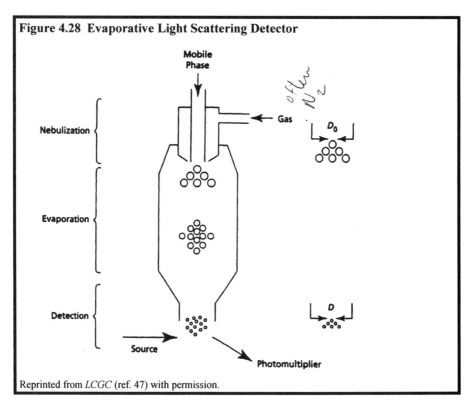

Figure 4.28 Evaporative Light Scattering Detector

Reprinted from *LCGC* (ref. 47) with permission.

The droplet size decreases throughout the evaporative light scattering process but it is important to attain constant droplet size. The larger the droplet, the higher the required evaporation temperature but the more significant the scattered light intensity. Lower temperatures in the evaporation step allow detection of less volatile solutes and also enhance crystallization, which increases sensitivity (47).

In evaporative light scattering detectors, the relationship of concentration to peak area is not linear but rather a logarithmic function:

$$\log A = b \log m + \log a$$

Eq. 4.7

$10^{-6}/10^{-3}$

where A is the peak area, m is mass and a and b are coefficients related to droplet size, solute concentration and properties, flowrates of the gas and liquid, and temperature. The relationship is nonlinear because light scattering is related to particle size rather than to concentration. The mass range of sensitivity is about 10^3 (40). The limit of detection for galactose is $10\mu g/ml$ (46).

Evaporative light scattering does not require a high quality gas (usually nitrogen) or solvents; however, they must be free of particulate matter, including residue after evaporation (43). Because the major restriction on the solvent is that it be more volatile than the solutes, solvents like chloroform may be used successfully. Gradients are also compatible with the technique.

2. Applications

Evaporative light scattering is especially useful for carbohydrates and lipids which may be difficult to detect by other means. The compatibility with gradients allows flexibility for HPLC methods development. Fig. 4.29 illustrates the analysis of microgram amounts of sugars using an acetonitrile/water gradient. A comparison with RI response to the same conditions is also shown.

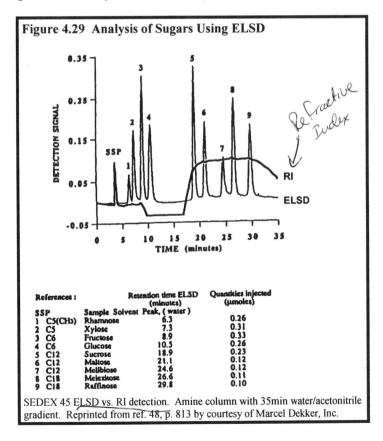

Figure 4.29 Analysis of Sugars Using ELSD

References :		Retention time ELSD (minutes)	Quantities injected (μmoles)
SSP	Sample Solvent Peak, (water)		
1 C5(CH3)	Rhamnose	6.3	0.26
2 C5	Xylose	7.3	0.31
3 C6	Fructose	8.9	0.33
4 C6	Glucose	10.5	0.26
5 C12	Sucrose	18.9	0.23
6 C12	Maltose	21.1	0.12
7 C12	Melibiose	24.6	0.12
8 C18	Melezitose	26.6	0.11
9 C18	Raffinose	29.8	0.10

SEDEX 45 ELSD vs. RI detection. Amine column with 35min water/acetonitrile gradient. Reprinted from ref. 48, p. 813 by courtesy of Marcel Dekker, Inc.

Chloroform and methanol can be used for analysis of phospholipids like those of soy bean lecithin or pig brain extract (49). The detection limit is about 20ng for phosphatidyl ethanolamine or phosphatidyl choline. Evaporative light scattering can also detect underivatized amino acids separated with a reversed phase gradient.

3. Advantages and Disadvantages

Evaporative light scattering is a universal detection method which can be used with gradients and is not temperature sensitive. Many solvents can be used; the main restriction is that they be more volatile than the solutes and free of particulate matter. Limits of detection are about 20ng. The nonlinear calibration curve (log/log) is the principle disadvantage, although computer programs can simplify calculations.

H. Mass Spectrometry

1. General

Mass spectrometry (MS) is an ideal detection mode in terms of its universality and selectivity (50). Mass spectrometry revolutionized gas chromatography (GC) of complex mixtures, especially in environmental and clinical areas, by providing resolution and identification of each of the components. Unfortunately, mass spectrometry is not as simply adapted to HPLC because the column effluent is a liquid and cannot be directly introduced into the vacuum environment of the analyzer. The key to the success of LC-MS is the interface between them which removes the bulk of the mobile phase from the sample. The most popular design is the solvent-assisted ionization interface, which nebulizes and vaporizes the mobile phase. This category includes thermospray, electrospray, and atmospheric pressure chemical ionization interfaces (50). Fig. 4.30 illustrates the molecular weight and polarity ranges for each.

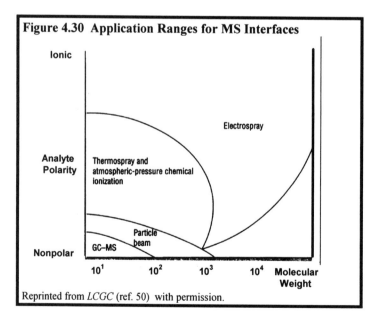

Figure 4.30 Application Ranges for MS Interfaces

Reprinted from *LCGC* (ref. 50) with permission.

The large molecular weight range of the electrospray interface gives it great utility in peptide and protein analysis; this interface is depicted in Fig. 4.31.

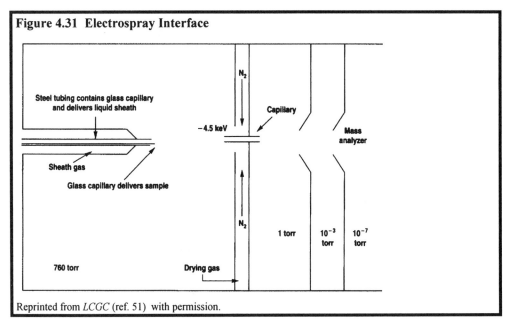

Figure 4.31 Electrospray Interface

Reprinted from *LCGC* (ref. 51) with permission.

After the LC flow is introduced into the electrospray probe, the solvent and solute molecules are ionized by the gain or loss of protons and the LC mobile phase is evaporated with nitrogen (51). As the solvent in the droplets evaporates, the ions come closer together and are eventually ejected into the gas phase as ions, which are then drawn into the mass analyzer. The addition of multiple charges to a molecule produces an envelope of peaks corresponding to various mass-to-charge ratios for each molecule (51).

2. Mass Spectrometer

A mass spectrometer is composed of an inlet system/ion source, a mass analyzer and a detector. Typical ion sources and mass analyzers are listed in Table 4.7.

Table 4.7 Ion Sources and Mass Analyzers

Ion Sources	Mass Analyzers
Electron Ionization (EI)	Sector (magnetic and electric, single/multiple stage)
Chemical Ionization (CI)	
Fast Atom Bombardment (FAB)	Quadrupole (single and triple)
Atmospheric Pressure Ionization (API)	Ion Trap
(electrospray, thermospray, etc.)	Fourier Transform Ion Cyclotron Resonance
Matrix-Assisted Laser Desorption Ionization	(FT-ICR)
(MALDI)	Time-of-Flight (TOF)

The ion source converts the sample into charged species by various means, such as electron impact, chemical ionization, fast atom bombardment, laser desorption, or the spray techniques. These charged species enter a mass analyzer which sorts and resolves ions with different mass to charge (m/z) ratios. The most commonly used analyzers for LC-MS have a quadrupole design which is relatively inexpensive and has good sensitivity. A quadrupole MS is composed of four voltage-carrying rods. Under conditions which maintain a constant ion velocity through the rods, a set of complex oscillations are created which allow only ions of a particular m/z ratio to traverse the entire path and be detected. These conditions are systematically altered to monitor new m/z ratios. Quadrupole mass spectrometers have an upper mass range of 1000 - 4000 Da (51). This range limitation is not a major problem if an electrospray interface is used because it produces ions with multiple charges; therefore, the mass of detectable molecules can actually reach 100,000 Da (51). Mass accuracy is excellent with quadrupole instruments, but can be improved with the use of magnetic sector (52) or time-of-flight analyzers. Ion traps allow MS-$(MS)^n$ experiments and operate in a manner similar to that of a quadrupole - with varying electric fields "trapping" only certain m/z ions. With the greater mass accuracy and sensitivity of MS/MS, it is possible to detect mass changes of $\pm$ 1Da, such as in amidation (53).

3. Applications

Mass spectrometry is nearly universal for compounds which ionize under the experimental conditions. For example, a protein may bear a proton for every 5 - 17 amino acids. An electrospray mass spectrum of myoglobin is shown in Fig. 4.32.

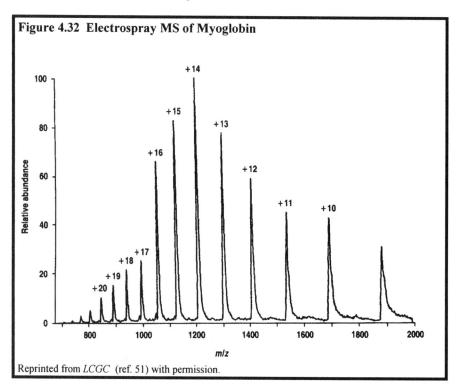

Figure 4.32 Electrospray MS of Myoglobin

Reprinted from *LCGC* (ref. 51) with permission.

As seen in this spectrum, the envelope of charges exhibits a characteristic pattern of electrospray ionization, with the lower m/z peaks closer together as more charges are added. Each peak represents an ion of $m/z = [M + z\ H^+]/z$, where M is the molecular weight of the compound and z is the number of protons added upon ionization. An algorithm to determine the molecular weight of a compound by choosing adjacent peaks is usually included in the mass spectrometer software.

Mass spectrometry provides molecular weight determination of biopolymers much more accurately than electrophoretic gels. LC-MS has additionally provided a deeper knowledge of glycoproteins, especially when total ion current plots and contour plots (m/z vs. time) have been correlated (54). LC-MS is very useful for identification of drugs and metabolites which can utilize either thermospray or electrospray interfaces due to the smaller molecular weights of these molecules (55). High femtomole amounts of some compounds can be detected. The sensitivity of the mass spectrometer is dramatically improved and quantitation is possible when one or two ions are chosen for monitoring [selected ion monitoring (SIM) or single ion recording (SIR)].

Mass spectrometry has found wide usage in biological applications, some of which are listed in Table 4.8.

Table 4.8 Applications of LC-MS in Biomolecule Separations

Molecular weight determination
Structure determination
Peptide mapping/sequencing
Determination of post-translational modifications
Metabolism/pharmokinetic studies
Carbohydrate characterization
Nucleic acid characterization
Phospholipid characterization

An example of mapping protein digests is shown in Fig. 4.33. This tryptic digest is shown to be incomplete by the molecular weights of the fragments. The peak at 24.69min still has a terminal lysine, whereas that at 25.03min has had the lysine cleaved.

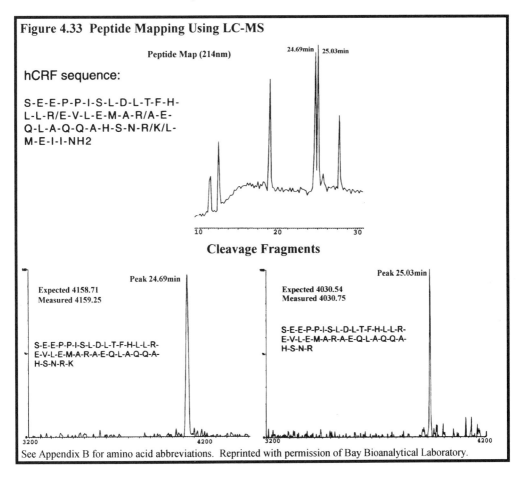

Figure 4.33 Peptide Mapping Using LC-MS

Peptide Map (214nm)

24.69min   25.03min

hCRF sequence:

S-E-E-P-P-I-S-L-D-L-T-F-H-
L-L-R/E-V-L-E-M-A-R/A-E-
Q-L-A-Q-Q-A-H-S-N-R/K/L-
M-E-I-I-NH2

Cleavage Fragments

Peak 24.69min

Expected 4158.71
Measured 4159.25

S-E-E-P-P-I-S-L-D-L-T-F-H-L-L-R-
E-V-L-E-M-A-R-A-E-Q-L-A-Q-Q-A-
H-S-N-R-K

Peak 25.03min

Expected 4030.54
Measured 4030.75

S-E-E-P-P-I-S-L-D-L-T-F-H-L-L-R-
E-V-L-E-M-A-R-A-E-Q-L-A-Q-Q-A-
H-S-N-R

See Appendix B for amino acid abbreviations. Reprinted with permission of Bay Bioanalytical Laboratory.

4. Advantages and Disadvantages

Mass spectrometry gives molecular weight and structural information for individual chromatographic peaks to allow identification. Electrospray ionization paired with a quadrupole analyzer has low noise (dependent on mobile phase composition), mass accuracy of 0.01%, and sub-picomole detection limits, in addition to a mass range of greater than 100,000 Da (51).

The maximum flowrate, however, for electrospray interfaces is often 200μl/min, although some instruments allow as high as 0.5 ml/min. The ideal is usually 1 - 20μl/min (51). For this reason, microbore or capillary columns are usually implemented in LC-MS, or the effluent stream is split. A 1mm ID microbore column is typically operated at 50μl/min, allowing direct introduction of the flow into the electrospray probe. The smaller columns also permit analysis of quantity-limited samples due to the smaller injection volumes needed.

An additional consideration for the interfacing of LC with MS is the choice of mobile phase. The mobile phase must be volatile so that it can be removed before the molecules enter the mass analyzer; therefore, reversed phase is often the method of choice. Table 4.9 lists volatile mobile phase additives that are compatible with MS; Table 4.10 list incompatible additives. TFA has signal suppressing effects on the electrospray process because it ion-pairs strongly with basic analytes, effectively rendering them neutral (56). This phenomenon can be alleviated by a "TFA Fix", which adds a solution of 75% propionic acid/ 25% 2-propanol to the column effluent through postcolumn addition, thus disrupting the strong ion-pairing (56, 57). Signal enhancement is 10 - 50 times.

Table 4.9 Volatile Mobile Phase Additives for MS	
cations	**anions**
ammonium	formate
diethylamine (DEA)	acetate
triethylamine (TEA)	carbonate
	trifluoroacetate (TFA)
	heptafluoroacetate (HFBA)

Table 4.10 Incompatible Mobile Phase additives for MS	
cations	**anions**
sodium	chloride
potassium	phosphate
most metals	sulfate
	sulfonic acids (e.g. ion-pairing agents)

Destruction of the sample by mass spectrometry is a serious disadvantage, particularly when sample is limited. In such cases, it is best to perform all nondestructive methods, such as NMR, first.

I. Comparison of Detectors

Table 4.11 lists comparative characteristics of the major kinds of HPLC detectors. Because the selectivity and sensitivity are different for each, no one detector is ideal, and selection must be determined by the samples, conditions, and goals of the analysis. Instruments are continually being improved in terms of performance and detection limits. The approximate values in this table should only be used as guidelines; specifications of current instruments should be consulted for absolute detection limits and sensitivity.

Table 4.11 HPLC Detectors

	Universality	Structural Information	Sensitivity	Limits of Detection (g)	Linear Dynamic Range	Temperature Sensitivity	Gradient Compatible	Cost
UV/VIS, fixed λ	med	low	high	10^{-10} to 10^{-11}	10^5	low	yes	low
Photodiode Array	med	med	high	3ppb	>2 AU	low	yes	med*
Refractive Index	high	low	low	10^{-7}	10^4	high	no	low
Fluorescence	low	low	high	10^{-12}	10^3	low	yes	low
Electrochemical	low	low	high	10^{-11}	10^6	high	no	low
Conductivity	low	low	high	10^{-11}	10^4	high	no	low
Radioactivity	low	low	med	<100 DPM	4×10^6DPM	low	yes	med
Multi-Angle Laser Light Scattering	med	high	low	10^{-7}		low	yes	med*
Evaporative Light Scattering	high	low	med	10^{-8}	10^3	low	yes	med
Mass Spectrometry	high	high	med	10^{-10}		low	yes	high*

* generally requires sophisticated software. Data from ref. 1, 6, 55 and manufacturers' literature.

III. POST-COLUMN REACTION

Sometimes a molecule can only be differentiated from other compounds in the sample or detected at trace levels by changing its properties through derivatization (17, 27). This is a common tactic in amino acid analysis because many amino acids have poor detectability with standard detectors. Derivatization can occur before (pre-column) or after (post-column) the separation on the analytical column. Post-column reaction (PCR) is often used to avoid separation of extra peaks caused by incomplete derivatization. Post-column derivatization reagents are usually pumped into a tee and/or reactor where they are mixed with the column effluent (58, 59). If it is a time-dependent reaction, the reactor requires enough volume to allow derivatization, with a geometry that minimizes band spreading - usually a capillary column or packed column with nonreactive, nonporous beads (60).

All amino acids except tryptophan, tyrosine, and phenylalanine must be derivatized before they can be effectively detected by UV or fluorescence (61, 62). Similarly, small peptides need to be derivatized if they do not contain these amino acid residues (15, 27). Three popular methods of converting amines into fluorescent derivatives utilize o-phthalaldehyde (OPA) for amino acids and primary amines, fluorescamine for peptides (63), and naphthalenedialdehyde-cyanide (NDA) for amines and peptides (64). All these reagents react with primary amines; however, OPA does not react with the N-terminus of peptides, limiting its utility to peptides containing lysine residues (27).

The OPA reaction, as seen in Fig. 4.34, is a simple reaction which yields fluorescent derivatives for amino acids. The fluorescamine reaction, also shown in Fig. 4.33, requires an organic solvent to dissolve the reagent and an alkali buffer to maximize the conversion (27). NDA produces stable derivatives which can be detected by fluorescence, electrochemical, absorbance or chemiluminescence detection (64). The reaction requires 30 minutes for completion; therefore, precolumn derivatization methods using automation are effective (65). Excess NDA reagent does not have to be removed before injection (65).

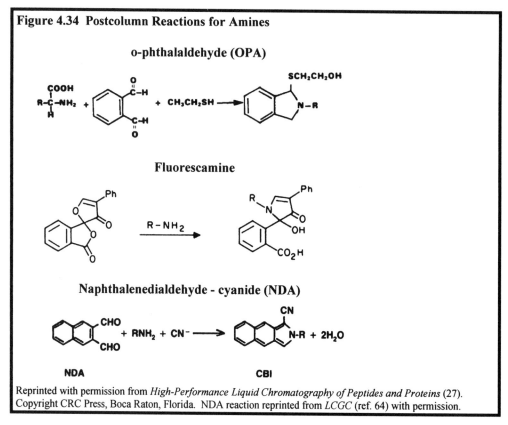

Figure 4.34 Postcolumn Reactions for Amines

o-phthalaldehyde (OPA)

Fluorescamine

Naphthalenedialdehyde - cyanide (NDA)

Reprinted with permission from *High-Performance Liquid Chromatography of Peptides and Proteins* (27). Copyright CRC Press, Boca Raton, Florida. NDA reaction reprinted from *LCGC* (ref. 64) with permission.

The use of fluorescamine allows detection sensitivity in the low picomole to mid-femtomole range (63). Fig. 4.35 illustrates the resolution and detection of peptides from a single cluster of sinus gland neurosecretory nerve endings using postcolumn fluorescamine derivatization.

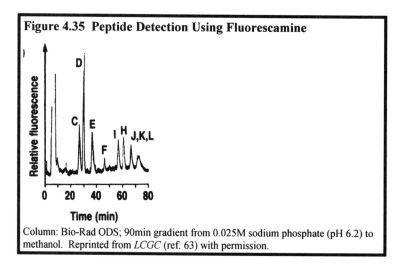

Figure 4.35 Peptide Detection Using Fluorescamine

Column: Bio-Rad ODS; 90min gradient from 0.025M sodium phosphate (pH 6.2) to methanol. Reprinted from *LCGC* (ref. 63) with permission.

One other type of PCR used for biomolecules is postcolumn enzymatic reaction (60). In these procedures, substrates and any other required reagents are mixed with the column effluent; by selective detection of the reaction products using fluorescence or visible absorption, trace amounts of enzymes in biological samples can be assayed. This method has been used to measure isoenzyme levels of creatine kinase and lactate dehydrogenase in clinical samples (60).

IV. REFERENCES

1. L.R. Snyder and J.J. Kirkland, *Introduction to Modern Liquid Chromatography, 2nd Edition*, John Wiley & Sons, New York (1979) Chap.4.
2. H. Poppe, "Column Liquid Chromatography" in *Chromatography, 5th Edition* (ed. E. Heftmann), Elsevier Science Publishers, Amsterdam (1992) A151.
3. E.L. Johnson and R. Stevenson, *Basic Liquid Chromatography*, Varian (1978) Chap. 11.
4. E.S. Yeung, *LC-GC* 7 (1989) 118.
5. J.B. Li, *LC-GC* 10 (1992) 856.
6. S.R. Bakalyar, C. Phipps, B. Spruce, and K. Olsen, *J. Chromatogr. A,* 762 (1997) 167.
7. *HP 1050 Series Multiple Wavelength Detector,* Hewlett-Packard.
8. R.L. Stevenson, "UV-VIS Absorption Detectors for HPLC" in *Liquid Chromatography Detectors* (ed. T.M. Vickrey), Marcel Dekker, New York (1983) 51.
9. *LC at Work* 111, Varian.
10. C. Seaver and P. Sadek, *LC-GC* 12 (1994) 742.
11. N.-I. Jang and P.R. Brown, *LC-GC* 10 (1992) 526
12. *HP 1100 Series Variable Wavelength Detector*, Hewlett-Packard, 1996.
13. M.V. Pickering, *LC-GC* 8 (1990) 846.
14. *HP 1100 Series Diode-Array Detector*, Hewlett-Packard, 1996.
15. I.S. Krull, J.R. Mazzeo, R. Mhatre, M.E. Szule, J.T. Stults, and J.H. Bourell, "Detection and Identification in Biochromatography" in *High Performance Liquid Chromatography: Principles and Methods in Biotechnology* (ed. E.D. Katz) John Wiley & Sons, Chichester (1996) 163.
16. D. Vost, W.B. Gratzer, R.A. Cox, and P. Dory, *Biopolymers* 1 (1963) 193.
17. I.S. Krull, M.E. Szulc, and S.-L. Wu, *LC-GC* 11 (1993) 350.
18. A.L. Lehninger, *Biochemistry*, Worth Publishers, New York, 1970.
19. M. Meys and S. Cohen, *LC-GC* 9 (1991) 422.
20. J.B. Wilson, "Separation of Human Hemoglobin Variants by HPLC" in *HPLC of Biological Macromolecules* (ed. K.M. Gooding and F.E. Regnier) Marcel Dekker, New York (1990) 457.
21. J. Del Rios, *Amer. Lab.* 17, June 1985, p.10.
22. *Waters 410 Differential Refractive Index Detector*, Waters, 1995.
23. *HP 1047A Refractive Index Detector*, Hewlett-Packard, 1993.
24. *The Optilab DSP*, Wyatt Technology, 1995.
25. *Varian 9075 Fluorescence Detector*, Varian, 1996.
26. *Waters 474 Scanning Fluorescence Detector*, Waters, 1994.
27. C.T. Wehr, "Post Column Reaction Systems for Fluorescence Detection of Polypeptides" in *High-Performance Liquid Chromatography of Peptides and Proteins* (ed. C.T. Mant and R.S. Hodges), CRC Press, Boca Raton (1991) 579.
28. R.E. Shoup, "Liquid Chromatography/Electrochemistry" in *High-Performance Liquid Chromatography, Advances and Perspectives* (ed. Cs. Horvath) Academic Press, Orlando (1986) 91.
29. M.E. Hall, B.J. Hoffer and G.A. Gerhardt, *LC-GC* 7 (1989) 258.
30. H.F. Walton, "Ion-Exchange Chromatography" in *Chromatography, 5th Edition* (ed. E. Heftmann), Elsevier Science Publishers, Amsterdam (1992) A227.
31. I.S. Krull, C.M. Selavka, M. Lookabaugh, and W.R. Childress, *LC-GC* 7 (1989) 758.
32. K. Lockhart, C. Nguyen, and M. Lee, "Detection Alternatives for HPLC Analysis of LHRH and LHRH Analogs" in *High-Performance Liquid Chromatography of Peptides and Proteins* (ed. C.T. Mant and R.S. Hodges), CRC Press, Boca Raton (1991) 571.
33. *Radiomatic ™ Flow Scintillation Analyzers*, Packard Instrument Company, 1996.

34. *Radio-HPLC Flow Cells*, Packard Instrument Company, 1996.
35. R.D. Wiehle and J.L. Wittliff, "Assessment of Steroid Receptor Polymorphism by High-Performance Ion-Exchange Chromatography" in *High-Performance Liquid Chromatography of Peptides and Proteins* (ed. C.T. Mant and R.S. Hodges), CRC Press, Boca Raton (1991) 255.
36. S..M. Hyder, J. Dong, P. Folk, and J.L. Wittliff, "High-Performance Hydrophobic Interaction Chromatography of a Labile Regulatory Protein" in *High-Performance Liquid Chromatography of Peptides and Proteins* (ed. C.T. Mant and R.S. Hodges), CRC Press, Boca Raton (1991) 451.
37. D.M. Boyle, R.D. Wiehle, N.A. Shahabi, and J.L. Wittliff, *J. Chromatogr.* 327 (1985) 369.
38. I.S. Krull, R. Mhatre and J. Cunniff, *LC-GC* 13 (1995) 30.
39. H.H. Stuting, I.S. Krull, R. Mhatre, S.C. Krzysko, and H.G. Barth, *LC-GC* 7 (1989) 402.
40. P.J. Wyatt, *LC-GC* 15 (1997) 160.
41. M. Kunitani, S. Wolfe, S. Rana, C. Apicella, V. Levi, and G. Dollinger, *J. Pharm. Biomed. Anal.* 16 (1997) 573.
42. G. Dollinger, B. Cunico, M. Kunitani, D. Johnson, and R. Jones, *J. Chromatogr.* 592 (1992) 215.
43. *The Absolute Detector for Macromolecular Analysis*, Wyatt Technology, 1996.
44. W. Kaye, *Anal. Chem.* 45 (1973) 221A
45. P.J. Wyatt and L.A. Papazian, *LC-GC* 11 (1993) 862.
46. "Evaporative Lght-Scattering Detector" in *Chromatography*, Alltech Catalog 400, 1997, p.32.
47. M. Dreux, M. Lafosse, and L. Morin-Allory, *LC-GC Intl.* (March 1996) 149.
48. A. Clement, D. Yong, and C. Brechet, *J. Liq. Chromatogr.* 15 (1992) 805.
49. J. Becart, C. Chevalier, and JP. Blesse, *J. High Res. Chromatogr.* 13 (1990) 126.
50. D.A. Volmer and D.L. Vollmer, *LC-GC* 14 (1996) 236.
51. I.S. Krull, R. Mhatre, and J. Cunniff, *LC-GC* 12 (1994) 914.
52. B.L. Gillece-Castro and J.T. Stults, "Peptide Characterization in Mass Spectrometry" in *Meth. Enzymol.* 271 (1996) 427.
53. J.R. Yates, "Protein Structure Analysis by Mass Spectrometry", *Meth. Enzymol.* 271 (1996) 351.
54. J.A. Chakel, A. Apffel, and W.S. Hancock, *LC-GC* 13 (1995) 866.
55. P. Newton, *LC-GC* 8 (1990) 706.
56. F.E. Kuhlmann, A. Apffel, S.M. Fisher, G. Goldberg, and P. Goodley, *J. Am. Soc. Mass Spectrom.* 6 (1995) 1221.
57. W.S. Hancock, A. Apffel, J. Chakel, C. Souders, T.M. Timkulu, E. Pungor, Jr., and A.W. Guzzetta, "Reversed-Phase Peptide Mapping of Glycoproteins Using LC/EI-MS", *Meth. Enzymol.* 271 (1996) 403.
58. M.K. Freeman, S. Daunert, and L.G. Bachas, *LC-GC* 10 (1992) 112.
59. W.J. Bachman and J.T. Stewart, *LC-GC* 7 (1989) 38.
60. T.D. Schlabach, S.H. Chang, K.M. Gooding, and F.E. Regnier, *J. Chromatogr.* 134 (1979) 91.
61. C.T. Mant, N.E. Zhou, and R.S. Hodges, "Amino Acids and Peptides" in *Chromatography, 5th Edition* (ed. E. Heftmann), Elsevier Science Publishers, Amsterdam (1992) B75.
62. C. Lazure, J.A. Rochemont, N.G. Seidah, and M. Chretien, "Amino Acids in Protein Sequence Analysis" in *HPLC of Biological Macromolecules* (ed. K.M. Gooding and F.E. Regnier), Marcel Dekker, New York (1990) 263.
63. R. Newcomb, *LC-GC* 10 (1992) 34.
64. S.M. Lunte and O.S. Wong, *LC-GC* 7 (1989) 908.
65. F. Lai and T. Sheehan, *BioTechniques* 14 (1993) 642.

CHAPTER 5
HPLC PACKINGS

The "heart" of an HPLC is the column; it alone provides the potential for a separation with given efficiency. It is therefore essential to understand the general physical and chemical characteristics of the support materials employed as column packings to utilize them effectively for HPLC. These column packings or supports may be used intact, as in the case of bare silica or crosslinked polystyrene. More often, however, supports are derivatized with a stationary phase with chemical characteristics like hydrophobicity or ionic charge. The stationary phase separates molecules by their unique affinities for the particular functionality.

I. PHYSICAL CHARACTERISTICS

A. Theoretical Considerations

As described in Chapter 2, the efficiency or plate height of a column (H) can be expressed in terms of the linear velocity (u):

$$H = Au^{0.33} + \frac{B}{u} + Cu \qquad\qquad \text{Eq. 5.1}$$

where A is the contribution of eddy diffusion, B is the contribution due to axial diffusion, and C is the broadening attributable to mass transfer. All of these constants are highly dependent on the physical characteristics of the column packing material and the column bed. The relative contributions of the terms of Eq. 5.1 can be understood by examining the plot in Fig. 5.1.

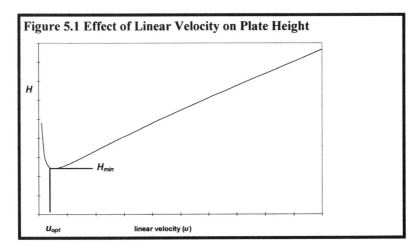

Figure 5.1 Effect of Linear Velocity on Plate Height

The plot of linear velocity (u) vs. plate height (H) has a minimum (H_{min}) at a velocity (u_{opt}) where optimum efficiency and resolution are achieved. The eddy diffusion (A) contribution primarily defines H_{min}. At linear velocities lower than the optimum (u_{opt}), the effects of axial diffusion (B) increase the peak width. At higher velocities, the resistance to mass transfer (C) is predominant.

The four major physical parameters of support particles, shape, particle diameter, pore diameter and porosity, as illustrated in Fig. 5.2, are primarily responsible for the potential efficiency of a well-packed column. Each of these characteristics has benefits and disadvantages which must be understood for intelligent column selection.

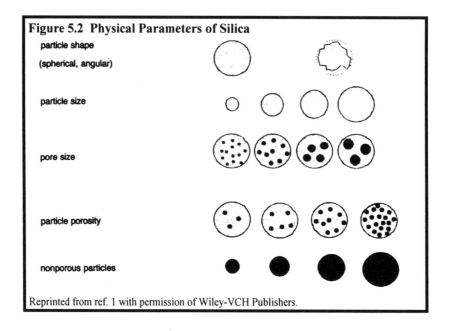

Figure 5.2 Physical Parameters of Silica

particle shape

(spherical, angular)

particle size

pore size

particle porosity

nonporous particles

Reprinted from ref. 1 with permission of Wiley-VCH Publishers.

B. Particle Shape

There are two classes of particles by shape - spherical and non-spherical (irregular or angular). Angular particles are the least expensive packings because of the relative ease by which they can be manufactured. The distribution of their particle diameters is generally much broader than it is for spherical particles, frequently including very small particles called fines. Angular particles often provide acceptable performance for preparative applications in large columns, but they are rarely used for analytical HPLC. They cannot be packed as well as spherical particles; therefore, columns have poorer efficiency and often have increased operating pressures due to the presence of fines. For these reasons, angular packings will not be discussed further.

The potential advantages of spherical particles over angular are fairly apparent. A bed of spherical particles can be more consistently packed due to the homogeneity of the support. In such a column, paths of flow are more uniform, reducing eddy diffusion (the A term in Eq. 5.1) and, therefore, band broadening. The more uniform interstitial spaces lead to reduced operating pressures and increased retention or bed capacity. The process for manufacturing uniform spherical particles has been refined over the last 15 years and the quality is very good. However, it is not perfect, as shown in the electron micrographs in Fig. 5.3.

Figure 5.3 TEM Photographs of Glass and Silica

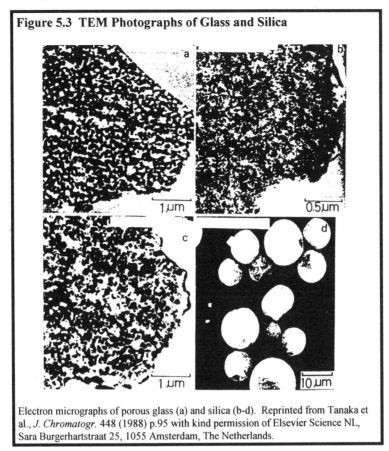

Electron micrographs of porous glass (a) and silica (b-d). Reprinted from Tanaka et al., *J. Chromatogr.* 448 (1988) p.95 with kind permission of Elsevier Science NL, Sara Burgerhartstraat 25, 1055 Amsterdam, The Netherlands.

Photo (a) depicts glass beads with very uniform pores; the remaining photos are spherical silica with controlled pores at different magnifications, as indicated. It can be seen that there is some irregularity to both the spheres and the pore structure.

C. Particle Diameter

1. Effect on Efficiency

In Chapter 2, it was theoretically determined that the particle diameter (d_p) is one of the most important physical parameters of the packing in its contribution to plate height (*see* Fig. 2.16, Eq. 2.27). The experimental plots of plate height vs. flowrate for 5µm, 12µm and 15µm reversed phase supports shown in Fig. 5.4 confirm the conclusions obtained theoretically.

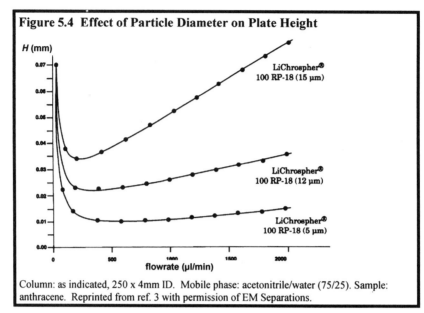

Figure 5.4 Effect of Particle Diameter on Plate Height

Column: as indicated, 250 x 4mm ID. Mobile phase: acetonitrile/water (75/25). Sample: anthracene. Reprinted from ref. 3 with permission of EM Separations.

At any given flowrate, the absolute plate height of the 15μm packing is higher than that of the 12μm and the 5μm packings, demonstrating the higher efficiency (and therefore greater resolution) of packings with smaller particle diameters. The advantages of small diameter particles result from the dependence of the A and C terms of Eq. 5.1 on particle diameter, as was seen in Eq. 2.27. A more striking aspect of these plots is the difference in their slopes. The linear portion of the plot of the 15μm packing exhibits a steep slope with an optimum flowrate of about 0.2 ml/min. The plot of the 5μm packing, on the other hand, exhibits a relatively shallow slope with no distinct minimum. The practical significance is that columns with small particle diameters permit faster flowrates to be used with only a slight loss of resolution. The effect of particle diameter on column efficiency is further demonstrated in Fig. 5.5 for a mixture of proteins run with a gradient on three reversed phase columns. Although the retention times of the major peaks are the same, the peak widths are much narrower on the 5μm support.

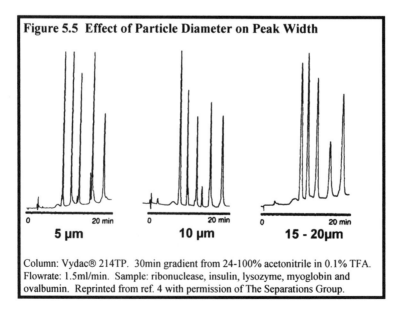

Figure 5.5 Effect of Particle Diameter on Peak Width

Column: Vydac® 214TP. 30min gradient from 24-100% acetonitrile in 0.1% TFA. Flowrate: 1.5ml/min. Sample: ribonuclease, insulin, lysozyme, myoglobin and ovalbumin. Reprinted from ref. 4 with permission of The Separations Group.

2. Effect on Pressure

It is clear from the examples and discussion why there has been such an effort made to manufacture smaller and smaller particles. Unfortunately, the price to pay for the high efficiency and fast flow afforded by small particles is an increased pressure drop across the column. The pressure drop can be defined by the following equation (5):

$$\Delta P = \frac{150\,\eta\,L\,F}{d_p^{\,2}\,d_c^{\,2}} \qquad\qquad \text{Eq. 5.2}$$

where η is the mobile phase viscosity, L is the column length, F is the flowrate, d_p is the particle diameter and d_c is the column diameter. It can be seen that the pressure drop increases with the inverse of the square of the particle diameter. Therefore, changing from a 15μm to a 5μm particle diameter creates a 9-fold increase in column pressure at the same flowrate. The theoretical relationship between particle diameter and pressure is diagrammed in Fig. 5.6.

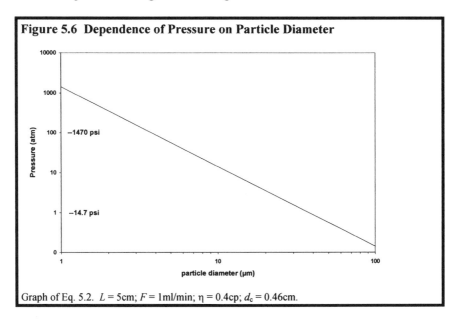

Figure 5.6 Dependence of Pressure on Particle Diameter

Graph of Eq. 5.2. L = 5cm; F = 1ml/min; η = 0.4cp; d_c = 0.46cm.

It is obvious that the current trend towards small particles can generate very high pressures, pushing a pump to its limits and straining seals and pulse dampers. For this reason, such columns are often made quite short (3 - 5cm) to keep ΔP manageably small, yet maintain satisfactory resolution. It should be noted that the column length used for Fig. 5.6 is only 5cm.

In recent years, the trend has been to reduce particle diameter to increase resolution. Columns with 5μm particles are still most popular but 1.5 - 2.5μm column packings are promoted, especially for use at high flowrates. These have proven effective for some applications; however, there are two major disadvantages. The primary one is the increased pressure generated by such particles. This limitation has been diminished by using short columns and elevated temperatures which reduce the viscosity. Small columns with 1 - 2μm particles are very sensitive to extracolumn volume, similar to microbore and capillary columns, necessitating the use of low volume injectors and flow cells. A secondary problem is associated with column frits (6). The standard mesh sizes are 0.5μm and 2μm. If the column packing includes particles which are close to the size of the holes in the frit, plugging will result. Smaller frit porosities inherently generate higher pressures and can plug more easily with sample or mobile phase impurities. A compromise of efficiency and usability regarding particle size may be 3 - 4μm (6).

D. Porosity

The porosity of a support has two aspects: the pore diameter and the abundance of pores. These characteristics are illustrated in Fig. 5.2. Pore diameters can be generally grouped into three size ranges: small (5 - 10nm), medium (10 - 30nm) and large ($\geq$ 50nm). A high abundance of pores will result in large pore volumes (V_i) which are a requirement for effective size exclusion chromatography. Unfortunately, high porosity yields concomitant fragility because of the lack of silica density in the structure, which is why the high porosity silicas used for size exclusion chromatography have lower pressure limits than supports used for other modes. Porous particles can be made in varying particle diameters, from 2μm or less for highest resolution to 100μm or greater for preparative separations, where low cost and minimum pressure drop are important considerations.

Most commonly used HPLC columns contain porous packings; however, nonporous varieties have been developed in particles as small as 1.5μm (7, 8). Nonporous particles eliminate band broadening caused by poor mass transfer into the stagnant mobile phase within the pores, but removal of pores also results in diminished surface area and thus, loading capacity. The loading capacity of a packing is greatly increased by the introduction of pores because they substantially augment the surface area. Table 5.1 lists typical surface areas for particles with various pore diameters (1). Pore diameter values are in nanometers, but are frequently given in literature in Ångstroms (10^{-1}nm).

Table 5.1 Effect of Pore Diameter on Surface Area	
Pore Diameter (nm)	Surface Area (m^2/g)
10	250
30	100
100	20
400	5 - 10

Loading capacity is directly related to surface area for small molecules which have total access to the pores. For proteins and other macromolecules, the available surface area will be maximal only if the pores in the particle are large enough to permit permeation; otherwise, the proteins will only have contact with the outside of the particles. The permeation of a macromolecule, which will be described fully in the chapter on size exclusion chromatography (SEC), is related to its size, which is in turn dependent on its shape and molecular weight. Highly folded or globular proteins resemble spheres. Randomly coiled proteins are more extended or rod shaped, rotating fully in solution and carving out a large hydrodynamic volume. Nominal values for the diameters of extended and globular proteins are provided in Table 5.2 (9).

Table 5.2 Protein Diameter vs. Molecular Weight		
	Hydrodynamic Diameter	
Molecular Weight (kD)	Random Coil (nm)	Globular (nm)
1	2.6	1.6
10	8.2	3.5
100	25.8	7.6
1000	81.6	16.3

These values, combined with the information in Table 5.1, provide guidelines as to the appropriate pore diameter for total penetration of proteins of different molecular weights and shapes. It can be seen that the common practice of using supports with pores that are at least 30nm for protein analysis should allow some permeation of most globular proteins under 1000kD or random coil molecules of 100kD; however, the actual limits of molecular weights may be somewhat lower than these data imply.

In practical terms, choosing a pore size which restricts the entry of macromolecules will result in a measurable reduction of binding capacity for excluded molecules, but choosing too large a pore will result in much lower potential surface area. Maximal loading is obtained with pores which are somewhat larger than the solute. If the dimension of a molecule is very close in size to the pore, diffusion is restricted (10, 11). The problem is most severe when the solute only has access to 10 - 20% of the pore volume (K_D of 0.1 - 0.2 in size exclusion chromatography, *see* chapter 6) (11). The phenomenon of reduced capacity due to pore diameter for peptides and proteins is shown experimentally in Fig. 5.7.

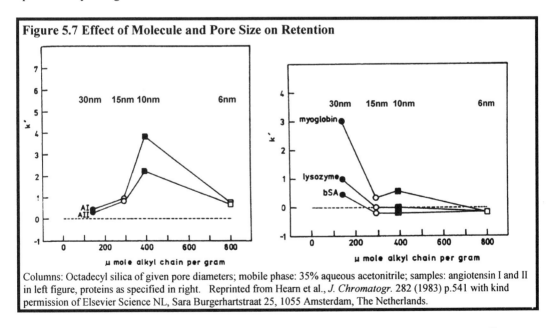

Figure 5.7 Effect of Molecule and Pore Size on Retention

Columns: Octadecyl silica of given pore diameters; mobile phase: 35% aqueous acetonitrile; samples: angiotensin I and II in left figure, proteins as specified in right. Reprinted from Hearn et al., *J. Chromatogr.* 282 (1983) p.541 with kind permission of Elsevier Science NL, Sara Burgerhartstraat 25, 1055 Amsterdam, The Netherlands.

Maximum retention of angiotensins I and II, which are about 1kD, occurs on the 10nm pore diameter support. Retention is reduced on the 6nm pores due to restriction of penetration. The proteins, to the contrary, exhibit highest capacity on the 30nm support into which they can all penetrate; the smaller pores have low capacity due to exclusion of the proteins.

II. CHEMICAL CHARACTERISTICS

A. Silica Supports

1. Composition

The silica gels used for HPLC are synthesized from pure sodium silicate and an acid (13) or tetraethoxysilane to achieve high purity spherical particles and tightly controlled porosity and particle diameter. The two primary methods of silica preparation result in substantially different surface and physical properties (14). Sol-gel silica supports are formed by aggregating silica sols; they often result in type B silica which generally has lower surface area, porosity and surface reactivity than type A supports (15). Sol-gel synthesis may sometimes result in Type A silica (16). Sil-gel silicas are made by gelling soluble silicates, producing type A silica which is a more acidic support, generally with higher surface area and porosity than type B silica. Some of the properties of the two kinds of silica used for HPLC are listed in Table 5.3. These characteristics are not absolute because they are predetermined by manufacturing processes like heating and rehydroxylation, as well as impurities in the silica (15).

Table 5.3 Silica Characteristics by Type

Type Designation	Sil-gel Type A	Sol-gel Type B
synthesis	gelling soluble silicates	aggregating silica sols
surface area	high	moderate
porosity	high	moderate
pore walls	variable	thick
silanols	isolated or unbonded	associated or bonded
pH stability	moderate	good
purity	may contain metal impurities	high
solubility	metals may increase stability	moderate

From ref. 14 -16.

The structure of fused silica contains several kinds of bonds, as shown in Fig. 5.8, each of which has specific chemical properties. The siloxane bonds (Si–O–Si) form a crosslinking network, imparting rigidity and mechanical strength to the particle.

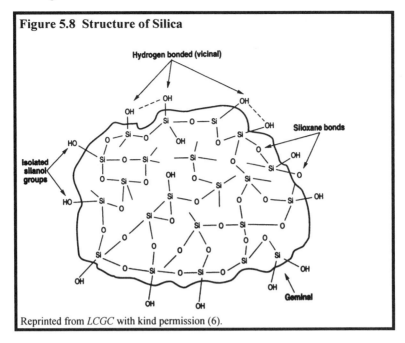

Figure 5.8 Structure of Silica

Reprinted from *LCGC* with kind permission (6).

The surface silanol groups are of three species: isolated, geminal and vicinal, which are further diagrammed in Fig. 5.9.

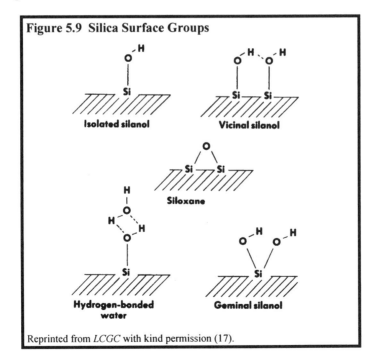

Figure 5.9 Silica Surface Groups

Reprinted from *LCGC* with kind permission (17).

Each of these, in addition to the siloxane moiety, has unique chemical characteristics and reactivity. Geminal silanols are the most reactive; isolated silanols are primarily responsible for the adsorption of basic molecules (15). Those silanols which are found in proximity to others may undergo hydrogen bonding, as illustrated in the figures for the vicinal silanol, and some silanols will be hydrated. The presence of isolated silanols on Type A silica, as well as the frequent presence of metal impurities, limit its utility for basic analytes or those which interact with metals. Type B silica affords a relatively inert surface to these molecules.

Silica has a number of advantages as a chromatographic support. Silica particles have good mechanical stability and their synthesis allows finely controlled porosity and particle diameters to be achieved. Porosity can range from low to high (0 - 80% of the particle volume), to provide silicas with different mechanical stabilities and pore volumes. Commercially available silica includes pore diameters from 5 - 100nm which are compatible with the requirements of biomolecules, allowing maximal surface areas (20 - 500m^2/g) and capacities dependent on the size of the solute molecule. Silica can be manufactured in a wide range of particle diameters appropriate for high efficiency analytical work (≤10μm) or for preparative work (>10μm). Perhaps most importantly, the surface of silica can be readily and reproducibly modified by reaction with organosilane modifiers, as described below.

A representative list of suppliers of silica-based HPLC supports for biopolymers is seen in Table 5.4. The many mergers and acquisitions of HPLC column manufacturers which have occurred recently may make the sources of particular columns confusing. In this table, the companies in the left column have acquired those in the right.

Table 5.4 Selected Sources of Silica-based Supports	
MANUFACTURER	**ASSOCIATED OR MERGED COMPANIES**
Alltech	Exmere Ltd.
Bio-Rad	
Beckman	
E. Merck	
Hewlett Packard	Rockland Technology
Higgins Analytical	
Keystone	
Macherey Nagel	
MICRA Scientific	SynChrom
Perkin-Elmer	Applied Biosystems, Perseptive Biosystems
Poly LC	
Separations Group	
Supelco	
Toso Haas	
Thermo Quest	Hypersil
Waters	Phase Separations, YMC

2. Derivatization

There are several reasons why the surface of silica is usually modified with a bonded phase before use as an HPLC column packing. First, the bonded phase imparts a selective chemistry for the specific interactions characteristic of each mode of chromatography. Second, nonspecific interactions of the surface silanols are reduced or eliminated by the derivatizing process. Finally, the bonded phase sometimes increases the stability of the silica backbone by shielding it from caustic agents in the mobile phase.

There are three primary categories of bonded phases based on their linkage to the silica support, as illustrated in Fig. 5.10 (1).

Figure 5.10 Bonded Phases

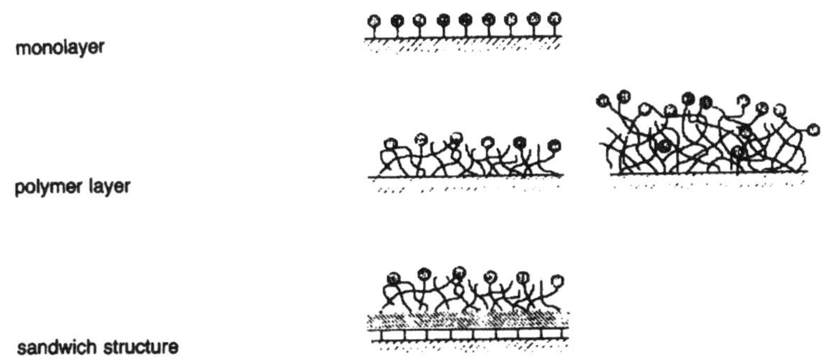

monolayer

polymer layer

sandwich structure

Reprinted from ref. 1 with permission of Wiley-VCH Publishers.

The most chemically simple bonded phase is the monolayer or so-called brush coating. This bonded phase is attached directly to the silica through a monofunctional silane which protrudes, not unlike the bristles of a brush. The intended interactions occur at the end of the chain farthest from the silica matrix.

A di- or trifunctional silane produces a layer which is slightly more complex because it possesses some crosslinking between the ligands. This polymerization creates molecular heterogeneity at the silica surface, as can be seen in Fig. 5.10. In addition to hydrophobic properties, the polymeric bonded phase structure provides a spatial display of interactive sites, giving it some shape selectivity, especially for polynuclear hydrocarbons (18). Sometimes polymeric layers also increase retention or loading capacity for small molecules, if ligand density is high.

The sandwich structure, in a further step of complexity, employs a bonded phase constructed from multiple layers. An initial layer is bonded to the silica to provide reactive groups and/or impart chemical resistance and a second layer is added to provide the desired functionality (1, 19).

Silane chemistry is a very effective tool for modifying and enhancing the utility of the silica base support. For brush coatings, silica is easily derivatized with a monofunctional silane which possesses a leaving group X⁻ (i.e. Cl⁻ or OR⁻) to link the particular functional group (R) onto the free hydroxyls, as seen in Fig. 5.11a (20). By using a silane reagent with two or three leaving groups, trichlorosilane for example, a polymeric bonded phase is obtained, as shown in Fig. 5.11b.

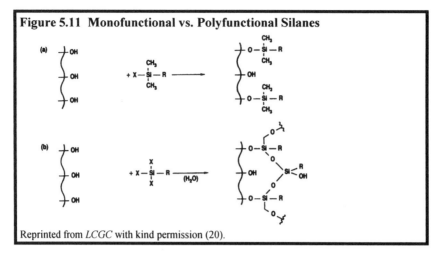

Figure 5.11 Monofunctional vs. Polyfunctional Silanes

Reprinted from *LCGC* with kind permission (20).

The synthesis of multilayer or sandwich bonded phases which contain the functional groups in the outermost layer is more complex. They may be constructed by using a silane with a reactive group for the initial layer and subsequently derivatizing it with a polymeric amine or alcohol which provides functional and/or crosslinking groups (19). Another scheme for the synthesis of multilayer bonded phases adsorbs a polyethyleneimine layer to the silica; this is later crosslinked with a compound containing nucleophilic leaving groups and various hydrophobic functionalities (21). Sandwich bonded phases are most frequently employed in ion exchange and hydrophobic interaction chromatography of proteins where surfaces must be biocompatible and nondenaturing, as well as capable of causing separation.

More details about specific bonded phases will be found in the chapters which describe the individual modes of HPLC. Examples of functional (R) groups used in several popular methods of chromatography are shown in Table 5.5.

Table 5.5 Functional Groups Used for Specific Chromatographic Modes

R Group	Mode of Chromatography
C4 (Butyl) C8 (Octyl) C18 (ODS, octadecyl)	Reversed Phase
DEAE (diethylaminoethanol) PEI (polyethyleneimine) QAE (quaternaryaminoethyl)	Anion Exchange
CM (carboxymethyl) SP (sulfopropyl)	Cation Exchange
Glycerylpropyl	Size Exclusion
Propyl (C3) Phenyl	Hydrophobic Interaction

3. Limitations of Silica

Silica supports have several noteworthy disadvantages despite their great utility in HPLC. Silica has been known to dissolve as the pH increases above pH 7 and is generally not used above pH 8. Fig. 5.12 illustrates the rapid dissolution of one silica as the pH was increased. In actuality, some silica is more stable than in this case because the solubility is dependent on a myriad of factors including composition, impurities (especially metals), particle diameter, the manufacturing process and hydration (23). Recent studies by Kirkland suggest that the relevant factors of silica composition on solubility include surface area, pore wall thickness and impurity levels (24). For example, a sol-gel silica exhibited a significantly slower rate of dissolution than a sil-gel support, as seen in Fig. 5.13.

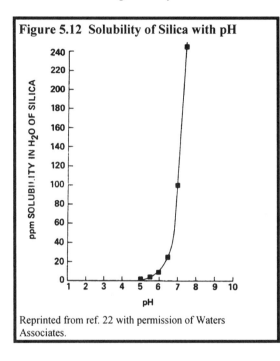

Figure 5.12 Solubility of Silica with pH

Reprinted from ref. 22 with permission of Waters Associates.

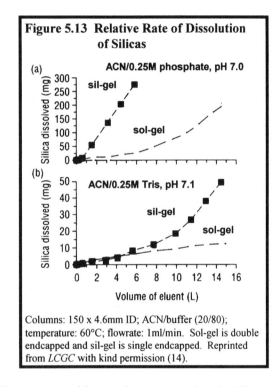

Figure 5.13 Relative Rate of Dissolution of Silicas

Columns: 150 x 4.6mm ID; ACN/buffer (20/80); temperature: 60°C; flowrate: 1ml/min. Sol-gel is double endcapped and sil-gel is single endcapped. Reprinted from *LCGC* with kind permission (14).

The rate of dissolution varies between silicas with different compositions and structures. Certain silicas have been seen to be stable to pH 10, probably because of their specific impurities or method of synthesis (24). The nature of the bonded phase can further improve the stability of the support by protecting the silica from hydrolysis by means of a hydrophobic or polymeric barrier (14, 24). Generally, bonded phases enhance the stability of silica supports.

Fig. 5.13 also illustrates that the buffer and operating conditions radically affect the dissolution rate. Phosphate causes rapid dissolution on both types of silica, whereas in the organic buffer, Tris, they are m ore stable. Note that the scale in the lower graph is a factor of six lower than that of the upper one. Other conditions which enhance silica stability are the use of temperatures under 40°C and ionic strengths less than 50mM (14).

In weakly acidic to neutral environments (pH 4 - 7), the silanol group ionizes: $SiOH \rightarrow SiO^-$, imparting a negative charge to the silica surface and, in effect, creating a cation exchange matrix. The hydrogen bonds formed by exposed silanols with basic and other molecules are usually unpredictable and irreproducible. The silanol linkage between the silane and the silica is susceptible to hydrolysis under strongly acidic conditions, further limiting the range of useful pH.

4. Strategies for Improving the Stability of Silica-Based Supports

Because of the limitations described above for silica supports, manufacturers have striven to develop methods to stabilize this matrix. One approach has been to coat silica with zirconium or its derivatives, which improves its stability by reducing dissolution at alkaline pH (25). The zirconyl-clad silica is further derivatized to provide functionality, as above. ZORBAX® GFC supports are stabilized in this way.

Polymerized bonded phases improve the stability of silica packing materials. If a crosslinked silane is used, a two point attachment of the R groups to the silica backbone is formed so that if one of the hydrolytically-sensitive bonds is attacked and broken, the moiety still remains attached, available for chromatography (26). This effect is similarly achieved by the polymeric bonded phases shown in Fig. 5.11b. Hydrolysis of the silanol bond is an equilibrium reaction; therefore, the silanol generated from hydrolysis may reattach if it is appropriately positioned.

Another method to improve stability of reversed phase supports, illustrated in Fig. 5.14, utilizes a monofunctional silane that has two relatively bulky and hydrophobic protective groups (labeled R) to derivatize the silica (20).

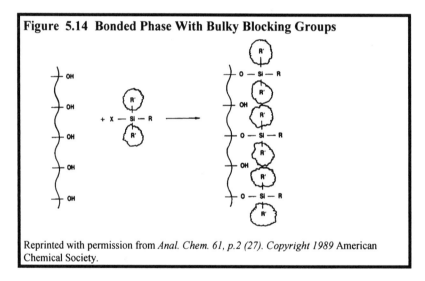

Figure 5.14 Bonded Phase With Bulky Blocking Groups

Reprinted with permission from *Anal. Chem. 61, p.2 (27). Copyright 1989* American Chemical Society.

These groups inhibit the access of water to the phase attachment point and thus hinder the degradation of the siloxane bond of silica. Using densely reacted, endcapped, long chain alkyl bonded phases also creates a hydrophobic barrier to hydrolysis (14).

B. Polymer-based Supports

1. Composition

Polymeric supports are also used as an HPLC matrix. These range from soft to rigid and hydrophilic to hydrophobic, depending on the chemical composition. Carbohydrate gels were the first media to be successfully used for protein purification (28) and the continued popularity of Sephadex and related materials attest to their utility. One type of carbohydrate matrix used for chromatography at moderate pressures is agarose, which is usually modified to increase rigidity or to modify physical or chemical properties. Agarose gels are very hydrophilic but not very robust to pressure. Their porosity properties are related to the solvent and the operating pressure.

Poly(styrene-divinylbenzene) (PSDVB) supports can be formulated to be very rigid with maximum operating pressures of 5000psi (29). However, they are extremely hydrophobic and must be modified before use with biopolymers, which generally call for a hydrophilic surface with the ability to wet. Bare polystyrene is very hydrophobic and can sometimes be used without modification as a reversed phase matrix. PSDVB polymers are very robust and compatible with many solvents including acids or bases.

A family of polymeric supports which combine hydrophilicity and pressure stability are those based on methacrylate (30-31). This technology has been extensively studied and enhanced at the Institute of Macromolecular Chemistry in the Czech Republic in collaboration with several Czech companies. Recent developments in the synthesis of methacrylate supports have increased the hydrophilicity as well as the stability (31).

Some polymeric supports experience bed instability caused by a change of solvent under pressure; the subsequent swelling or contraction can alter the bed volume, leading to channeling, cracking or other catastrophic failures of the matrix. A major advantage of many polymer-based supports is that they can be cleaned with base. Table 5.6 lists some properties of polymeric supports and selected suppliers.

Table 5.6 Properties and Suppliers of Polymeric LC Supports

	Polystyrene	Methacrylate	Agarose
hydrophobicity	very	fairly	hydrophilic
pressure stability	5000psi	4300psi	220psi
pore structure	permanent	permanent	defined by solvent and pressure
pH range	1 - 13	2 - 12	3 - 12
durability	high	good	moderate
Sources	Bio-Rad	Bio-Rad	Amersham Pharmacia
	Hamilton	E. Merck	Bio-Rad
	Polymer Labs	Labio	
	Perseptive Biosystems	Polymer Labs	
	Amersham Pharmacia	Showa Denko	
		Toso Haas	

Information from ref. 29, 31 and 32.

2. Derivatization

Polymerized gels are usually chemically modified with functional groups like those listed in Table 5.5 before use as HPLC supports. The process for synthesis is dependent on both the chemical composition of the polymeric base and the desired bonded phase (30). The synthesis is often more complex than the silane chemistry used for silica derivatization but most of the bonded phases are not as prone to hydrolysis. Methacrylate and carbohydrate gels are usually reacted through the hydroxyl or epoxy groups (33). Carboxylic acid functionalities can often be incorporated during synthesis of the support matrix. Polystyrene-based matrices require more vigorous conditions for derivatization, such as boiling with fuming sulfuric acid to produce a cation-exchange support or a two stage reaction using chloromethyl ether to allow attachment of other ligands like an amine for an anion-exchange bonded phase (34). If PSDVB supports are not totally covered with bonded phase or a hydrophilic layer, proteins and peptides are likely to exhibit hydrophobic interactions in addition to the expected associations.

The actual attachment of the functional group to the support can provide specific spatial characteristics for the attached ligand (35-36), as illustrated in Fig. 5.15 for affinity supports.

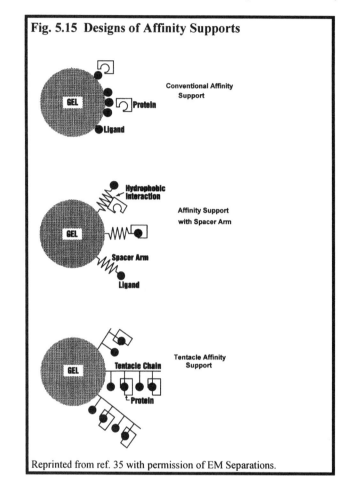

Fig. 5.15 Designs of Affinity Supports

Conventional Affinity Support

Protein

Ligand

Hydrophobic Interaction

Affinity Support with Spacer Arm

Spacer Arm

Ligand

Tentacle Chain

Tentacle Affinity Support

Protein

GEL

Reprinted from ref. 35 with permission of EM Separations.

The top panel represents a "classical" gel and its direct attachment of ligand. As shown, not every ligand attached to the gel is able to interact with and bind a protein because the ligand density is high and the binding of one protein blocks the binding of another. In the late 1960's, this problem of interference was reduced by the use of a "spacer arm" (37). This linker positioned the ligand 5 - 10Å from the gel matrix like quills on a porcupine (Fig. 5.15, center). Unfortunately, the spacer arms are hydrophobic and capable of providing sites for interactions other than those designed for the protein and ligand.

A further development in the late 1980's by E. Merck was the "tentacle affinity support" (36). In this variation, a flexible chain made of a polyelectrolyte extends from the gel (Fig. 5.15, lower panel). This "tentacle", which serves as a connector to attach various ligands, has a number of advantages over the other two designs of affinity support. The available ligand density of the tentacle support is substantially higher. The tentacle is hydrophilic and virtually eliminates any nonspecific interactions. The tentacle mobility permits rapid binding kinetics, while reducing the tendency for certain labile proteins to denature, as is known to occur on more rigid supports. The stereochemistry of the ligand can also be tailored to increase its selectivity. If the ligand itself is expensive, as in the case of antigens used in immunoaffinity chromatography, utilization of ligand can be optimized by substantially lowering the ligand density on the gel so that the steric hindrance to subsequent protein binding is virtually eliminated. This protocol may also result in low binding capacity. Tentacle phases are commercially available from E. Merck for ion-exchange, metal chelation, and thiophilic adsorption chromatography, as well as affinity.

C. Comparison of Silica- and Polymer-based Supports

Silica- and polymer-based supports are both used for HPLC of biomolecules. The choice is dependent on the requirements of the specific analysis. Silica-based supports are most often used for reversed phase chromatography due to their excellent efficiency and selectivity. The ability of the unreacted silanols to wet yields a different environment than the hydrophobic surface of polystyrene. Both silica- and polymer-based supports are used for ion-exchange and hydrophobic interaction chromatography of proteins and peptides; generally, the chemistry used for these packings totally covers the support matrix and renders it inconsequential. Size exclusion chromatography is still frequently carried out on carbohydrate gels like agarose but HPLC often implements silica-based supports with glyceryl bonded phases. High pH can be used to clean polymeric columns - a major benefit in pharmaceutical purification methods. Table 5.7 lists some of the advantages and disadvantages of each matrix. The polymeric supports in this table only represent the rigid polystyrene and methacrylate categories which are stable at typical HPLC pressures. Variations in each category of the specific composition or the porosity will effect the characteristics.

Table 5.7 Silica- vs. Polymer-based Supports	
ADVANTAGES	
Silica	**Polymer**
easily derivatized	pH stable, even in acid and base
rigid	very durable
high pressure stability (> 10,000psi)	
DISADVANTAGES	
limited pH range	limited pressure stability ($\leq$ 5000psi)
silanol effects	matrix may be hydrophobic

III. STRATEGIES FOR IMPROVING MASS TRANSFER

A. Nonporous Supports

As described in Chapter 2 on chromatographic theory, factors which impede mass transfer in either the mobile phase or the stationary phase reduce column efficiency by increasing the C term of Eq. 5.1 (or Eq. 2.27). The use of nonporous particles is one solution to the problem because it eliminates the C term altogether (7, 8). The protein analysis in Fig. 5.16 shows that very fast separations with excellent resolution can be achieved on nonporous supports (38). An elevated temperature was used to reduce the pressure at the high flowrate on this 2µm support.

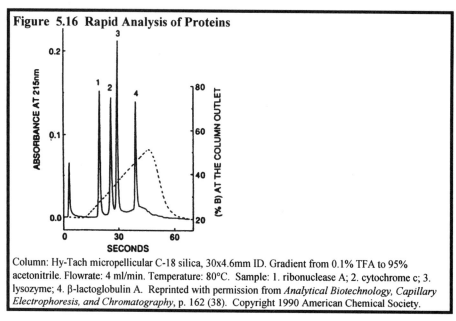

Figure 5.16 Rapid Analysis of Proteins

Column: Hy-Tach micropellicular C-18 silica, 30x4.6mm ID. Gradient from 0.1% TFA to 95% acetonitrile. Flowrate: 4 ml/min. Temperature: 80°C. Sample: 1. ribonuclease A; 2. cytochrome c; 3. lysozyme; 4. β-lactoglobulin A. Reprinted with permission from *Analytical Biotechnology, Capillary Electrophoresis, and Chromatography*, p. 162 (38). Copyright 1990 American Chemical Society.

For analytical applications or small samples, nonporous supports provide excellent resolution; however, for larger masses they are unsatisfactory because of their low surface area and loading capacity. When very small particles are used, the loss of capacity is regained somewhat, but it does not approach that of porous particles. Another characteristic of nonporous supports is that lower solvent strengths are required for elution. The extremely low organic solvent concentrations plus the high sensitivity of the separations to small changes in mobile phase sometimes present difficulties in optimizing separations.

B. Large Pores

Particles with large pores (≥ 100nm), also minimize mass transfer problems because they permit ready access for high molecular weight proteins; however, this is also achieved at the expense of capacity (10, 39). Fig. 5.17 illustrates the excellent resolution of proteins that can be achieved on a 400nm reversed phase support (40).

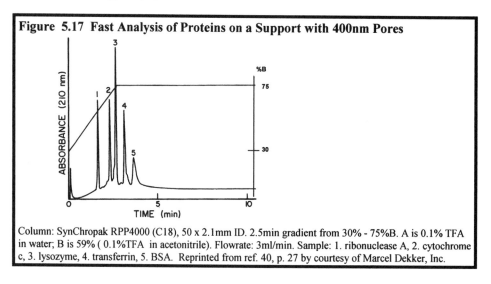

Figure 5.17 Fast Analysis of Proteins on a Support with 400nm Pores

Column: SynChropak RPP4000 (C18), 50 x 2.1mm ID. 2.5min gradient from 30% - 75%B. A is 0.1% TFA in water; B is 59% (0.1%TFA in acetonitrile). Flowrate: 3ml/min. Sample: 1. ribonuclease A, 2. cytochrome c, 3. lysozyme, 4. transferrin, 5. BSA. Reprinted from ref. 40, p. 27 by courtesy of Marcel Dekker, Inc.

C. Perfusion Supports

Considerations about reducing mass transfer without an inordinate decrease in loading capacity led to the development of "perfusion" chromatography by Perceptive Biosystems, Inc. (41). The "perfusive" particle has a series of primary and secondary channels running through it, as shown in Fig. 5.18.

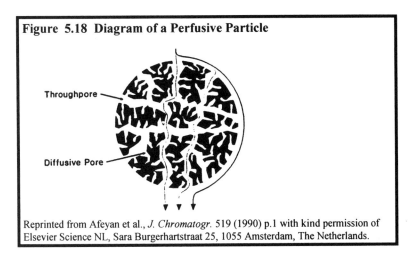

Figure 5.18 Diagram of a Perfusive Particle

Throughpore

Diffusive Pore

Reprinted from Afeyan et al., *J. Chromatogr.* 519 (1990) p.1 with kind permission of Elsevier Science NL, Sara Burgerhartstraat 25, 1055 Amsterdam, The Netherlands.

The primary pore, called a throughpore, is large enough to allow macromolecules to pass through, with little, if any, restriction. The secondary pores are diffuse and shallow. These secondary pores help regain some of the capacity lost to the throughpores, yet their shallow depths minimize the stagnant mobile phase effects typical of deep pores.

A comparison of bandspreading as a function of flow velocity for a packing with 30nm pores and the perfusion support is shown in Fig. 5.19. Note that the bandspreading is given in terms of reduced plate height in this graph; therefore, the actual plate height would be this value multiplied by the particle diameter (*see* Eq. 2.28).

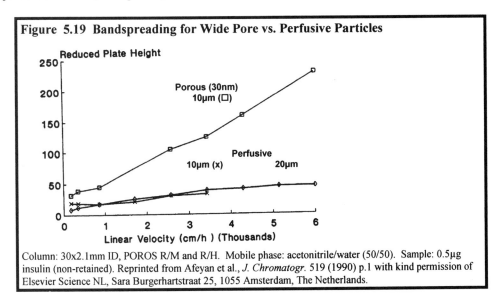

Figure 5.19 Bandspreading for Wide Pore vs. Perfusive Particles

Reduced Plate Height

Porous (30nm)
10µm (□)

Perfusive

10µm (x)   20µm

Linear Velocity (cm/h) (Thousands)

Column: 30x2.1mm ID, POROS R/M and R/H. Mobile phase: acetonitrile/water (50/50). Sample: 0.5µg insulin (non-retained). Reprinted from Afeyan et al., *J. Chromatogr.* 519 (1990) p.1 with kind permission of Elsevier Science NL, Sara Burgerhartstraat 25, 1055 Amsterdam, The Netherlands.

Little change in efficiency is observed with the perfusion packing over a fivefold range of flow, in marked contrast to the increase in plate height which occurs with the 30nm pore packing. The difference between efficiencies on the porous and perfusive columns is less than this for small molecule solutes (41). The minimal effects of increased flow velocity on the resolution of four proteins is further demonstrated in Fig. 5.20. The gradient slope was kept constant by decreasing the time by the same factor the flowrate was increased.

Figure 5.20 Effect of Flowrate on Resolution for a Perfusive Support

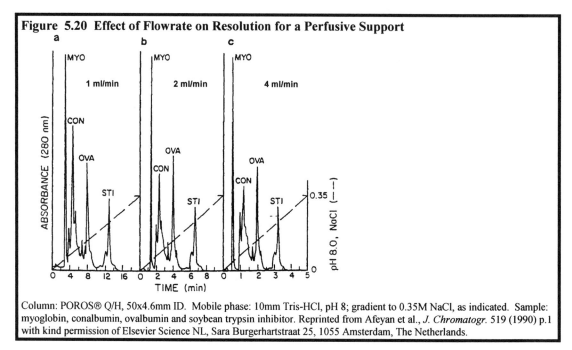

Column: POROS® Q/H, 50x4.6mm ID. Mobile phase: 10mm Tris-HCl, pH 8; gradient to 0.35M NaCl, as indicated. Sample: myoglobin, conalbumin, ovalbumin and soybean trypsin inhibitor. Reprinted from Afeyan et al., *J. Chromatogr.* 519 (1990) p.1 with kind permission of Elsevier Science NL, Sara Burgerhartstraat 25, 1055 Amsterdam, The Netherlands.

D. Comparison

No single category of support is superior in all aspects. Porous supports are most commonly used and the only choice for size exclusion chromatography. Nonporous supports yield excellent efficiency and ultrafast analyses for small samples. Table 5.7 lists some of the advantages and disadvantages of supports with different porosity characteristics.

Table 5.8 Comparison of Supports with Different Pore Characteristics

Porous 2 - 10µm	Perfusive 10 - 20µm	Nonporous 1.5 - 2.5µm
Advantages		
high efficiency suitable for size exclusion high loading	average efficiency good mass transfer low pressure	very high efficiency excellent mass transfer very fast analyses
Disadvantages		

mass transfer problems at
  high flowrates
	compromise of loading
	  and efficiency
		high pressure
		low loading
		very sensitive to extra
		  column volume

IV. NEW DEVELOPMENTS

Many of the negative aspects of HPLC columns such as high pressures and column instability would be alleviated if a solid bed could give the same performance as packed individual particles (42). Several research groups have addressed this issue and investigated the feasibility of solid bed columns from different perspectives.

A. Gel Beds

1. Compressed Gels

In an effort to obtain the performance and convenience of electrophoresis gels with HPLC columns, Hjerten and colleagues prepared acrylamide gels inside columns (43). The gels were later compressed at pressures greater than those used for operation. The reproducible process resulted in 3 - 4μm channels between the aggregated polymers which function as throughpores (44). These acrylamide gels have low cost, ease of preparation and good stability. They also possess high dynamic loading capacity with no loss at high flowrates, as shown in Fig. 5.21. Note that the x-axis is volume; therefore, the time of analysis will decrease as the flow is increased. Bio-Rad Laboratories has developed the technology as UNO™ columns.

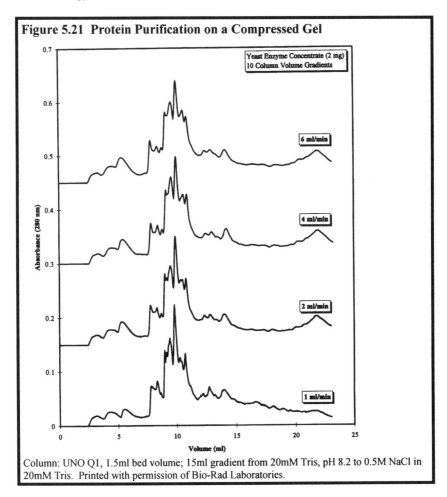

Figure 5.21 Protein Purification on a Compressed Gel

Column: UNO Q1, 1.5ml bed volume; 15ml gradient from 20mM Tris, pH 8.2 to 0.5M NaCl in 20mM Tris. Printed with permission of Bio-Rad Laboratories.

2. Molded Gels

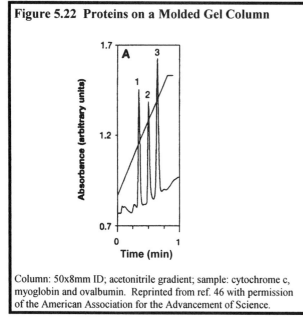

Figure 5.22 Proteins on a Molded Gel Column

Column: 50x8mm ID; acetonitrile gradient; sample: cytochrome c, myoglobin and ovalbumin. Reprinted from ref. 46 with permission of the American Association for the Advancement of Science.

Rigid continuous beds made of macroporous polymer were formed by Svec and colleagues by polymerization of methacrylate directly in columns (45-46). Derivatization was achieved by pumping in the appropriate reagents after the rod had been formed. The monoliths prepared by this process contain small diffusive pores and a large number of throughpores of 700 - 2000nm (46) which result in high efficiency. Fast flowrates, as seen in Fig. 5.22, can be used to analyze proteins with high resolution. ISCO is currently developing this technology.

B. Silica Rods

Silica has been shaped into a continuous bed as a result of research where rods were formed in molds by reaction of tetramethoxysilane and poly(ethylene oxide) (42). These rods were then packed into radial compression columns which conformed to their shapes. The continuous rods contain throughpores of about 1.7μm and mesopores of either 14 or 25nm. They have high porosity (~ 85%) and low pressure drop due to the rod structure. Derivatization is performed *in situ* on the column. The shallow mesopores eliminate most mass transfer problems and allow fast flowrates to be used effectively, as seen in Fig. 5.23 for insulin. These rods are being further developed by E. Merck.

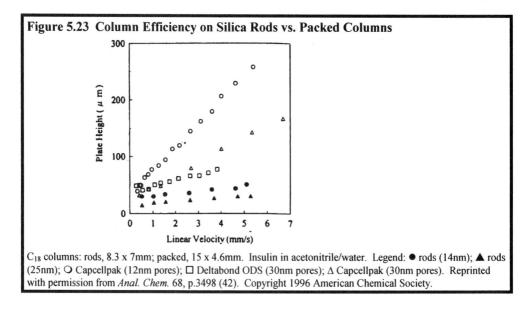

Figure 5.23 Column Efficiency on Silica Rods vs. Packed Columns

C$_{18}$ columns: rods, 8.3 x 7mm; packed, 15 x 4.6mm. Insulin in acetonitrile/water. Legend: ● rods (14nm); ▲ rods (25nm); O Capcellpak (12nm pores); □ Deltabond ODS (30nm pores); Δ Capcellpak (30nm pores). Reprinted with permission from *Anal. Chem.* 68, p.3498 (42). Copyright 1996 American Chemical Society.

For macromolecules, the 25nm pore consistently gives higher efficiency than the 14nm, similar to standard porous supports. At this time, the primary benefits of these columns appear to be for macromolecules, but advances in the technology may yield other advantages.

V. VALIDATED COLUMNS

An important issue in HPLC is column reproducibility and validation. HPLC columns have been notorious for variability between different manufacturers and even different batches. This has been due to batch-to-batch differences between silica lots, as well as manufacturing procedures. As HPLC has progressed from research applications to use as a critical analysis tool in manufacturing and quality control in the pharmaceutical industry, this situation has become intolerable. As a result, most column manufacturers are now striving for and achieving good reproducibility in many of their lines.

A. USP Designation

One attempt at standardization has been the use of the US Pharmacopoeia (USP) designations which list physical properties of HPLC supports (47). For example, USP L1 is octadecylsilane (ODS or C18) chemically bonded to porous silica or ceramic particles which are 3 - 10μm in diameter. USP designations allow broad categorization, but no guaranteed reproducibility due to the nonconformity of the HPLC industry. Validated methods must be carefully designed to be robust in terms of both the operating conditions and the column (48). It is best to indicate a general column description in a method so that a brand substitution can be made, if necessary. It is also imperative that a method be validated on multiple batches and preferably, a second brand of column. Many HPLC column suppliers provide lists of their columns which meet USP specifications.

B. Universal Column

Another proposed solution to validation has been the adoption of a "universal" column to be the benchmark of all reversed phase columns (49). Obviously, this is very controversial because one manufacturer would have the monopoly on this column. The proposal has generated a positive consequence in that many manufacturers are now using a battery of standard tests on each batch to test variations and conformity (50). The exact conditions for the evaluations are still being debated but, at present, it is fairly universally accepted that they include the categories in Table 5.9 (48). Such tests give good indications of metal ion content, efficiency, hydrophobicity, acidity, and stability.

> **Table 5.9 Vital Chromatographic Parameters for Reversed Phase Columns**
>
> Column-to-column reproducibility
> - plate number
> - peak symmetry
> - selectivity and adsorption phenomena: amines, acids and isomers
>
> Column lifetime data with aging at two pH values
> - loss of k'
> - loss of column performance
> - change in peak symmetry
> - change in hydrophobic retention
> - change in silanophilic retention
>
> Reprinted from <u>American Laboratory</u>, volume 29, number 8, page 40, 1997. Copyright 1997 by International
> Scientific Communications, Inc.

     All efforts thus far have been directed towards uniformity in small pore reversed phase columns. It is unlikely that these same efforts will be made for every variation of mode and bonded phase, but a series of identical tests of typically-run solutes would certainly aid in column selection.

VI. REFERENCES

1. K.K. Unger, K.D. Lork, and H.-J. Wirth, "Development of Advanced Silica-Based Packing Materials" in *HPLC of Proteins, Peptides and Polynucleotides* (M.T.W. Hearn, ed.), VCH, New York, 1991.
2. N. Tanaka, K. Hashidzume, M. Araki, H. Tsuchiya, A. Okuno, K. Iwaguchi, S. Ohnishi, and N. Takai, *J. Chromatogr.* 448 (1988) 95.
3. *LiChrospher & LiChroprep Sorbents Tailored for Cost Effective Chromatography*, EM Separations, Gibbstown.
4. *HPLC Columns and Separation Materials*, 1996-1997, The Separations Group, Hesperia, CA.
5. L.R. Snyder, "Theory of Chromatography" in *Chromatography, 5th Edition* (E. Heftmann, ed.), Elsevier, Amsterdam, 1992.
6. R.E. Majors, *LC-GC Current Issues in HPLC Technology* (1997) S8.
7. K. Kalghatgi and Cs. Horvath, *J. Chromatogr.* 398 (1987) 335.
8. K.K. Unger, G. Jilgs, J.N. Kinkel, and M.T.W. Hearn, *J. Chromatogr.* 359 (1986) 61.
9. G. Dollinger, B. Cunico, M. Kunitani, D. Johnson, and R. Jones, *J. Chromatogr.* 592 (1992) 215.
10. G. Vanecek and F.E. Regnier, *Anal. Biochem.* 109 (1980) 345.
11. E. Pfannkoch, K.C. Lu, F.E. Regnier, and H.G. Barth, *J. Chromatogr. Sci.* 18 (1980) 430.
12. M.T.W. Hearn and B. Grego, *J. Chromatogr.* 282 (1983) 541.
13. K.K. Unger, "Silica as a Support" in *HPLC of Biological Macromolecules: Methods and Applications* (K.M. Gooding and F.E. Regnier, eds.), Marcel Dekker, New York, 1990.
14. J.J. Kirkland, *LC-GC Current Issues in HPLC Technology* (1997) S46.
15. J. Kohler and J.J. Kirkland, *J. Chromatogr.* 385 (1987) 125.
16. J.J. Kirkland, personal communication.
17. P.J. van den Driest, H.J. Ritchie, and S. Rose, *LC-GC* 6 (1988) 124.
18. L.C. Sander and S.A. Wise, *LC-GC* 8 (1990) 378.
19. S.H. Chang, K.M. Gooding, and F.E. Regnier, *J. Chromatogr.* 120 (1976) 321.
20. J.L. Glajch and J.J. Kirkland, *LC-GC* 8 (1990) 140.
21. A.J. Alpert and F.E. Regnier, *J. Chromatogr.* 185 (1979) 375.
22. "Basic Liquid Chromatography Course Book," Waters Associates, Milford, p. CS-2A.

23. R.K. Iler, *The Chemistry of Silica*, John Wiley & Sons, New York, 1979, Chapter 1.
24. J.J. Kirkland, M.A. van Straten, and H.A. Claessens, *J. Chromatogr. A* 691 (1995) 3.
25. R.W. Stout and J.J. DeStefano, *J. Chromatogr.* 326 (1985) 63-78.
26. U. Esser and K.K. Unger, "Reversed-Phase Packings for the Separation of Peptides and Proteins by Means of Gradient Elution HPLC" in *High Performance Liquid Chromatography of Peptides and Proteins* (ed. C.T. Mant and R.S. Hodges), CRC Press, Boca Raton, 1991.
27. J.J. Kirkland, J.L. Glajch, and R.D. Farlee, *Anal. Chem.* 61 (1989) 2.
28. J. Porath and P. Flodin, *Nature (London)* 183 (1959) 1657.
29. *Hamilton HPLC Application Handbook*, Hamilton, Reno, 1993.
30. O. Mikes and J. Coupek, "Organic Supports" in *HPLC of Biological Macromolecules: Methods and Applications* (K.M. Gooding and F.E. Regnier, eds.), Marcel Dekker, New York, 1990.
31. S. Vozka, *LC-GC Current Issues in HPLC Technology* (1997) S56.
32. *BioDirectory '96*, Pharmacia Biotech, Uppsala, 1996.
33. F. Svec, "New Organic Polymer Support Materials" in *HPLC of Biological Macromolecules: Methods and Applications, 2nd Edition* (K.M. Gooding and F.E. Regnier, eds.), Marcel Dekker, New York, in press.
34. H.F. Walton, "Ion-Exchange Chromatography" in *Chromatography, 5th Edition* (E. Heftmann, ed.), Elsevier, Amsterdam, 1992.
35. *Tentacle Affinity Bioseparation Media*, EM Separations, Gibbstown, 1992.
36. W. Muller, *J. Chromatogr.* 510 (1990) 133.
37. P. Cuatrecasas, M. Wilchek, and C.B. Anfinsen, *Proc. Natl. Acad. Sci. USA* 37 (1968) 636.
38. K. Kalghatgi and Cs. Horvath, "Micropellicular Sorbents for Rapid Reversed-Phase Chromatography of Proteins and Peptides" in *Analytical Biotechnology, Capillary Electrophoresis and Chromatography* (ed. Cs. Horvath and J.G. Nikelly), American Chemical Society, (1990) 162.
39. M.P. Nowlan and K.M. Gooding, "High-Performance Ion-Exchange Chromatography of Proteins" in *High Performance Liquid Chromatography of Peptides and Proteins* (ed. C.T. Mant and R.S. Hodges), CRC Press, Boca Raton, 1991.
40. M. Kawakatsu and K.M. Gooding, *J. Liq. Chromatogr.* 16 (1993) 21.
41. N.B. Afeyan, N.F. Gordon, I. Mazsaroff, L. Varady, S.P. Fulton, Y.B. Yang, and F.E. Regnier, *J. Chromatogr.* 519 (1990) 1.
42. H. Minakuchi, K. Nakanishi, N. Soga, N. Ishizuka, and N. Tanaka, *Anal. Chem.* 68 (1996) 3498.
43. S. Hjerten, J.-L. Liao, and R. Zhang, *J. Chromatogr.* 473 (1989) 273.
44. J.-L. Liao, R. Zhang, and S. Hjerten, *J. Chromatogr.* 586 (1991) 21.
45. F. Svec and J.M.J. Frechet, *Anal. Chem.* 64 (1992) 820.
46. F. Svec and J.M.J. Frechet, *Science* 273 (1996) 205.
47. *Chromatographic Reagents Used in USP.NF and Pharmacopeial Forum*, United States Pharmacopeial, 1991.
48. K.M. Gooding and M.N. Schmuck, *Biotechniques* 11 (1991) 232.
49. G. Wieland, K. Cabrera, and W. Eymann, *LC-GC* 15 (1997) 98.
50. M.P. Henry, I. Birznieks, and M.W. Dong, *Amer. Lab.* 29, April 1997, 40.

CHAPTER 6
SIZE EXCLUSION CHROMATOGRAPHY

Size exclusion chromatography (SEC), as the name implies, separates molecules by selectively excluding them from a porous matrix based on their size. Historically, this type of chromatography has been called gel filtration (GFC), gel permeation (GPC), and steric exclusion (SEC), but size exclusion has become the accepted nomenclature, especially in HPLC. Supports used for SEC are highly porous, having carefully controlled pore diameters with narrow distribution. It is important that the surface of the support be totally noninteractive with the solutes. In the size exclusion process, molecules which are excluded from all pores go directly through the channels between the particles and elute first. Small molecules, which have total access to the pores, flow in and out as they migrate and elute last. Intermediate-sized molecules have partial access to the pores and elute between the included and excluded solutes. Fig. 6.1 diagrams this concept. If this support were designed for SEC, the pore diameters would be tightly controlled to have minimum variance.

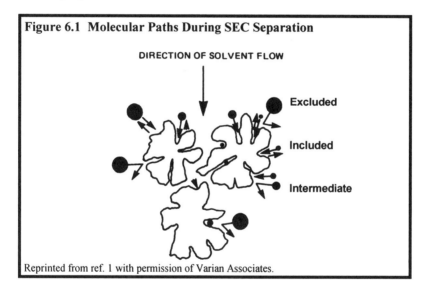

Figure 6.1 Molecular Paths During SEC Separation

Reprinted from ref. 1 with permission of Varian Associates.

I. THEORY

A. Retention Mechanism

The theory underlying the technique of SEC is the simplest of all chromatography because, in contrast to other methods of separation, retention is uniquely defined by the excluded and included volumes of the column (2-4). Following injection of sample onto the column, molecules diffuse in and out of the solvent-filled pores with the following results:

1. Molecules as large or larger than the pores are totally excluded from them and are the first to elute. They do so simultaneously at a specific column volume (V_o) called the exclusion volume or the void volume.
2. Molecules at or below a certain size which can totally penetrate the pores elute together unresolved in a volume called the mobile phase volume (V_M), total volume, included volume, or permeation volume. This peak should signify the end of the run.

3.   Molecules with intermediate sizes elute between the extremes of the excluded and included volumes. They can often be separated from other components in the mixture, but total resolution is limited to about ten species.

Fig. 6.2 whimsically represents the concept of retention order in SEC for analytes with vast differences in size. In this representation, objects of widely varying sizes - the planet Saturn, three mammals and various "bugs", have been fractionated by size. Saturn is excluded from the matrix and elutes first from the imaginary column in the excluded volume; the mammals are differentially separated at 13 to 32 min during the run, and the bugs elute simultaneously and without separation at the end of the run - in the inclusion volume - even though their sizes vary substantially.

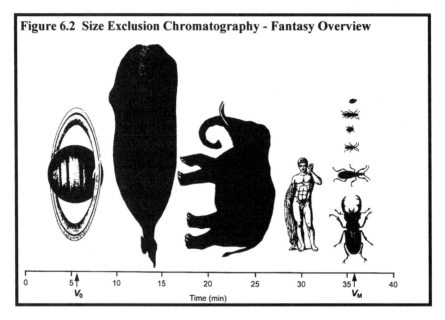

The relationship between the critical volumes in SEC is described by the following equation:

$$V_M = V_o + V_i \qquad\qquad \text{Eq. 6.1}$$

where V_o is the void volume (outside the pores), V_i is the internal volume (included within the pores), and V_M is the total volume of mobile phase in the column. The distribution coefficient is defined as:

$$K_D = \frac{V_R - V_o}{V_M - V_o} \qquad\qquad \text{Eq. 6.2}$$

where V_R is the retention or elution volume and K_D is the solute distribution coefficient. Combining the two equations yields:

$$V_R = V_o + K_D V_i \qquad\qquad \text{Eq. 6.3}$$

which illustrates that pore volume is a very important variable of retention in SEC. One measure of the porosity in SEC is the ratio of pore volume to the void volume (V_i/V_o). The larger this is, the more column volume is available for the separation. Given equal efficiency, a larger V_i/V_o will yield better resolution of more peaks (2, 4); however, supports with high V_i/V_o often have lower efficiency than those with lower pore volumes.

     The term "void volume" can be confusing if the column deadvolume is referred to as the "void". The void volume (V_o), as defined for size exclusion chromatography, is not the same as the deadvolume of a column, which is actually the total volume of mobile phase (V_M). This problem in nomenclature occurred because HPLC originally dealt only with small molecules and exclusion from pores was irrelevant. Techniques of size exclusion were used by a totally different group of scientists. The problems in terminology began when HPLC of proteins and macromolecules introduced exclusion into the mechanism of separation and the nomenclature had to be redefined.

B. Molecular Weight

     The retention volume (V_R) or retention time (t_R) in SEC is linearly related to the log of the molecular weight for a series of molecules of similar shape. The graph of log molecular weight vs. retention time, retention volume, or distribution coefficient (K_D) is known as the calibration curve of the column. Fig. 6.3 shows an example of a size exclusion chromatogram with a calibration curve corresponding to the column.

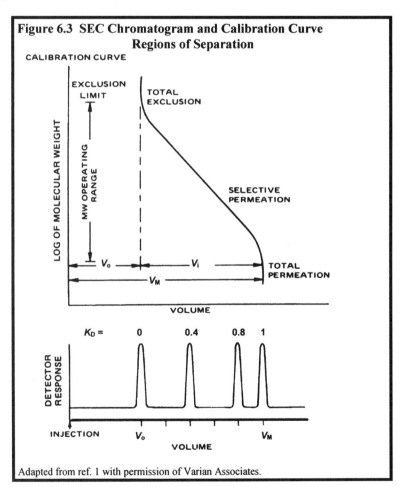

Figure 6.3 SEC Chromatogram and Calibration Curve
Regions of Separation

Adapted from ref. 1 with permission of Varian Associates.

A size exclusion calibration curve is usually linear at K_D values of about 0.2 - 0.8 with sharp slopes near the exclusion limits. Under ideal SEC conditions, all molecules have distribution coefficients between 0 and 1. Small totally included molecules elute at a K_D of one and large excluded molecules elute at a K_D of zero. This is in contrast to other methods of separation where distribution coefficients are usually larger, and can be so high that they require gradient elution.

If separations are performed under constant flow conditions, molecules will elute at characteristic volumes and therefore, at predictable times after sample injection. The retention of sample peaks or areas of biological activity can be directly compared with the retention of standard molecules of known molecular weights (M) to estimate the size of the unknowns. A variety of well-behaved marker proteins, such as those listed in Table 6.1, are used for calibration. Protein molecular weight is usually denoted in Daltons (Da) or kilodaltons (kD).

Table 6.1 Molecular Weight Standards for SEC	
STANDARDS	***M* (kD)**
Thyroglobulin	669
Ferritin	440
Catalase	232
IgG	150
Lactate dehydrogenase (LDH)	109
Amylase	100
Bovine serum albumin (BSA)	67
Ovalbumin	44
Chymotrypsinogen A	25
Ribonuclease (RNase)	13.7
α-endorphin	1.5
2'-AMP	0.4

These molecules fractionate reproducibly by SEC under nondenaturing conditions like 0.05 - 0.2M phosphate buffer, pH 6 - 7. The smaller proteins in this table (<70kD) and/or those composed of a single chain are also suitable for calibration with denaturing conditions like 5 - 6M guanidinium hydrochloride or 0.1% SDS in 0.2M sodium phosphate. It is standard practice to run a calibration curve of standards on each new column or column series before running unknown samples. Calibration will indicate if any standards interact with a support under the chosen conditions.

C. Molecular Shape

Separation in SEC is actually based on molecular size, which is determined not only by molecular weight but also by molecular shape. Because separations are occurring completely in solution, molecules are free to tumble. Hence, molecular "size" must take this into account and can be approximated by the radius of gyration (r). A compact globular protein and a rod-shaped protein of the same molecular mass "carve out" the surrounding solvent very differently; that is, their hydrodynamic volumes vary greatly. This being so, a close correlation between molecular weight and the elution volume would not be expected unless the shapes were similar. Because separation is based on differences in hydrodynamic volumes, two molecules of very similar molecular weights but different shapes may very well be separated by SEC. Fig. 6.4 demonstrates how the shape of a molecule influences the Stokes radius observed during tumbling in solution.

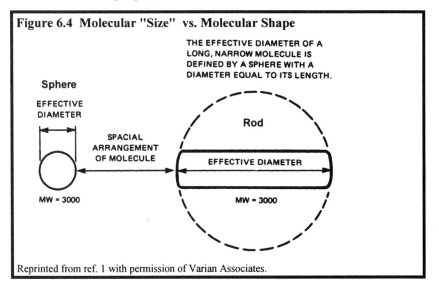

Figure 6.4 Molecular "Size" vs. Molecular Shape

Reprinted from ref. 1 with permission of Varian Associates.

The two molecules in this figure are each 3000 Da but, based on their shapes, they exhibit distinct hydrodynamic volumes as they diffuse and spin in solution. This tumbling creates very different effective diameters. Table 6.2 lists the molecular weights and gyration radii for seven selected macromolecules (2, 5). Note that the muscle protein myosin is approximately twice the molecular weight of the globular protein catalase, yet it generates a radius more than ten times greater.

Table 6.2. Molecular Weight Correlation With Molecular Volume		
	M	*r* (Å)
Serum albumin (globular, solid spheres)	66,000	29.8[a]
Catalase (globular)	225,000	39.8[a]
Myosin (rodlike)	493,000	468
Polystyrene fraction (flexible coil)	3.2×10^6	494
DNA (rodlike)	4×10^6	1170
Bushy stunt virus (globular)	10.6×10^6	120[a]
Tobacco mosaic virus (rigid rod)	39×10^6	924

a: obtained from x-ray scattering; all other *r* values are from light scattering. Data from Tanford, ref. 5.

There is a relationship between molecular volume, as measured by Stokes radius, and retention in SEC, irregardless of molecular weight, such that smaller molecular volumes result in higher retention This is shown in Fig. 6.5 for a mixture of proteins and other biological macromolecules on each of two size exclusion columns. The Stokes radius, which is measured by diffusion, is plotted against σ which is equivalent to K_D (6). In this study, it was found that there is even better correlation with a molecular radius determined by viscosity (6).

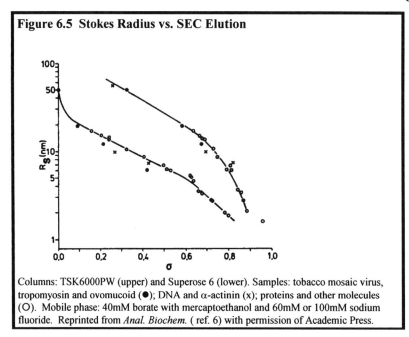

Figure 6.5 Stokes Radius vs. SEC Elution

Columns: TSK6000PW (upper) and Superose 6 (lower). Samples: tobacco mosaic virus, tropomyosin and ovomucoid (●); DNA and α-actinin (x); proteins and other molecules (O). Mobile phase: 40mM borate with mercaptoethanol and 60mM or 100mM sodium fluoride. Reprinted from *Anal. Biochem.* (ref. 6) with permission of Academic Press.

The importance of shape and volume in SEC retention is further illustrated in Fig. 6.6, where linear sulfonated polystyrenes (SPS) and globular proteins of varying molecular weights produce two calibration curves which are definitely distinct. A linear relationship exists between log molecular weight and elution for each molecular series. The more compact proteins occupy less volume and therefore elute later, at higher K_D, than polystyrenes of corresponding molecular weight. This strongly emphasizes the need for using standards of the same shape as the samples when generating a calibration curve.

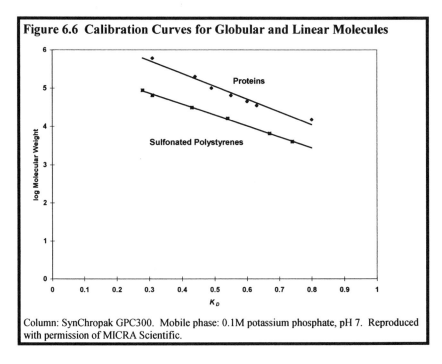

Figure 6.6 Calibration Curves for Globular and Linear Molecules

Column: SynChropak GPC300. Mobile phase: 0.1M potassium phosphate, pH 7. Reproduced with permission of MICRA Scientific.

The extended structures of the sulfonated polystyrenes give them volumes similar to those of proteins of much larger molecular weights. As seen in Fig. 6.5, plotting the Stokes radius instead of the molecular weight normalizes these differences and captures a much more accurate picture of molecular size than molecular weight does. Despite the increased accuracy, it is common practice to use calibration curves of molecules similar in shape to the sample for molecular weight estimation, because it is much easier than using the Stokes radius which is frequently unknown.

An important consequence of the shape effect in SEC is seen when surfactants or denaturants are used for protein analysis. Calibration curves of proteins using 0.1% sodium dodecylsulfate (SDS), urea or another surfactant in the mobile phase more nearly resemble those for linear polymers than globular proteins, as seen in Fig. 6.7.

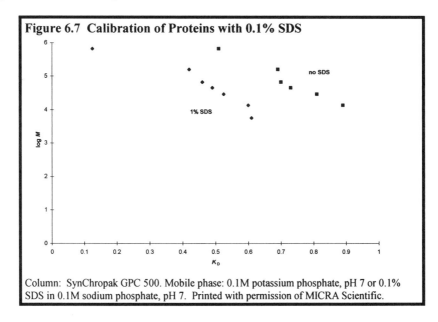

Figure 6.7 Calibration of Proteins with 0.1% SDS

Column: SynChropak GPC 500. Mobile phase: 0.1M potassium phosphate, pH 7 or 0.1% SDS in 0.1M sodium phosphate, pH 7. Printed with permission of MICRA Scientific.

II. SUPPORTS

A. General

An absolute requirement of size exclusion supports is noninteraction with the solutes; otherwise, true separation by size cannot occur. Supports must also have sufficient porosity to allow separation and narrow pore distribution to limit the range of included molecules. Table 6.3 lists general requirements for an SEC support.

Table 6.3 Characteristics of an Ideal SEC Support

- Inert, Hydrophilic Surface
- Uniform Pore Diameter
- Large Pore Volume
- Wide Choice of Pore Diameters
- Chemically Stable - permits use of denaturants, strong detergents and wide pH range
- Incompressible, mechanically stable

Regardless of type or brand, SEC column packing materials are 5 - 50μm in diameter with controlled porosity comprised of 10 - 400nm pores. Most SEC supports for HPLC are 5 - 10μm in diameter which is 100 -1000 times larger than the molecules to be separated. For low pressure preparative applications, the packing is 1000 - 10000 times larger than the solutes.

B. Carbohydrate Gels

Size exclusion chromatography became a viable technique for protein separation around 1960 when carbohydrate gel matrices were developed by Porath and Flodin (7) and Mosbach and Hjerten (8, 9). These carbohydrate gels are hydrophilic with pore characteristics related to crosslinking and the mobile phase environment. The pore volumes are fairly large with V_i/V_o ranges of 1.5 - 2.5 (4). They are soft gels which are compressible when pressure is applied. Soft gels are unsuitable for HPLC and only used with gravity or peristaltic pumps as the driving force. Crosslinking with acrylamide or agarose has been used to stabilize the gels to pressure and the deforming effects of solvents (10). Some stabilized gels (Superose and Superdex) are able to withstand moderate pressures, such as those generated by FPLC systems. Carbohydrate matrices are of three general types - dextran, agarose and polyacrylamide - which have been modified chemically to increase their utility. These basic structures are shown in Fig. 6.8.

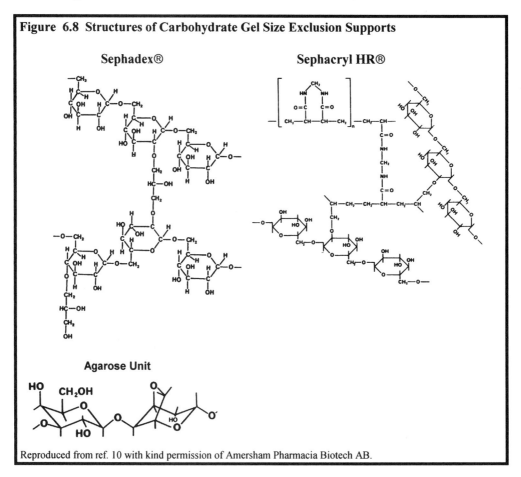

Figure 6.8 Structures of Carbohydrate Gel Size Exclusion Supports

Sephadex®

Sephacryl HR®

Agarose Unit

Reproduced from ref. 10 with kind permission of Amersham Pharmacia Biotech AB.

Some examples of commercially available SEC packing materials for low pressure applications are listed in Table 6.4.

Table 6.4 Comparison of Low Pressure SEC Columns

Product	Gel	Composition	Molecular Weight	min d_p (µm)	Manufacturer
Bio-Gel	P-2	polyacrylamide	100-1800	<45	Bio-Rad
	P-4	polyacrylamide	800-4000	<45	
	P-6	polyacrylamide	1000-6000	<45	
	P-10	polyacrylamide	1500-20000	45-90	
	P-30	polyacrylamide	2500-40000	45-90	
	P-60	polyacrylamide	3000-60000	45-90	
	P-100	polyacrylamide	$5000-10^5$	45-90	
	A-0.5m	agarose	$<10^4 - 5\times10^5$	38-75	
	A-1.5m	agarose	$<10^4 - 1.5\times10^6$	38-75	
	A-5.0m	agarose	$10^4 - 5\times10^6$	38-75	
	A-15m	agarose	$4\times10^4 - 15\times10^6$	38-75	
Sephadex	G-10	dextran	<700	55-166	Amersham Pharmacia Biotech
	G-15	dextran	<1500	60-181	
	G-25	dextran	1000-5000	17-69	
	G-50	dextran	1500-30000	20-80	
	G-75	dextran	3000-70000	23-92	
	G-100	dextran	$4000-1.5\times10^5$	26-103	
Sepharose	6B	agarose	$10^4 - 4\times10^6$	40-165	Amersham Pharmacia Biotech
	4B	agarose	$6\times10^4 - 2\times10^7$	40-165	
	2B	agarose	$7\times10^4 - 4\times10^7$	60-200	
	CL-6B	crosslinked agarose	$10^4 - 4\times10^6$	40-165	
	CL-4B	crosslinked agarose	$6\times10^4 - 2\times10^7$	40-165	
	CL-2B	crosslinked agarose	$7\times10^4 - 4\times10^7$	60-200	
Sephacryl HR	S-100	dextran/bis-acrylamide	$1000-10^5$	25-75	Amersham Pharmacia Biotech
	S-200	dextran/bis-acrylamide	$5000-2.5\times10^5$	25-75	
	S-300	dextran/bis-acrylamide	$10^4 - 1.5\times10^6$	25-75	
	S-400	dextran/bis-acrylamide	$2\times10^4 - 8\times10^6$	25-75	

Data from manufacturer's catalogs.

The manufacturer's estimates of effective fractionation ranges, as listed in Table 6.4, provide a useful guide to choosing an appropriate gel. Matrices with very large separation ranges are appropriate for separating molecules of vastly different sizes, but the broad pore distribution yields reduced resolution between given pairs of molecular weights. As a rule, the gel with the narrowest working range which encompasses the molecular weights of interest should be chosen. This can also be described as the support whose calibration curve has the slope with the lowest absolute value. For example, a mixture of bovine serum albumin (BSA) (67kD) and ovalbumin (44kD) could be fractionated on Biogel® P-100, Sephacryl® S200 or Sephadex® G75. On Bio-Gel® P-100, BSA and ovalbumin fall into the size range most effectively resolved - in the middle of the separation range (K_D = 0.2 - 0.8) or where the solutes elute on the linear portion of the calibration curve. Sephadex® G75 would appear to be a possible choice for this separation but it has the drawback that the stated working range (3 - 70kD) may result in the elution of BSA in the void volume with only ovalbumin being included. On both Bio-Gel® P100 and Sephadex® G75, BSA would not be fractionated from any larger molecules, such as BSA aggregates, which are present to some extent in all preparations. On Sephacryl® S200, aggregates would be fractionated. Additionally, solutes which elute at $0 < K_D < 0.2$ on any of the columns may experience bandspreading due to poor diffusion, as discussed in Chapter 5. It is obvious that there are many factors which must be considered when choosing a column for SEC.

C. High Performance Supports

1. Supports

Size exclusion supports designed for HPLC are generally based on silica that has been modified with a carbohydrate bonded phase. The silica must be highly porous and have narrow pore distribution to be suitable for SEC. Because the porosity is usually greater than that of supports used for reversed phase or other modes of chromatography, the pressure limits are sometimes reduced by a factor of two. The particle diameters for SEC packings used for HPLC are generally in the range of 5 - 10μm, which are 5 - 50 times smaller than the low pressure packings described in Table 6.4. The use of finer particles increases resolution dramatically, improving the mobile phase mass transfer and column efficiency. Because diffusion distances are shorter, equilibrium conditions inside and outside the particles can be nearly maintained so that flow rates can be faster. The pressure required to force this flow is also increased substantially. The pore volumes of rigid supports are lower than the carbohydrate gels, with V_i/V_o ranging from 0.8 - 1.5, but they are not variable with the mobile phase as those of the gels are. High performance rigid gels also have smaller pore volumes than the soft gels because of the crosslinking required to give mechanical stability.

2. Bonded Phase

A neutral bonded phase must be applied to silica before it can be used for SEC of biological macromolecules which are usually adsorbed on silanols. Basic amino acid residues of proteins adsorb to the free silanols, frequently denaturing proteins (11). The requirements of neutrality and hydrophilicity are encompassed by a carbohydrate layer. Glycerylpropyl or a related structure is most frequently used (12, 13). It is synthesized by the reaction of γ-glycidoxypropylsilane to produce a diol bonded phase, as seen in Fig. 6.9. Such silica-based supports are subject to the restrictions of pH discussed in Chapter 5.

Figure 6.9 Diol Bonded Phase

Silica-Si-O-Si-CH₂-CH₂-CH₂-O-CH₂-CHOH-CH₂OH

Table 6.5 lists some high performance size exclusion columns. The manufacturer's specifications can be used to select the appropriate column for a particular application. This table serves as a beginning for making comparisons to similar products. The smaller particle size provides greater resolution at higher pressures but except for Superose®, Superdex®, and the TSK PW series, these size exclusion matrices are all based on silica and, therefore, have a limited range of working pH. Fortunately, this is not generally a problem because most protein separations by SEC are performed at pH near neutrality. The fractionation range required for most separations is fairly well covered by the three major exclusion limits shown (e.g. 60kD, 300kD and 1000kD cutoffs).

Table 6.5 Comparison of HPLC Size Exclusion Columns

Manufacturer	Product	Pore Diameter (nm)	Particle (µm)	Exclusion Limit Protein *M*	Optimal Range	pH Range
Bio-Rad	Bio-Sil SEC 125	12.5	5	100kD	5 - 100kD	2.0 - 8.0
(silica-based diol)	Bio-Sil SEC 250	25	5	300kD	10 - 300kD	
	Bio-Sil SEC 400	40	5	1000kD	20 - 1000kD	
MICRA	SynChropak GPC Peptide	5	5	35kD	1 - 35kD	2.0 - 8.0
(SynChrom)	SynChropak GPC100	10	5	500kD	5 - 160kD	
(silica-based diol)	SynChropak GPC300	30	5	2000kD	10 - 500kD	
	SynChropak GPC500	50	7	5000kD	40 - 1000kD	
	SynChropak GPC1000	100	7	10000kD	40 - 10000kD	
	SynChropak GPC4000	400	10	>10000kD		
Amersham Pharmacia	Superose 6		13	40000kD	5 - 5000kD	1.0 - 14
(semi-rigid agarose)	Superose 12		10	2000kD	1 - 300kD	
	Superdex 75		11-15	100kD	3 - 70kD	3 - 12
	Superdex 200		11-15	1300kD	10 - 600kD	
Hewlett Packard	ZORBAX GF-250	15	5	400kD	4 - 400kD	2.5 - 8.5
(Rockland Technologies)	ZORBAX GF-450	30	6	1000kD	10 - 1000kD	
(DuPont)						
(silica-based diol, stabilized)						
TosoHaas	TSK 2000SW	12.5	10	60kD	10 - 60kD	3.0 - 7.0
(silica-based diol)	TSK 2000SWXL	12.5	5	60kD	10 - 60kD	
	TSK 3000SW	25	10	300kD	10 - 300kD	
	TSK 3000SWXL	25	5	300kD	10 - 300kD	
	TSK 4000SW	40	10	7000kD	30 - 1000kD	
	TSK 4000SWXL	40	5	7000kD	30 - 1000kD	
(hydrophilic polymer)	TSK G3000PWXL	20	6	800kD	0.5 - 800kD	2 - 12
	TSK G4000PWXL	50	10	1500kD	10 - 1500kD	
	TSK G5000PWXL	100	10	10000kD	<10000kD	
	TSK G6000PWXL	>100	13	50000kD	<200000kD	

Data from manufacturer's catalogs.

The major manufacturers of SEC columns continue to refine and develop new and improved packings which have fewer secondary interactions, more tightly controlled pore diameters, greater rigidity and chemical stability, and compatibility with faster flow rates. Zorbax® GF 250 and 450 are silica-based materials stabilized by a zirconia cladding process which extends the usable pH range to 2.5 - 8.5 and minimizes ionic and hydrophobic interactions with the sample (12). The bonded phase of this support contains diol functionalities

III. OPERATION

There are significant advantages of size exclusion over other chromatographic methods. Separations are isocratic and rapid, usually in 30 minutes or less. Aside from solubilizing the sample in the mobile phase and making logical choices of a column, method development is minimal. By choosing physiological conditions for separation, denaturation can be avoided, with excellent recoveries - often exceeding 90% of the quantity loaded. An understanding of the effects of the operational variables on resolution is a key factor in setting up an analysis.

A. Mobile Phase

In size exclusion chromatography, it is essential that samples remain soluble throughout the separation and be reasonably low in viscosity to reduce peak broadening and optimize efficiency.

Although size exclusion matrices are designed to minimize physical or chemical interactions (thus distinguishing it from all other types of chromatography where specific interactions are the basis of separation), most size exclusion supports are weakly anionic and slightly hydrophobic, possessing functional groups which can lead to sample interaction or nonideal separation with certain solutes or mobile phases (14). Such interactions may create deviations from ideal behavior, generally resulting in cationic or hydrophobic solutes being retained longer on the column than would otherwise be expected. Conversely, a negatively charged molecule may elute early due to repulsion from the matrix, which is known as ion-exclusion. Such potentially unwanted interactions with the support are difficult to predict and may result in less reproducible separations than those based strictly on size. Ionic interactions can generally be minimized with the addition of 100 - 300mM salt and operation at pH 6.5 - 7.0; however, high concentrations of salt can cause hydrophobic interactions of proteins with the bonded phase. Undesirable hydrophobic interactions can be minimized by adding glycol or alcohol (e.g. 5 - 10% methanol, isopropanol or glycerol) or simply reducing the mobile phase ionic strength. Clearly, one must strike some degree of balance here because the remedy for one type of interaction may contribute to the other. The manufacturer's recommendations about mobile phase should usually be heeded because they are optimized for the physical and chemical characteristics of the specific support (15).

At times, interactions of the sample and the support can be exploited to effect isocratic separations simultaneously by both size and charge or hydrophobicity. This type of separation has the misnomer, "nonideal SEC". Additives like butylated hydroxyanisole (BHA) or 2,6 di(t-butyl) hydroxytoluene (BHT) often have hydrophobic interactions with size exclusion supports and elute after the excluded volume. If hydrophobic molecules bind to a sample, they may cause the resultant separation to be based on both size and hydrophobicity.

B. Loading

SEC has very low loading capacity compared to interactive modes of chromatography. The volume loaded should not generally exceed 2% of the column volume or bandspreading may occur (2). Mass loads of more than 1 - 2mg of protein will broaden peaks on a 300 x 7.8mm ID column. These loading characteristics for SEC are generally independent of the brand of column because the separation occurs in the liquid volume rather than on the surface. Loading capacity is also related to the number of components in the sample and whether they are resolved (15). Table 6.5 lists some general loading guidelines for various sizes of size exclusion columns.

Table 6.6 Maximum Volume and Mass Loading on SEC Columns		
Column ID (mm)	Sample Volume (μl)	Sample Mass (mg)
2.1	2	0.050-0.100
4.6	8	0.200-0.400
7.8	30	1-2
10.0	50	2-4
21.2	200	8-16

These values assume a length of 25 cm. Maximum volumes and masses are loads where bandspreading may occur for a protein of about 45 kD molecular weight. Actual loading will vary with solute, mobile phase, and the number of sample components.

C. Flowrate

Because the solutes in SEC are macromolecules which at least partially penetrate the support, the diffusion coefficients can cause significant band spreading due to mass transfer into the pores and eddy diffusion. The net result is that low flowrates are necessary to obtain optimum resolution;

however, this is not a severe limitation because the runs are isocratic and generally fast. The effect of flowrate on peak width increases with the size of the solute, as seen in Fig. 6.10 for amylase (100kD), BSA (67kD), ovalbumin (44kD), carbonic anhydrase (29kD), and glycyltyrosine, a small dipeptide.

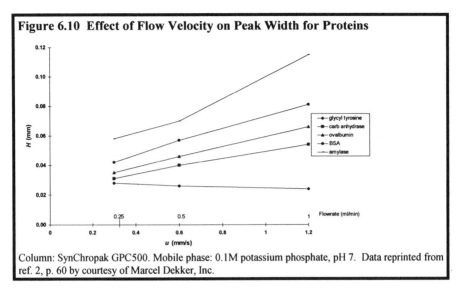

Figure 6.10 Effect of Flow Velocity on Peak Width for Proteins

Column: SynChropak GPC500. Mobile phase: 0.1M potassium phosphate, pH 7. Data reprinted from ref. 2, p. 60 by courtesy of Marcel Dekker, Inc.

Although there is a considerable difference in efficiency with linear velocity for the large proteins, it is often best to use a moderate flowrate, such as 1 ml/min on a 7.8mm ID column when developing a method. If more resolution is required, the flowrate can be lowered.

D. Solute Probes

A variety of common biochemicals have been utilized as probes to examine the behavior of size exclusion column packings and to determine the empirically important column parameters. When operating at neutral pH, column void volume (V_o) can often be determined with DNA from calf thymus. If the mobile phase does not contain adequate salt, a slight underestimate of void volume may occur for the highly anionic phosphate groups of the nucleic acid due to ion exclusion from the weakly acidic surfaces of silica-based matrices. Nucleosides and glycyltyrosine are convenient markers for mobile phase volume (V_M) or internal volume (V_i) because they usually have no interaction with the supports (14). The positively-charged amino acids, arginine and lysine, and negatively charged glutamic acid or citrate serve as excellent probes of electrostatic interactions and the damping effects exerted by different concentrations of various buffer salts (2, 14). Similarly, matrix hydrophobicity and the counter effects of glycols and alcohol can be assessed with phenethyl- or benzyl alcohol (13). Columns should be calibrated with appropriate standards. For protein analysis, proteins such as those listed in Table 6.1 can be used. For linear polymers, sulfonated polystyrenes or dextrans are suitable.

E. Optimization

Size exclusion chromatography can be optimized by a variety of strategies. The number of effective plates (N) or the column efficiency can be increased by reducing the mobile phase velocity, increasing the length of the column, or reducing the particle size. Pore diameter and pore volume determine the separation in SEC; therefore, these characteristics should be reviewed intelligently when choosing a column or changing columns to improve a separation. Calibration curves and standard chromatograms can give good indications of the potential for resolution of sample components with specific molecular weights and/or molecular sizes. Although generally a minor factor, reduction of buffer viscosity can sometimes improve a separation. This can be accomplished by increasing the temperature of the run, if sample stability permits, or by substituting an additive like glycerol with an appropriate alcohol. Maximum efficiency will be achieved in SEC if the sample is made up in the mobile phase. Table 6.7 lists some ways to increase resolution in SEC.

Table 6.7 Optimization of Resolution in SEC

- Decrease flow rate (F) or linear velocity (u).
- Prepare sample in mobile phase.
- Increase column length (L).
- Decrease particle diameter (d_p).
- Increase pore volume (V_i).
- Optimize pore diameter so that $0.2 \leq K_D \leq 0.8$.
- Decrease mobile phase viscosity (η).

IV. APPLICATIONS

SEC has had wide utility in protein purification methods. It is used to approximate the molecular weight or size of a protein, to prefractionate complex mixtures at early stages, and to desalt. Desalting is the separation of macromolecules from small molecules, including salts, on a size exclusion column with very small pores, so that all the large molecules elute in the void volume. It can separate the macromolecules from the small molecules while concomitantly exchanging the sample solvent for the mobile phase. By utilizing the final formulation buffer as the running buffer, "desalting" can, in fact, be a buffer exchange, and the final step in the purification process of a drug or biologic. SEC has been used to determine aggregation states and to eliminate aggregates from a monomer. Similarly, it can be used in soluble-phase ligand binding studies because a receptor bound with ligands can often be separated from free receptor and then readily identified and quantitated (16). Table 6.8 lists some common applications of SEC.

Table 6.8 Common Applications of SEC

- Prefractionation of complex samples
- Estimation of molecular weight
- Desalting (buffer exchange)
- Determination or elimination of protein aggregates
- Soluble ligand binding studies

A. Proteins

To compare the effect of pore diameter on protein resolution in SEC, Fig. 6.11 shows the UV elution patterns for a test mixture of seven proteins and sodium azide on supports with 15nm and 30nm pores in a neutral phosphate buffer. The runs were completed in less than 15 minutes, with the actual separation of the proteins occurring in about half the run time.

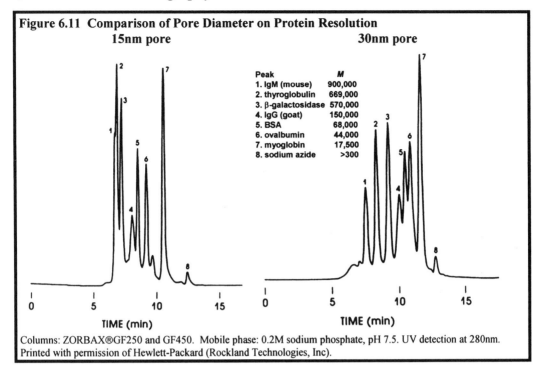

Figure 6.11 Comparison of Pore Diameter on Protein Resolution

Columns: ZORBAX®GF250 and GF450. Mobile phase: 0.2M sodium phosphate, pH 7.5. UV detection at 280nm. Printed with permission of Hewlett-Packard (Rockland Technologies, Inc).

The smaller pore size (15nm) of ZORBAX® GF-250 resulted in a finer resolution of smaller components than that observed on the GF-450, whereas the 30nm pore diameter resolved the larger molecules better. The calibration curves corresponding to these columns, shown in Fig. 6.12, indicate that the selective fractionation range for ZORBAX® GF-250 runs from thyroglobulin to myoglobin and that for ZORBAX® GF-450 extends from IgM to BSA. Citing the example used earlier regarding the separation of BSA and ovalbumin, they are resolved better on ZORBAX® GF-250 because the slope of the calibration curve is less steep. The proteins elute on the linear portions of both curves. Aggregates would be better discerned on the ZORBAX® GF-450 where they would elute in the optimum range. A broader range of selective fractionation from 4000 to 1,000,000 would be obtained by running these two columns in series.

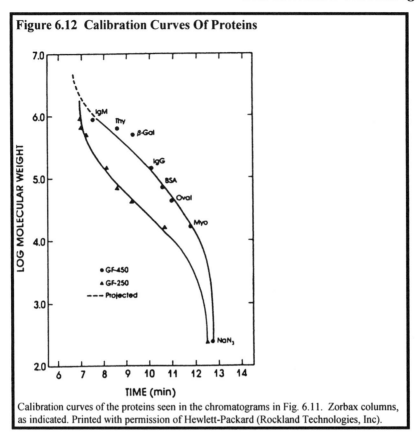

Figure 6.12 Calibration Curves Of Proteins

Calibration curves of the proteins seen in the chromatograms in Fig. 6.11. Zorbax columns, as indicated. Printed with permission of Hewlett-Packard (Rockland Technologies, Inc).

The chemical stability of high performance size exclusion supports permits reducing and denaturing conditions to be used without deterioration of the columns. Fig. 6.13 shows total resolution of immunoglobulin light (25kD) and heavy (50kD) chains, following disruption of interchain disulfide bonds with 5M guanidinium hydrochloride and dithiothreitol, followed by alkylation.

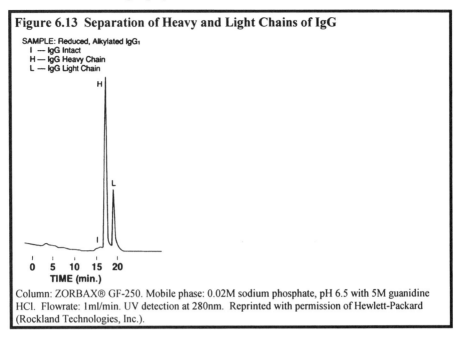

Figure 6.13 Separation of Heavy and Light Chains of IgG

SAMPLE: Reduced, Alkylated IgG₁
 I — IgG Intact
 H — IgG Heavy Chain
 L — IgG Light Chain

Column: ZORBAX® GF-250. Mobile phase: 0.02M sodium phosphate, pH 6.5 with 5M guanidine HCl. Flowrate: 1ml/min. UV detection at 280nm. Reprinted with permission of Hewlett-Packard (Rockland Technologies, Inc.).

Similarly, strong detergents can be used to disassociate hydrophobic complexes such as outer membrane proteins. The effects of the concentration of Triton X-100, a commonly used PEG-based nonionic detergent, on the aggregation state of bacteriorhodopsin is shown in Fig. 6.14 (17). Bacteriorhodopsin is a 25kD protein isolated from the membranes of *Halobacterium halobium,* which under physiological conditions exists primarily as an aggregate, with about 20% as a monomer. As with other hydrophobic proteins, detergents can disaggregate these preparations. With increasing concentrations of the detergent Triton™ X-100, the aggregates disassociated into monomers. Stronger detergents like SDS would give similar results. At detergent concentrations above 1M, virtually all aggregates were converted to the corresponding monomer.

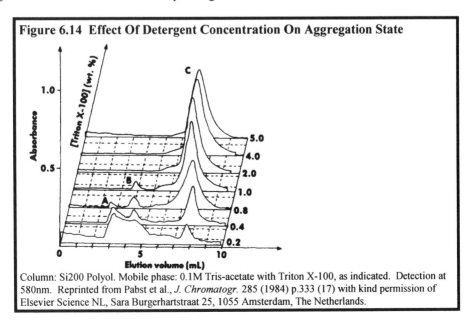

Figure 6.14 Effect Of Detergent Concentration On Aggregation State

Column: Si200 Polyol. Mobile phase: 0.1M Tris-acetate with Triton X-100, as indicated. Detection at 580nm. Reprinted from Pabst et al., *J. Chromatogr.* 285 (1984) p.333 (17) with kind permission of Elsevier Science NL, Sara Burgerhartstraat 25, 1055 Amsterdam, The Netherlands.

The use of detergents and chaotropes may not be possible if biological activity must be preserved. In such cases, great care should be taken to prevent aggregation during initial steps in isolation.

B. Peptides

The analysis of peptides by SEC is much more difficult than that of proteins for two reasons. First, peptides vary greatly in solubility and are often very hydrophobic or highly charged. These traits make it difficult to find mobile phase conditions where there is no interaction with the support for a mixture of peptides. Second, the shapes of peptides are not uniform and they vary from linear to semi-defined, such as helical, β-sheet, etc.

Efforts to find a universal SEC system for peptides have been largely unsuccessful; however, the addition of some organic solvent to the mobile phase often reduces interactions, as seen in Fig. 6.15. In 0.1M phosphate, several peptides, including the angiotensins, eluted substantially after the included volume (V_M). Most of the peptides eluted at the correct retention time after 35% methanol was added to the mobile phase although angiotensin III still interacted with the column. Both angiotensins contain predominantly hydrophobic amino acids. Angiotensin III has only one hydrophilic residue, arginine, which could have ionic interaction with the support. The use of 0.1% SDS in sodium phosphate yielded a linear calibration curve in this study, but peaks were broader in SDS than with small percentages of alcohol.

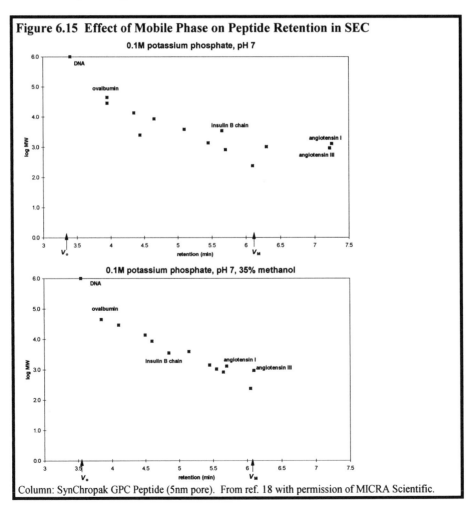

Figure 6.15 Effect of Mobile Phase on Peptide Retention in SEC

Column: SynChropak GPC Peptide (5nm pore). From ref. 18 with permission of MICRA Scientific.

V. CONCLUSIONS

A. Advantages and Disadvantages

Size exclusion chromatography is a simple isocratic method by which macromolecules can be separated on the basis of size. It is almost universally used as at least one step of protein purification. If used properly, it can both separate and yield structural information about sample components. Although simply implemented, it has comparatively low resolution and loading capacity. Table 6.9 and 6.10 list some of the advantages and disadvantages of the technique.

Table 6.9 Advantages of SEC

- Rapid separation - 30min or less
- Easy method development - simply choose buffer and fractionation range
- Nondenaturing under physiological conditions
- High recovery - often greater than 90%
- Reproducible and efficient

Table 6.10 Limitations of SEC

- Small sample volumes (2 - 5% of bed volume)
- Limited resolution - often 10 species or less
- Limited pH range of 2 - 8.5 for silica-based supports
- Nonideal behavior due to interaction with support must be eliminated, if used for estimation of molecular weight

---

B. Start-Up Procedures

RECOMMENDATIONS FOR SEC START-UP

1. Choose a column from those shown in Table 6.5 which has the appropriate molecular weight range. Consult calibration curves to ensure that the molecules will elute in the linear portion.
2. Use a column of 4.6 - 7.8mm ID if the sample concentrations fit into the guidelines shown in Table 6.6. These columns are less expensive, use less solvent and do not require special hardware.
3. Use a buffered mobile phase at pH 6 - 7 with 0.1 - 0.2M salt. It is often best and easiest to use the mobile phase recommended by the manufacturer or shown in the test chromatogram.
4. Use a moderate flowrate (0.5 ml/min for 4.6mm ID) and condition the column with at least ten column volumes of mobile phase. If the column was shipped in an organic solvent, wash with ten column volumes of distilled water or a miscible solvent before running the mobile phase.
5. Inject 1µl of a solution of AMP or glycyltyrosine in the mobile phase (approx. 1 mg/ml). Compare the theoretical plates with those achieved by the manufacturer. If they are not similar, try the measures suggested below in troubleshooting.
6. Inject a small amount of a dilute solution of DNA plus 1µl AMP to measure the V_M, V_i and V_o of the column. If DNA is offscale, dilute until slightly smaller than the AMP peak.
7. Calibrate the column with a series of standard proteins if protein samples will be run. Inject DNA and AMP, as in step 6 plus 1µl of each protein (approx. 1 mg/ml) individually so that all peaks are onscale. Graph log molecular weight (M) vs. retention time or K_D for the series. Wash the injector thoroughly with mobile phase between injections of different proteins.
8. Inject 5µl of the unknown sample. Adjust the injection volume to obtain suitable peak size but do not exceed the volume shown in the guidelines in Table 6.6.
9. Inject DNA and AMP occasionally to verify the void and included volumes and that efficiency is being maintained.
10. After the analyses are finished, wash the column in a solvent which will inhibit bacterial growth, usually 0.1% sodium azide or 10 - 20% aqueous methanol. Follow the recommendations of the column manufacturer.

C. Troubleshooting Procedures

TROUBLESHOOTING SEC COLUMNS

Symptom	Cause	Remedy
1 Poor plates on new column (small molecule)	sample	
	too concentrated	Dilute
	volume too large	Inject less
	mismatched solvent	Make sample in mobile phase
	old or contaminated	Make new sample
	dead volume	Minimize extra-column volume
2 Poor plates on used column (small molecule)	see #1 above	
	column fouling	Wash per manufacturer's recommendations. Organics will remove hydrophobic molecules. 0.1% TFA will remove some ionic molecules.
	column void	Replace column; use guard column in future; topping off may be temporary cure.
3 Poor plates for proteins	sample	as in #1 above
	specific protein problem	Run a standard protein to check column.
	flowrate too high	Reduce flowrate by 50%.
	mobile phase	
	column not conditioned	Run ten column volumes.
	inappropriate buffer	See #4 below
4 Retention greater than mobile phase volume $V_R > V_M$	solute-support interaction	Mobile phase must be buffered.
	ionic binding	Add salt (0.1 - 0.3M).
		Reduce pH, if solutes allow.
	hydrophobic binding	Reduce salt (> 0.05M).
		Add 5-10% organic or glycol.
5 Retention too low	ion exclusion	Add salt to mobile phase (> 0.05M).

VI. REFERENCES

1. E.L. Johnson and R.L. Stevenson, "Exclusion Chromatography" in *Basic Liquid Chromatography*, Varian, 1978.
2. K.M. Gooding and F.E. Regnier, "Size Exclusion Chromatography" in *HPLC of Biological Macromolecules: Methods and Applications* (K.M. Gooding and F.E. Regnier, eds.), Marcel Dekker, New York, 1990.
3. W.W. Yau, J.J. Kirkland, and D.D. Bly, *Modern Size Exclusion Chromatography*, John Wiley and Sons, New York, 1979.
4. L. Hagel and J.C. Janson, "Size-Exclusion Chromatography" in *Chromatography, 5th Edition* (E. Heftmann, ed.), Elsevier, Amsterdam, 1992.
5. C. Tanford, *Physical Chemistry of Macromolecules*, Wiley, New York, 1961, Ch. 3 and 5.
6. M. Potschka, *Anal. Biochem.* 162 (1987) 47.
7. J. Porath and P. Flodin, *Nature (London)*, 183 (1959) 1657.
8. S. Hjerten and R. Mosbach, *Anal. Biochem.* 3 (1962) 109.
9. S. Hjerten, *Arch. Biochem. Biophys.* 99 (1962) 466.
10. *Gel Filtration Principles and Methods, 6th Edition*, Pharmacia Biotech.
11. K.K. Unger, "Silica as a Support" in *HPLC of Biological Macromolecules: Methods and Applications* (K.M. Gooding and F.E. Regnier, eds.), Marcel Dekker, New York, 1990.

12. R.W. Stout and J.J. DeStefano, *J. Chromatogr.* 326 (1985) 63.

13. F.E. Regnier and R. Noel, *J. Chromatogr. Sci.* 14 (1976) 316.

14. E. Pfannkoch, K.C. Lu, F.E. Regnier, and H.G. Barth, *J. Chromatogr. Sci.* 18 (1980) 430.

15. *Zorbax Product Bulletin 96051TB*, Rockland Technologies, Inc., 1996.

16. N. Sato, S.M. Hyder, L. Chang, A. Thais, and J.L. Wittliff, *J. Chromatogr.* 359 (1986) 475.

17. R. Pabst, T. Nawroth, and K. Dose, *J. Chromatogr.* 285 (1984) 333.

18. H. Freiser, M.N. Schmuck, and K.M. Gooding, presented at *HPLC '92,* Baltimore.

CHAPTER 7
REVERSED-PHASE CHROMATOGRAPHY

I. INTRODUCTION

Reversed-phase chromatography (RPC) has become one of the most widely used and most powerful tools of separation in biochemistry. Its rapid development over the last twenty years has been driven in part by the demands of biotechnology and the recognition that hydrophobicity provides a selectivity distinct from the mechanisms of separation modes based on size or charge. Unique aspects of peptide and protein structure, unexploited by other methods, drive the highly selective binding to (and elution from) the reversed-phase matrix, through multiple weak van der Waals-type interactions. The resolving power of this technique is remarkable, permitting, in many cases, total separation of nearly identical molecules. Examples of separations of peptides which differ by only one amino acid residue abound in the literature.

The chromatographic separation of biomolecules by differences in hydrophobicity evolved from work performed in the 1950's with soft gels that were modified noncovalently with alkane chains (1). The term "reversed-phase" was coined to contrast the technique with normal phase chromatography. In normal phase or "straight" phase chromatography such as paper, thin-layer or column chromatography on bare silica or a very polar and hydrophilic stationary phase, molecules elute in order of increasing polarity. The initial mobile phase is a nonpolar solvent like hexane, with stronger, more polar solvents like methylene chloride, propanol, or methanol used to effect elution. The opposite characteristics for solvents are required for RPC. The stationary phase is nonpolar and hydrophobic; the initial mobile phase is polar. Examples of initial or weak solvents which cause hydrophobic binding are water and buffers. Strong solvents that effect elution are more nonpolar, to include methanol, isopropanol and acetonitrile. Solutes elute in increasing order of hydrophobicity or in decreasing order of net charge, degree of ionization, and ability to participate in hydrogen bonding. The elution profile of small molecules separated in RPC would resemble the mirror image of the profile obtained with normal phase chromatography.

The relevance of hydrophobic interactions in chromatography is not surprising when one considers the significant role they play in many important biological processes, as briefly summarized in Table 7.1.

Table 7.1   Biologically Important Hydrophobic Interactions

- Orientation of cell membrane proteins in phospholipid bilayers
- Assembly of multi-subunit proteins, enzymes and organelle complexes
- Enzyme-substrate interactions
- Antibody-antigen binding
- Cell receptor and ligand interactions

Almost all peptides contain some hydrophobic amino acid residues (tryptophan, phenylalanine, leucine, isoleucine, methionine, valine, tyrosine, alanine, and proline). Despite their being water-soluble, most biologically active proteins have hydrophobic patches as part of their surface topography; therefore, most peptides and proteins bind to some extent to reversed-phase supports.

II. MECHANISM

The behavior of peptides in reversed-phase chromatography is frequently different from that of small molecules, although, mechanistically, the hydrophobic interactions are identical. Mixtures of small molecules can often be separated isocratically in a classical chromatographic process where they continuously partition between the mobile and stationary phases as they traverse the column. In contrast, polypeptides display strong, and frequently multi-site, interactions with the stationary phase so that gradient elution is usually required and isocratic elution of a mixture of analytes is often impossible. In gradient elution RPC, a polypeptide is adsorbed onto the column at low solvent strength and remains essentially fixed, until the concentration of the organic modifier in the mobile phase reaches a discreet solvent strength, at which the peptide desorbs and elutes from the column; at higher solvent strengths, it may not elute due to silanophilic interaction (2). The peptide only elutes in this "window" of organic concentration.

The binding of a protein to the reversed-phase support is a more complex consideration. A typical protein in solution can be viewed as a three dimensional, irregularly shaped object with many surfaces that can be distinguished by their shape and "chromatographic" chemistry. These surfaces may be flat, involuted, or protruding from the surrounding areas, and can be neutral, charged, hydrophilic, or hydrophobic. Thus, a protein has the ability to bind via a variety of sites and chromatographic mechanisms. In RPC, it is the predominating hydrophobic surfaces of the protein that interact with the matrix. The subtle differences in the contact areas permit the chromatographic separation of biomolecules. The binding of proteins to the RPC matrix is also influenced by conformational changes induced by the chromatographic process itself. For example, mobile phase or stationary phase interactions may produce changes in protein tertiary structure at locations away from the contact area which may alter binding strength or create new binding surfaces. The binding is driven thermodynamically by the protein attempting to reduce its surface tension and attain its lowest free energy by escaping the water environment and binding to the bonded phase.

III. COLUMNS

A. Stationary Phase

1. Support Matrix

The contributions to band broadening which were presented in the chapters on theory and column packings are all applicable to RPC. Because of the crucial role which particle size plays in resolution, 2 - 5µm particles are commonly used for analytical applications and 15 - 30µm particles for preparative separations. Increased capacity is attained by using porous particles to increase the surface area available for binding (3). The optimum pore diameter is dictated by the particular sample, with 8 - 12nm used for analysis of small molecules and 30 - 100nm being large enough for most peptides and proteins.

The support matrix can be silica, polystyrenedivinyl benzene (PSDVB) or polystyrene. Silica is the most popular and widely used support (3), but for applications in strong acid or base, polystyrene is used successfully as a stationary phase (4). Silica and most polymeric supports are derivatized with a suitable hydrophobic ligand. High quality reversed phase supports are available from all of the vendors listed in Table 5.4.

2. Bonded Phase

A variety of hydrophobic ligands varying from weakly nonpolar to very hydrophobic are used as stationary phases for RPC. Common ligands are listed in Table 7.2:

```
┌─────────────────────────────────────────────────┐
│ Table 7.2  Reversed-Phase Ligands                 │
│                                                   │
│  •  Trimethyl                                     │
│  •  Butyl (C4)                                    │
│  •  Octyl (C8)                                    │
│  •  Octadecyl (C18, ODS)                          │
│  •  Phenyl                                        │
│  •  Diphenyl                                      │
│  •  Cyano                                         │
└─────────────────────────────────────────────────┘
```

Most silica-based reversed-phase supports are synthesized with silanes, as discussed in Chapter 5. Monofunctional silanes yield very reproducible bonded phases, but the resultant supports always contain some residual silanols which are sterically hindered from reacting; additionally, monofunctional silanes are subject to acid hydrolysis. Polyfunctional silanes yield more stable bonded phases but are somewhat less reproducible because polymerization is dependent on the water content of both the reaction solvents and the silica. These synthetic methods also leave unreacted silanols. Many of the residual silanols can be derivatized by endcapping, a procedure of silanization with a small reactant like trimethylchlorosilane, which can access the free silanols.

The reversed-phase supports designed for peptide analysis include both endcapped and nonendcapped, monomeric and polymeric supports. Most have good selectivity for peptides using gradient elution. Despite having similar physical properties (e.g. 30nm, C-18), products from different sources will not be identical. Although the synthesis of reversed-phase supports appears to be trivial, it is in fact, an art as well as a science. The specific silica matrix, chemicals, pre- and post-treatments, and the experience of the synthetic chemist all influence the quality of the support.

An important subset of reversed-phase supports is the group termed "base-deactivated" (5, 6). Many reversed-phase supports, including some with standard endcapping, exhibit tailing for cationic solutes, like drugs. The addition of amines to the mobile phase to eliminate silanol interactions is generally unsatisfactory due to operational difficulties, including poor reproducibility. Base-deactivated supports synthesized from low reactivity silica (Type B, see Chapter 5) and special endcapping techniques to deactivate the unreacted silanols allow amines or other weak bases to be run on the columns without tailing or additives (5). These supports are especially useful for the analysis of drugs and other basic molecules. Usually they have 10nm pores and are thus not generally suited for peptide analysis.

A few simple guidelines aid in selecting the best ligand for a specific analysis. Octyl (C8) and octadecyl (C18) ligands are the most commonly used for RPC and are usually the best initial choice when developing a method for analyzing small molecules or peptides. Both are quite durable and they partially shield the silica backbone from degradation, particularly in highly aqueous environments. Octadecyl ligands are the most hydrophobic, but in aqueous environments with only a small amount of organic modifier, the alkyl chains tend to withdraw from the mobile phase and "fold in " on themselves towards the hydrophobic backbone, thus reducing their hydrophobic character (7). The C18 ligand may also bind too tightly to certain hydrophobic proteins, making elution difficult and potentially denaturing. The folding and the tight binding have led to the popularity of C8 and C4 ligands for protein and peptide analysis. The differences in hydrophobicity between the ligand chains are recognized by small molecules whose retention is related to chain length (8). For peptides and proteins using gradient elution, however, little difference in selectivity between C18, C8 and C4 is observed (9).

Less nonpolar bonded phases, like trimethyl or butyl, permit elution of strongly hydrophobic biomolecules with only modest concentrations of organic modifiers (10). Phenyl and diphenyl columns provide hydrophobicity similar to a C5 or C6 ligand and sometimes have a particular selectivity due to π - π interactions for solutes which contain a lot of aromatic amino acids. Cyano groups, the most polar ligands for RPC, are used to separate highly hydrophobic peptides, particularly fragments generated from cyanogen bromide (CNBr) cleavage. Studies of the effects of ligand density on the resolution of small molecules and on stability have concluded that high density bonded phases are most stable (11).

B. Column Size

Small molecules analyzed isocratically by RPC exhibit the typical effect of increased plates and resolution with column length. For gradient elution, however, column length plays a surprisingly small role in the resolution of peptides (12), especially when gradients are steep. Frequently, a 5cm column will give equal or better resolution than one that is 25cm, along with lower solvent consumption and shorter analysis time. Sometimes longer column lengths will improve the resolution of complex mixtures of peptides.

Over the past few years, columns with very small diameters have become popular for RPC of peptides, especially when only minute quantities of sample are available. Columns with 2.1mm and 1mm ID are routinely used for LC-MS, with packed capillaries gaining in popularity (9, 10). Columns with diameters of 1mm or less generally require special hardware to minimize extra-column bandspreading, as discussed in Chapters 3 and 4.

IV. OPERATION

A. Mobile Phase

1. pH

As discussed previously, most reversed-phase columns contain residual silanols which may interact with any positively charged amino acid residues in peptides or proteins. The population of free silanols is likely to increase over the lifetime of a column, due to gradual hydrolysis of the silane bonded phase. One way to negate the unwanted interactions of peptides with free silanols is to operate at low pH (< 4) where the silanols are not charged. The addition of acid to the mobile phase, frequently at levels of 0.1%, achieves this. Another option is to raise the pH so that all the cationic amino acids lose their charges (pH >11). This basic pH region is not compatible with silica-based columns because these are not stable, but it is with polymer-based supports (13). Operation at high pH is a principle application for polymer-based columns in peptide analysis.

The pH dependence of the selectivity of reversed-phase columns for peptides is due to the ionic amino acids whose charges depend on their pI values, neighboring residues, and the mobile phase pH. Charged amino acids are very hydrophilic but they become hydrophobic when neutralized. Operation at neutral pH frequently yields totally different resolution than that at low pH; however, at neutral pH, it is often necessary to include an additive like sodium perchlorate in the mobile phase, as well as a buffer, to minimize silanol interactions (14).

2. Ion-Pair Agents

a. Purpose

Ion-pairing agents are used to alter the ionic or hydrophobic characteristics of the support or the solute. For small cationic solutes, ion-pair agents may be necessary to prevent interaction with residual silanols. This is demonstrated in Fig. 7.1a where a mixture of drugs is analyzed on a standard reversed-phase column. The residual silanols present on this reversed-phase support caused the basic drugs to exhibit tailing and adsorption. When 0.008M triethylamine (TEA) was added to the mobile phase (Fig. 7.1b), the peaks became sharper with no tailing.

Figure 7.1 Effect of Ion-Pairing Agent on Drug Retention

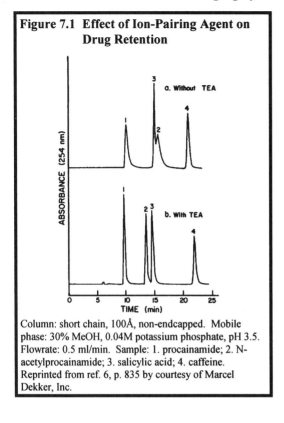

Column: short chain, 100Å, non-endcapped. Mobile phase: 30% MeOH, 0.04M potassium phosphate, pH 3.5. Flowrate: 0.5 ml/min. Sample: 1. procainamide; 2. N-acetylprocainamide; 3. salicylic acid; 4. caffeine. Reprinted from ref. 6, p. 835 by courtesy of Marcel Dekker, Inc.

Figure 7.2 Effect of Ion-Pair Agent on a Base-Deactivated Support

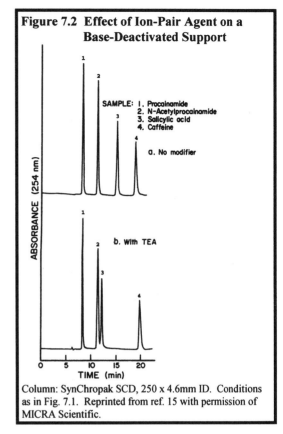

Column: SynChropak SCD, 250 x 4.6mm ID. Conditions as in Fig. 7.1. Reprinted from ref. 15 with permission of MICRA Scientific.

Base-deactivated columns produce good peak shapes for small cationic analytes without the addition of an ion-pair agent, as seen in Fig. 7.2a. In this example, addition of the ion-pair agent did not change the shape or retention times of the amine peaks but it did alter the selectivity for salicylic acid, as seen in Fig. 7.2b.

b. Mechanism of Action

The charge of a polypeptide is dependent on pH and its amino acid composition. At low pH, peptides are likely to be cationic; under alkaline conditions, anionic; and at neutral pH, possess mixed-charges with zwitterions. The hydrophilicity induced by charged ionic groups reduces the affinity of the polypeptide for the reversed-phase matrix. Ion-pairing agents, which are small molecules that always possess a charged group and often have a hydrophobic functionality, are commonly used to neutralize the ionic groups of a peptide and sometimes concomitantly to increase its hydrophobicity (16, 17). Two models of ion-pairing have been proposed (16):

Model 1: Mobile Phase Ion-pair Formation

In this classical model, the ion-pairing agent binds to the charged solute in the mobile phase, forming a neutralized complex with fewer ionic groups (18, 19). This "ion-pair" more readily binds to the reversed-phase matrix because of its increased hydrophobicity. This mechanism is symbolized below:

$$S_M^+ + P_M^- \rightarrow \left(S^+P^- \right)_M$$

$$\left(S^+P^- \right)_M \rightarrow \left(S^+P^- \right)_S$$

where S_M^+ is the positively-charged solute in the mobile phase, P_M^- is the negatively-charged ion-pairing agent in the mobile phase and $(S^+P^-)_{M/S}$ is the neutralized solute/ion-pairing agent complex in the mobile phase or stationary phase

Model 2. Dynamic Ion Exchange

The other model of ion-pair formation is different in concept from the one described above. The ion-pairing agent is proposed to bind hydrophobically to the matrix as the first step (19, 20). This results in the ionic portion of the ion-pairing agent extending into the mobile phase and serving as a point for possible ion-exchange interaction with the charged solute. These interactions can be described by the following:

$$P_M^- \rightarrow P_S^-$$

$$S_M^+ + P_S^- \rightarrow \left(S^+P^- \right)_S$$

where P_M^- is the negatively-charged ion-pair agent in the mobile phase; P_S^- is the negatively-charged ion-pair agent bound hydrophobically to the stationary phase; and S_M^+ is the positively-charged solute in the mobile phase which binds to the negatively-charged ion-pair agent on the stationary phase to form a neutral complex.

Evidence supporting both mechanisms has been reported in the literature and the actual ion-pairing interaction may be a combination of both mechanisms. In practice, the classical Model 1 serves as a satisfactory paradigm for developing an ion-pair separation.

c. Ion-Pairing Compounds

Ion-pair agents can be either acids or bases; those commonly used for RPC are listed in Table 7.3. Some of these additives, such as TFA and TEAA, have the additional advantage of volatility.

Table 7.3 Common Ion-Pairing Agents	
ACIDS	**BASES**
Acetic	Triethylamine (TEA)
Formic	Tetramethylammonium (TMA)
Perchloric	Tetrabutylammonium (TBA)
Phosphoric	Triethylammonium acetate (TEAA)
Trifluoroacetic (TFA)	Nonylamine
Heptafluorobutyric (HFBA)	
Hexanesulfonic	
Heptanesulfonic	

The first four acids are polar and they thus form hydrophilic ion-pairs. These acids do not enhance the hydrophobic binding of peptides to the reversed-phase support and may even reduce binding because of the formation of hydrophilic ion-pairs (e.g. phosphoric acid). They primarily neutralize positive charges on the solutes. Trifluoroacetic and heptafluorobutyric acids are hydrophobic ion-pairing agents which increase the hydrophobicity, as well as reduce the positive charge of solutes and they thus promote binding to the support.

The bases listed in Table 7.3 have mainly been used in the analysis of small cationic molecules to reduce tailing or adsorption to reversed-phase supports that are not base-deactivated. They bind to both residual silanols and anionic groups of charged solutes. The most common method of reducing the silanol interactions of small molecules on standard reversed-phase columns which are not base-deactivated is the addition of an ion-pair base to the mobile phase. Such addition reduces tailing substantially and may even change selectivity. Ion-pair bases can be used for peptides also, especially when operation at neutral pH is desirable (21). The hydrophobicity of the resulting complex depends on the specific ion-pair agent. Obviously, tetrabutylammonium would yield a more hydrophobic complex than tetramethylammonium. RPC of oligonucleotides can be successfully implemented by using TEAA, TMA or TBA as the ion-pair agent, resulting in additional hydrophobicity related to the size of the oligonucleotide (22, 23).

As described in Chapter 5, the free silanol groups on silica supports can potentially create unwanted secondary ionic interactions. These sites can be neutralized by covalent endcapping with appropriate reagents by the column manufacturer; however, endcapping is not universally applied to columns designed for RPC of peptides, in part because of the standard use of ion-pair reagents. Additionally, endcapping agents are subject to hydrolysis, resulting in exposed silanols over time. The deleterious effects of the free silanols generated by hydrolysis of the bonded phase on the RPC of peptides are minimized by the routine use of ion-pair agents like TFA in the mobile phase. By reducing the pH of the mobile phase and complexing with amine functionalities, ion-pair agents minimize or eliminate interaction of peptides with silanols. The use of combinations of acid and base can also be effective in eliminating interactions due to silanols. An example of such a combination is TEA-phosphate used at 10 - 25mM at a pH below 4 or 5 (24).

Even with fully endcapped columns, TMA-chloride has been shown to increase recovery and improve peak shape by reducing tailing, particularly when separating a mixture of basic peptides. The use of such ion-pair agents has been referred to as "dynamic end capping". It should be noted that at neutral pH or higher, basic ion-pair agents can cause rapid degradation of silica columns.

d. Effect on Peptide Retention

Ion-pairing agents selectively affect the retention of peptides on reversed-phase matrices by changing their apparent hydrophobicities. It can be seen in Fig. 7.3 that the retention of peptides was different with each of three acids. Each peptide in the sample, as identified in Table 7.4, varied in hydrophobicity; S1 was identical to S2 except for a free terminal amino group.

Table 7.4 Structure of Model Peptides

Peptide	Sequence	Charge
S1	NH_2-Arg-Gly-Gly-Gly-Gly-Leu-Gly-Leu-Gly-Lys-amide	+3
S2	Ac-Arg-Gly-Gly-Gly-Gly-Leu-Gly-Leu-Gly-Lys-amide	+2
S3	Ac-Arg-Gly-Ala-Gly-Gly-Leu-Gly-Leu-Gly-Lys-amide	+2
S4	Ac-Arg-Gly-Val-Gly-Gly-Leu-Gly-Leu-Gly-Lys-amide	+2
S5	Ac-Arg-Gly-Val-Val-Gly-Leu-Gly-Leu-Gly-Lys-amide	+2
0	Ac-Thr-Asp-Leu-Leu-Gly-amide	0
1	Ac-Val-Ser-Lys-Thr-Glu-Thr-Ser-Gln-Val-Ala-Pro-Ala-amide	+1
2A	Ac-Arg-Gly-Ala-Gly-Gly-Leu-Gly-Leu-Gly-Lys-amide	+2
4	Ac-Ser-Asp-Gln-Glu-Lys-Arg-Lys-Gln-Ile-Ser-Val-Arg-Gly-Leu-amide	+4
6	Ac-Gly-Lys-Phe-Lys-Arg-Pro-Pro-Leu-Arg-Arg-Val-Arg-amide	+6

Reprinted with permission from *High-Performance Liquid Chromatography of Peptides and Proteins*, p.321 (16). Copyright 1991 CRC Press.

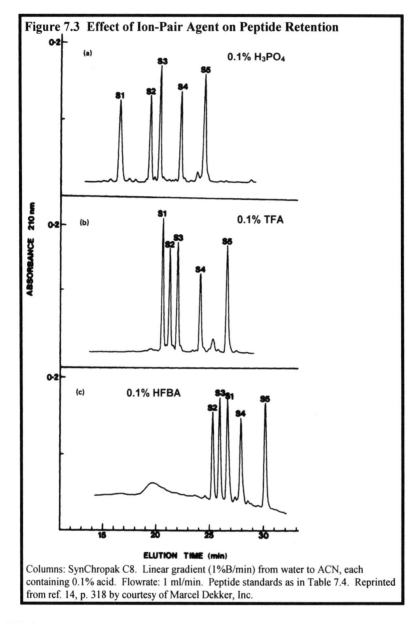

Figure 7.3 Effect of Ion-Pair Agent on Peptide Retention

Columns: SynChropak C8. Linear gradient (1%B/min) from water to ACN, each containing 0.1% acid. Flowrate: 1 ml/min. Peptide standards as in Table 7.4. Reprinted from ref. 14, p. 318 by courtesy of Marcel Dekker, Inc.

The hydrophilic ion-pair agent, phosphoric acid, induced rapid elution of the peptides. The more hydrophobic ion-pair agents, TFA and HFBA, increased the hydrophobicity and therefore the retention of the peptides. With HFBA, the relative hydrophobicity of S1 was increased due to its extra amine group.

Jagged baselines are frequently observed for HFBA which is often contaminated with other isomers. Only ion-pair agents of high quality should be used for RPC or their impurities will bind to the support and cause baseline irregularities.

The concentration of ion-pairing agent is another factor in the retention and selectivity of peptide separations. In Fig. 7.4, the separations of six peptides on a C18 column were compared when the concentration of TFA was varied from 0.02 - 0.8% (25).

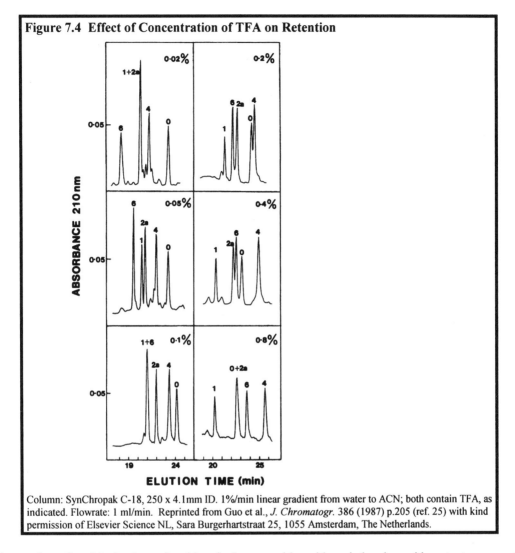

Figure 7.4 Effect of Concentration of TFA on Retention

Column: SynChropak C-18, 250 x 4.1mm ID. 1%/min linear gradient from water to ACN; both contain TFA, as indicated. Flowrate: 1 ml/min. Reprinted from Guo et al., *J. Chromatogr.* 386 (1987) p.205 (ref. 25) with kind permission of Elsevier Science NL, Sara Burgerhartstraat 25, 1055 Amsterdam, The Netherlands.

The number of positively-charged residues in these peptides with varied amino acid content was equal to their designation, as seen in Table 7.4. Peptide 6 contained six basic amino acids and was therefore most sensitive to the hydrophobic effects of TFA. Its elution order changed from the first to the fourth as it became more hydrophobic at higher TFA concentrations. Peptide 0 had no basic groups and therefore showed little, if any, retention effects from TFA except for a slight decrease at high concentrations. Because the amino acid content varied substantially among the peptides, the effects of individual amino acids or combinations on retention may have been responsible for unexpected selectivity. For example, peptide 6 contained twelve amino acids with two proline residues, whereas peptide 0 had only five amino acids.

3. Organic Modifiers

Solvents used as the mobile phase for RPC are categorized as weak to strong depending on their ability to elute solutes from the bonded phase. Table 7.5 lists the relative strengths of some commonly used solvents.

Table 7.5 Commonly Used Solvents for RPC	
water containing salt or buffer	weak
water	
methanol	
acetonitrile	
isopropanol	strong

The specific organic solvent used in a reversed-phase procedure exerts a substantial influence on the resulting separation. Fig. 7.5 shows the differences in retention of model peptides of the sequence Ac-Gly-X-X-$(Leu)_3$-$(Lys)_2$-amide with the indicated amino acids as "X", when three solvents, isopropanol (IPA), acetonitrile (ACN) and methanol (MeOH), were used (26). Amino acid abbreviations can be found in Appendix B.

Figure 7.5 Effect of Organic on Peptide Retention

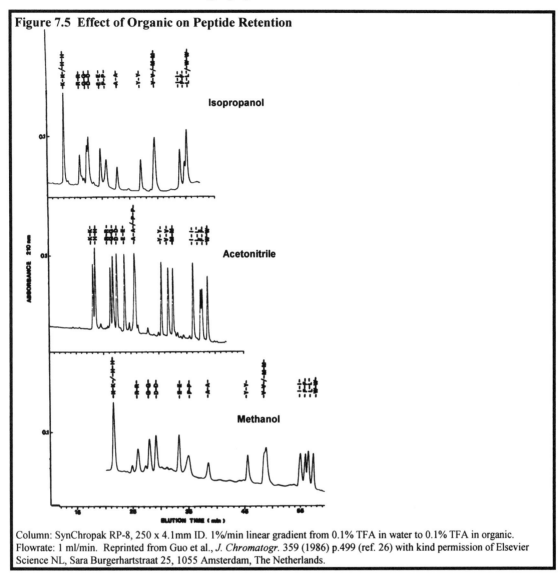

Column: SynChropak RP-8, 250 x 4.1mm ID. 1%/min linear gradient from 0.1% TFA in water to 0.1% TFA in organic. Flowrate: 1 ml/min. Reprinted from Guo et al., *J. Chromatogr.* 359 (1986) p.499 (ref. 26) with kind permission of Elsevier Science NL, Sara Burgerhartstraat 25, 1055 Amsterdam, The Netherlands.

In this study, gradients (0 - 70%B) of 0.1% trifluoroacetic acid (TFA) in water to the indicated organic solvent were used on a C8 column. The relative solvent strengths are in the order of IPA > ACN >> MeOH, but the best separation was observed with acetonitrile where the peaks were sharp and fairly well resolved. Isopropanol is a stronger solvent, as indicated by the shorter time required to elute all twelve peptides; however, its high viscosity caused some band broadening due to the concomitant decrease in the diffusion coefficients. Methanol is the weakest solvent, increasing the time required for peptide elution. The most hydrophobic peptides required about 50% more time for elution with methanol than with either of the stronger solvents. The substantial band broadening is a consequence of the longer retention and the increased viscosity caused by the extensive hydrogen bonding that occurs between methanol and water molecules. The high viscosity also increased the operating pressure of the system. The data in Fig. 7.5 demonstrate why acetonitrile is often the preferred solvent for RPC of peptides. Acetonitrile has an additional advantage for the analysis of proteins and peptides in that it is transparent at 210nm, a wavelength used to detect peptide bonds, whereas isopropanol and methanol have significant absorbance there.

Although TFA displays little UV absorption at 210nm, the absorption changes as the concentration of acetonitrile or another organic reagent is increased. This is due to disruption of the electron interactions, which changes the spectrum in the 190 - 250nm region. The resulting shifts in baseline can be minimized by working at 214 - 216nm instead of 210nm, or by using less TFA in the organic mobile phase than in the aqueous. It is also important to use high quality TFA because some impurities of TFA have absorbance in this region.

Other considerations for selecting organic modifiers include the cost of the purified solvent, its intrinsic toxicity, the risk it may pose to laboratory personnel, and the cost of proper storage and disposal. Because acetonitrile is more toxic and expensive than isopropanol, isopropanol is sometimes substituted if the resolution is adequate. Feldhoff developed methods using a mixture of isopropanol and ethanol, which is nontoxic but yielded resolution similar to acetonitrile (27).

4. Surfactants

There are some peptides and proteins which are most effectively analyzed in the presence of surfactant for solubilization or prevention of aggregation. A variety of surfactants can be added to the mobile phase to reduce solvent viscosity and surface tension. They reduce band widths because they increase both eluent strength and diffusion coefficients. Despite these benefits, detergents should only be used with caution in RPC because they bind very strongly to hydrophobic ligands as well as to proteins, and are difficult, if not impossible, to totally remove from RPC columns. When detergents like sodium dodecylsulfate (SDS) bind to proteins and columns, they alter their chromatographic properties. The solubility of detergents is also affected by the concentration of salts and organic solvents; therefore, care must be taken to avoid precipitation. For these reasons, detergents should be used in RPC only when absolutely necessary and the column should thereafter be dedicated to use with the specific detergent. If detergents are present as impurities in the sample, inverse gradients are sometimes effective for removing them (28).

B. Flowrate

The effects of isocratic flowrate on plate height (*H*) are related to the solute molecular weight, as seen for benzyl alcohol, met-enkephalin, and lysozyme in Fig. 7.6 (29). These curves resemble those observed during SEC of proteins (Fig. 6.10) (30).

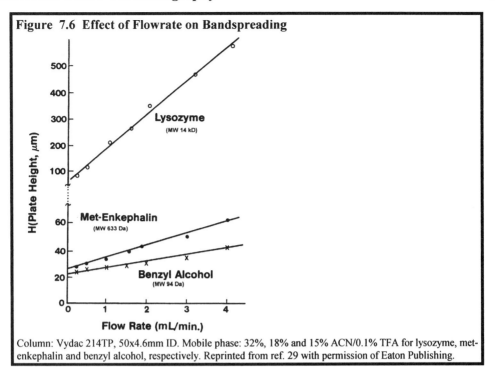

Figure 7.6 Effect of Flowrate on Bandspreading

Column: Vydac 214TP, 50x4.6mm ID. Mobile phase: 32%, 18% and 15% ACN/0.1% TFA for lysozyme, met-enkephalin and benzyl alcohol, respectively. Reprinted from ref. 29 with permission of Eaton Publishing.

Under isocratic conditions, flowrate has only a modest effect on the plate height of small molecules like an alcohol or a small peptide. For larger polypeptides and proteins, flowrate plays a significant role in isocratic resolution and indicates the importance of running at relatively slow rates of flow. Isocratic methods are usually impractical for polypeptides due to multipoint adsorption, necessitating gradient elution which also permits the effective use of higher flowrates.

When using gradient elution for peptide analysis, the rate of change of the gradient plays a greater role in determining the resolution than does flowrate. This is seen in Fig. 7.7 where two decapeptides were separated using four gradients of varying steepness with a tenfold range of flowrates (31).

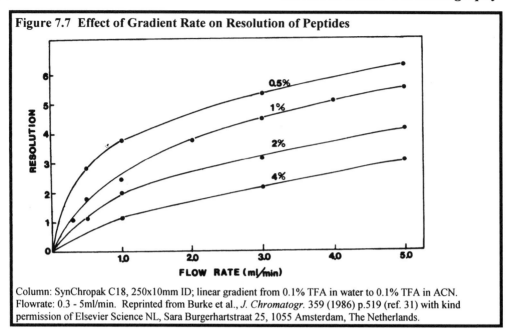

Figure 7.7 Effect of Gradient Rate on Resolution of Peptides

Column: SynChropak C18, 250x10mm ID; linear gradient from 0.1% TFA in water to 0.1% TFA in ACN.
Flowrate: 0.3 - 5ml/min. Reprinted from Burke et al., *J. Chromatogr.* 359 (1986) p.519 (ref. 31) with kind
permission of Elsevier Science NL, Sara Burgerhartstraat 25, 1055 Amsterdam, The Netherlands.

Resolution increased with flowrate or with a lower gradient rate (longer gradients). The effects on
resolution caused by flowrate and gradient slope are accompanied by differences in sensitivity, as
illustrated in Fig. 7.8.

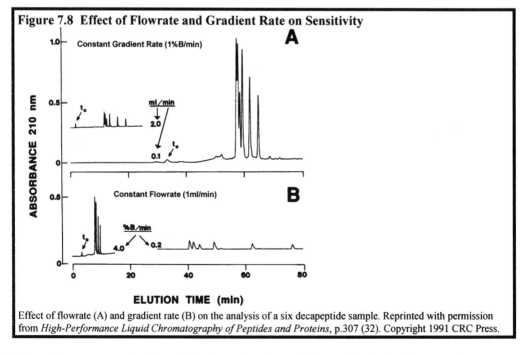

Figure 7.8 Effect of Flowrate and Gradient Rate on Sensitivity

Effect of flowrate (A) and gradient rate (B) on the analysis of a six decapeptide sample. Reprinted with permission
from *High-Performance Liquid Chromatography of Peptides and Proteins*, p.307 (32). Copyright 1991 CRC Press.

The faster flowrate caused earlier elution but reduced sensitivity. With the shallower gradients,
retention (k) increased to improve the resolution, but sensitivity decreased because the peaks
broadened. Conversely, the steeper gradient improved the sharpness of the eluting peaks, and thus
sensitivity, but peaks eluted earlier and some resolution was lost.

C. Temperature

Increasing the temperature in RPC is another means of improving the resolution. Efficiency is increased due to reduction of the viscosity of the mobile phase and increase of the diffusion coefficient. Temperatures over ambient should be used with caution for biological macromolecules, however, because many are denatured at elevated temperatures. This denaturation may be irreversible or irreproducible. High temperatures also increase column degradation by hydrolysis of bonded phase and silica (11, 12). One series of applications that routinely implements high temperatures for reversed-phase separations of peptides and proteins are those using 1.5 - 2µm nonporous particles (33, 34), as was seen in Fig. 5.15. For these columns, there is a major benefit of pressure reduction so that high flowrates and fast analyses (45s in Fig. 5.15) can be achieved. The greater stability of some nonporous supports and the short run times partially reduce the deleterious effects on the columns. Denaturing HPLC routinely uses elevated temperatures for the separation of oligonucleotides to inhibit formation of secondary structures (23). This will be further discussed in Section V.D.

D. Loading

There are two kinds of chromatographic loading capacity. Dynamic loading is the load at which the peak width of a solute increases by a specified percentage (often 50 - 100%) using standard analysis conditions. Absolute capacity is the maximum amount of a solute which can be loaded on a column under mass overload using displacement chromatography (35). As was discussed in Chapter 5, loading depends on the available surface area and thus, the pore diameter and solute size. Loading capacities for reversed-phase columns are lower than those for ion-exchange or hydrophobic interaction chromatography. Dynamic loading capacities for a 250 x 4.6mm ID column of 30nm C-18 are on the order of 5 - 10mg for a protein or 100 - 500µg for a peptide. Absolute loading is also fairly low, in the range of 20mg of protein per ml of support (35). Because RPC can denature proteins, artifacts due to the conversion to other forms can appear when loads are increased (36). Care must therefore be exercised when doing preparative chromatography of proteins by RPC. Various methods have been found to improve preparative RPC methods, including the use of radial compressed cartridges (37), discontinuous RPC (38), shallow gradients (39), and displacement chromatography (40).

V. APPLICATIONS

A. Small Molecules

Although the focus of this book is on HPLC of peptides, proteins, and other biological macromolecules, it is also appropriate to examine the chromatographic behavior of small molecules. Small molecules are usually used as test probes for column performance because their structures are simple, band broadening due to flow restrictions is minimal, and retention mechanisms are well defined. Small molecules are also frequently present in biological samples as excipients, substrates, or other significant species. Ideally in RPC, small molecules should elute in order of hydrophobicity or reversed order of hydrophilicity. Retention is related to the hydrophobicity of the bonded phase which is primarily based on the length of the ligand chain and the density of the coating. Secondary interactions may occur due to the chemical structure of the bonded phase or interactions with the silica matrix (8, 41-44), as seen in Fig. 7.9 for the four supports described in Table 7.6 (41).

Table 7.6 Features of Reversed-Phase Supports Used in Fig. 7.9			
	C-18	Silica Pretreatment	Endcapping
C_{18}	X		
A-C_{18}	X	acid	
C_{18}-S	X		X
A-C_{18}-S	X	acid	X

The treatment and bonding of the silica, as well as its composition, are responsible for the presence or absence of primary and secondary interactions. Acid treatment will tend to remove metal impurities and endcapping will block residual silanols. These interactions have been primarily studied with small molecules where they can be isolated, but they are also observed when biopolymers are analyzed.

Figure 7.9 Solute Interactions on a Reversed-Phase Column

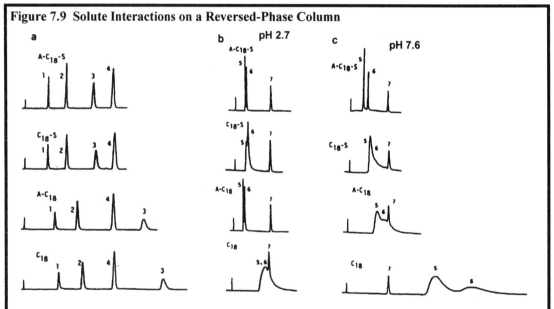

Column as indicated. Mobile phase: a. 20% methanol (MeOH); b. 40% buffered MeOH, pH 2.7; c. 40% buffered MeOH, pH 7.6. Solutes: 1. theobromine, 2. theophylline, 3. caffeine, 4. phenol, 5. procainamide, 6. N-acetylprocainamide, 7. benzyl alcohol. Reprinted from the *Journal of Chromatographic Science* by permission of Preston Publications, a Division of Preston Industries, Inc.

In a study by Dr. Tanaka and associates, the mechanisms of specific and nonspecific interactions with reversed-phase bonded phases were defined and evaluation methods were developed to measure the interactions and compare columns (41, 42). To evaluate hydrophobicity, amyl- and butylbenzene were run so that the only structural difference was a methylene group. Shape selectivity was determined with triphenylene and o-terphenyl, as recommended by Sanders (43).

The three types of nonspecific interactions with reversed-phase columns are either caused by silanols (hydrogen bonding or ion-exchange) or by metal impurities (chelate formation). The presence of any of these will result in a mixed mode separation. Hydrogen bonding was tested with caffeine and phenol, as seen in Fig. 7.9a. Fig. 7.9b and c illustrate the effect of pH on ion-exchange properties, as evaluated with procainamide, N-acetylprocainamide and benzyl alcohol, similar probes to those used to evaluate base deactivation (6). Metal impurities were detected with two tests. For high amounts of metal impurities, quinizarin was used; for lower amounts, 8-hydroxyquinoline was run. Using these or similar tests, all columns can be compared. Table 7.7 lists the probes used to detect interaction mechanisms (41).

Table 7.7 Probes for Interaction on RPC Columns	
hydrophobicity	amyl- and butylbenzene
shape selectivity	triphenylene and o-terphenyl
hydrogen bonding	caffeine and phenol
ion-exchange	procainamide, N-acetylprocainamide and benzyl alcohol
metal impurities	quinizarin or 8-hydroxyquinoline

The frequent presence of nonspecific interactions on reversed-phase columns means that probes to measure deadtime (t_0) must be chosen carefully and tested to insure that there is no interaction. The best compounds to use for t_0 determination are hydrophilic so that they will not bind to the bonded phase, and negatively charged to eliminate any possible adsorption to silanols. They should also be small enough to totally penetrate the pores. Nucleotides like UTP or CTP are good choices for markers of t_0 because they embody those qualities and are soluble in water or buffer mobile phases as well as organic solvents; however, anionic compounds like these may experience ion-exclusion if there is no salt in the mobile phase.

B. Peptide Retention

1. Amino Acid Composition

Given the growing importance of RPC in the separation of biomolecules, in particular the peptides and proteins derived from the burgeoning biotechnology industry, substantial effort has gone into developing models to provide a deeper understanding of this mode of chromatography. Such a model, if it could accurately predict the behavior of peptides and proteins in RPC, would have invaluable utility in the selection of buffers, organic modifiers, and elution conditions.

Early models of peptide retention were based on the retentive behavior of individual amino acids (45). The models regarded the chromatographic behavior of a peptide as the "sum of its (amino acid) parts". Initially, the degree to which individual amino acids would partition in various organic/buffer solvents was examined (46). This approach was expanded to measure the retention of amino acids or small model peptides on reversed-phase columns under standardized conditions (47). These methods produced a set of partitioning and retention coefficients for each amino acid and solvent that could then be used to help predict the retention of larger peptides. Peptides with fewer than twenty residues, with polarities that span the entire hydrophilic to hydrophobic range have been tested and shown to elute essentially as predicted; however the theory is inadequate for predicting the behavior of larger peptides (17).

An advanced chromatographic theory was presented by Robert Hodges and his colleagues who studied the following model peptide (26):

$$Ac\text{-}Gly\text{-}X\text{-}X\text{-}(Leu)_3\text{-}(Lys)_2\text{-}amide$$

The amino and carboxyl termini of the purified synthetic peptides were blocked by acetylation and amidation to eliminate any ionic binding by the end groups. The neutral hydrophobic core $(Leu)_3$ was flanked by two basic residues $(Lys)_2$ and two variable residues (X-X). Peptides were synthesized with the variable residues being substituted with each common amino acid; then retention times were measured at pH 2 and pH 7. The set of retention coefficients obtained from these data and from the unblocked termini are shown in Table 7.8 (26).

Table 7.8 Retention Times for Model Peptides

Amino Acid Residue*	Retention Coefficient (min)	
	pH 2.0	pH 7.0
Trp	8.8	9.5
Phe	8.1	9.0
Leu	8.1	9.0
Ile	7.4	8.3
Met	5.5	6.0
Val	5.0	5.7
Tyr	4.5	4.6
Cys	2.6	2.6
Pro	2.0	2.2
Ala	2.0	2.2
Glu	1.1	-1.3
Thr	0.6	0.3
Asp	0.2	-2.6
Gln	0.0	0.0
Ser	-0.2	-0.5
Gly	-0.2	-0.2
Arg	-0.6	0.9
Asn	-0.6	-0.8
His	-2.1	2.2
Lys	-2.1	-0.2
α-Amino	-6.9	-2.4
α-COOH	-0.8	-5.2

* abbreviations in Appendix B.
Column: SynChropak C18, 250x4.1mm ID. At pH2: linear gradient (1%B/min) from 0.1% TFA in water to 0.1% TFA in ACN. At pH7: linear gradient (1.67%/min) 10mM ammonium phosphate/0.1M sodium perchlorate to 0.1M sodium perchlorate in 60% ACN. Flowrate: 1ml/min. Reprinted from Guo et al., *J. Chromatogr.* 359 (1986) p.499 (ref. 26) with kind permission of Elsevier Science NL, Sara Burgerhartstraat 25, 1055 Amsterdam, The Netherlands.

The amino acids in the study have been arranged in decreasing order of hydrophobicity, as indicated by their time of retention by RPC at pH 2. Glutamine (Gln) was chosen as the "neutral point" and assigned a value of zero; therefore, hydrophilic amino acids have negative values under these specific conditions. The effect of the protonation of the amino acids on hydrophobicity was evaluated with aspartic and glutamic acids. At pH 2, the carboxyl groups were titrated and the peptides were retained. At pH 7, these amino acids were anionic and hydrophilic, with negative retention coefficients. Similar effects were seen with the basic amino acids which were less hydrophobic at pH 2, where they were more highly ionized than at neutral pH. It is important to remember that the retention coefficients in Table 7.8 are dependent on the particular chromatographic parameters used and will vary with the column, pH, organic modifier, or other operational parameter.

When this model was applied to fifty-eight peptides composed of 16 or fewer amino acids, the predicted vs. observed retention time had a linear relationship as shown in Fig. 7.10. The slope of this plot is approximately one, indicating a good correlation between observation and theory.

Unfortunately, like the simpler theories, this method has difficulty predicting the behavior of peptides larger than 15 - 20 residues.

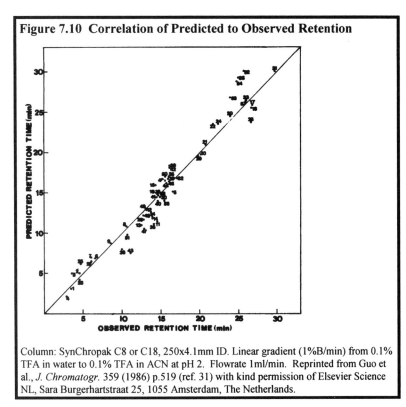

Figure 7.10 Correlation of Predicted to Observed Retention

Column: SynChropak C8 or C18, 250x4.1mm ID. Linear gradient (1%B/min) from 0.1% TFA in water to 0.1% TFA in ACN at pH 2. Flowrate 1ml/min. Reprinted from Guo et al., *J. Chromatogr.* 359 (1986) p.519 (ref. 31) with kind permission of Elsevier Science NL, Sara Burgerhartstraat 25, 1055 Amsterdam, The Netherlands.

The effects of different amino acid compositions on peptide retention has also been examined using four polymer series of different compositions, as detailed in Table 7.9 (48).

Table 7.9 Polymeric Peptide Structures

Designation	Structure	n	residues
X	Ac-(GLGAKGAGVG)$_n$-amide	1 - 5	10 - 50
G	Ac-(GKGLG)$_n$-amide	1 - 10	˙ 5 - 50
A	Ac-(LGLKA)$_n$-amide	1 - 10	5 - 50
L	Ac-(LGLKL)$_n$-amide	1 - 4	5 - 20

Correlation to the retention model for these classes of peptides rich in either leucine (L), alanine (A), glycine (G), or random amino acids (X), is shown in Fig. 7.11 (48).

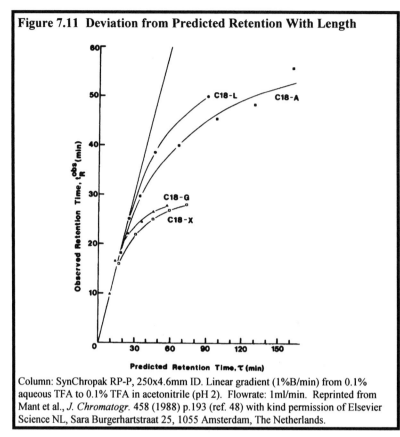

Figure 7.11 Deviation from Predicted Retention With Length

Column: SynChropak RP-P, 250x4.6mm ID. Linear gradient (1%B/min) from 0.1% aqueous TFA to 0.1% TFA in acetonitrile (pH 2). Flowrate: 1ml/min. Reprinted from Mant et al., *J. Chromatogr.* 458 (1988) p.193 (ref. 48) with kind permission of Elsevier Science NL, Sara Burgerhartstraat 25, 1055 Amsterdam, The Netherlands.

As the peptides increased in length, there was a significant decrease in the observed hydrophobicity, as indicated by the fall-off of the plots from linearity. These deviations were attributed to nearest neighbor, positional, and conformational effects. Clearly, a large and difficult combinatorial problem exists in trying to factor these considerations of peptide sequence into the model.

2. Positional Effects of Amino Acids

As the length of peptides increases, a point is reached where secondary and tertiary structures are formed. Just as point mutations and single amino acid substitutions can destroy the biological activity of a protein or peptide, they also play a crucial role in chromatographic behavior. This idea was tested in an elegant way with sequence specific peptide analogs. Dr. Houghten and colleagues at Scripps Institute generated all possible single amino acid substitutions of a 13-mer parent peptide with the following sequence (49):

$$Y\text{-}P\text{-}Y\text{-}D\text{-}V\text{-}P\text{-}D\text{-}Y\text{-}A\text{-}S\text{-}L\text{-}R\text{-}S$$
$$\#1 \qquad\qquad\qquad\qquad\qquad\qquad \#13$$

Because there are twenty common amino acids and thirteen positions for substitution to occur, a total of 260 peptides were synthesized. The reversed-phase retention for each peptide was measured and a set of normalized coefficients obtained by subtracting the value of glycine as a standard. These data are given in Table 7.10 (49).

Table 7.10 Amino Acid Retention Coefficients

Amino Acid	Y Tyr	P Pro	Y Tyr	D Asp	V Val	P Pro	D Asp	Y Tyr	A Ala	S Ser	L Leu	R Arg	S Ser	Avg	Min.	Max
Lys	-3.21	-0.82	-2.70	-3.51	-3.11	-2.24	-3.20	-3.51	-1.41	-1.21	-1.71	-1.63	-1.51	-2.29	-3.51	-0.82
His	-2.81	-1.10	-3.50	-2.50	-2.87	-2.19	-1.80	-3.27	-1.60	-1.10	-1.35	-1.52	-2.22	-2.14	-3.50	-1.10
Gln	-1.20	-0.45	-1.10	-2.62	-1.21	0.08	-0.20	-1.96	0.10	-0.04	-0.83	-1.01	-0.42	-0.84	-2.62	0.10
Arg	-1.21	-1.80	-1.50	-0.80	-1.13	1.05	-2.60	-1.45	-1.50	-1.01	-0.58	-0.75	-1.10	-1.11	-2.60	1.05
Asn	-1.10	-0.35	-0.60	-1.60	-1.10	-1.31	0.10	-1.42	-0.30	-0.32	-0.42	-0.74	-0.58	-0.75	-1.60	0.10
Ser	0.60	-1.00	-1.00	-0.01	-0.32	-0.52	0.50	-0.86	-0.20	-0.81	-0.23	-0.24	-1.05	-0.40	-1.05	0.60
Asp	-0.80	0.00	0.20	-0.10	0.55	-1.03	1.80	-0.81	-0.81	0.23	0.64	1.00	-0.10	0.06	-1.03	1.80
Gly	0.00	0.00	0.00	0.00	0.00	0.00	0.00	0.00	0.00	0.00	0.00	0.00	0.00	0.00	0.00	0.00
Glu	-0.70	0.70	-0.40	0.55	1.60	1.48	1.00	-1.01	1.50	1.74	1.23	1.02	-0.36	0.64	-1.01	1.74
Ala	-0.30	0.30	1.80	0.42	1.34	-1.28	2.40	1.59	1.50	1.05	1.88	1.28	2.03	1.08	-1.28	2.40
Thr	-1.00	0.20	0.53	0.21	2.20	2.01	0.70	0.51	1.40	0.89	1.01	1.43	2.23	0.95	-1.00	2.23
Pro	0.66	2.10	3.75	1.42	3.18	0.97	2.30	3.39	1.70	2.21	2.32	2.65	3.42	2.31	0.66	3.75
Cys	1.00	2.00	3.50	1.85	4.22	3.68	3.70	4.34	2.48	2.71	3.21	3.98	3.74	3.11	1.00	4.34
Val	2.41	3.20	6.02	1.52	5.98	2.81	4.90	7.77	3.01	3.71	4.37	5.16	3.68	4.20	1.52	7.77
Tyr	3.20	3.90	6.30	2.75	6.42	5.33	5.30	7.79	3.52	3.76	4.78	5.67	7.33	5.08	2.75	7.79
Met	7.78	4.30	6.30	3.25	7.41	6.72	5.90	8.67	6.81	4.87	5.44	7.22	7.43	6.32	3.25	8.67
Ile	4.10	3.94	8.35	5.10	8.24	8.01	6.71	7.34	7.72	5.41	7.72	9.01	9.92	7.04	3.94	9.92
Leu	5.52	6.40	10.95	5.05	9.24	7.30	7.30	5.47	8.20	7.21	8.34	8.82	9.01	7.60	5.05	10.95
Phe	8.90	6.12	11.21	7.60	11.11	9.71	9.30	8.74	10.41	7.68	10.41	12.68	10.96	9.60	6.12	12.68
Trp	11.70	8.71	12.71	9.15	12.01	12.02	10.82	10.96	11.22	8.71	11.71	13.74	11.67	11.16	8.71	13.74
Min	-3.21	-1.80	-3.50	-3.51	-3.11	-2.24	-3.20	-3.51	-1.60	-1.21	-1.71	-1.63	-2.22	-2.29		
Max	11.70	8.71	12.71	9.15	12.01	12.02	10.82	10.96	11.22	8.71	11.71	13.74	11.67	11.16		
Range	14.91	10.51	16.21	12.66	15.12	14.26	14.02	14.47	12.82	9.92	13.42	15.37	13.89	13.45		
Ave	1.68	1.82	3.04	1.39	3.19	2.63	2.75	2.61	2.69	2.28	2.90	3.39	3.20	2.58		

Column: Vydac C18 TP, 250x4.6mm ID. 50min linear gradient from 0.1% TFA in 15% ACN to 0.1% TFA in 40% ACN. Flowrate: 1ml/min. Reprinted from Houghten et al., *J. Chromatogr.* 386 (1987) p.223 (ref. 49) with kind permission of Elsevier Science NL, Sara Burgerhartstraat 25, 1055 Amsterdam, The Netherlands.

If any particular position in the peptide were irrelevant to retention, then the minimum, maximum and range of retention (shown at the bottom of each column) would be identical. Likewise, if any specific amino acid were irrelevant to retention or if it had a finite effect, then the coefficients of a row would be identical. Neither is the case! The range data for each position in the peptide were plotted as a bar graph, shown in Fig. 7.12, to emphasize the variance.

Figure 7.12 Variations in Range for Amino Acid Substitutions

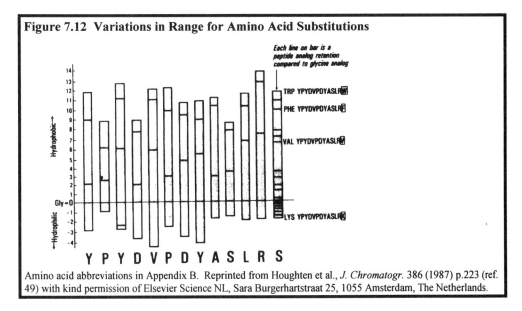

Amino acid abbreviations in Appendix B. Reprinted from Houghten et al., *J. Chromatogr.* 386 (1987) p.223 (ref. 49) with kind permission of Elsevier Science NL, Sara Burgerhartstraat 25, 1055 Amsterdam, The Netherlands.

Clearly, amino acid identity and position, as well as peptide sequence, all play extremely important roles in retention by RPC. Houghten further demonstrated the effects of amino acid position on reversed-phase retention in a study where a single glycine was inserted into a 13-mer peptide at 6 different positions to produce 14-mer peptides which had identical amino acid compositions (49). As seen in Fig. 7.13, the isomers differed enough in retention to be separated by RPC.

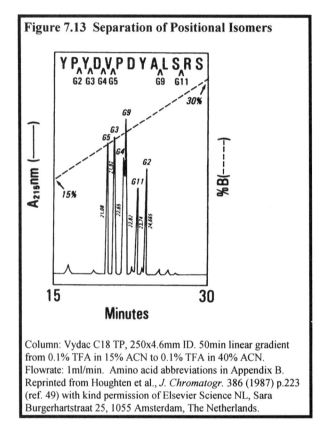

Figure 7.13 Separation of Positional Isomers

Column: Vydac C18 TP, 250x4.6mm ID. 50min linear gradient from 0.1% TFA in 15% ACN to 0.1% TFA in 40% ACN. Flowrate: 1ml/min. Amino acid abbreviations in Appendix B. Reprinted from Houghten et al., *J. Chromatogr.* 386 (1987) p.223 (ref. 49) with kind permission of Elsevier Science NL, Sara Burgerhartstraat 25, 1055 Amsterdam, The Netherlands.

3. Higher Orders of Structure of Larger Peptides

Most small peptides composed of 5 - 15 amino acids are assumed to exist in solution as a random coil. As the length of a peptide increases, opportunities exist for it to fold upon itself to assume its most stable state, forming more complex secondary structures, such as α-helices and pleated sheets. The most stable structure of a peptide is strongly influenced by the solvent and its immediate environment, which can include the stationary phase or other molecules in the mixture. This changing part of the surface of the molecule may or may not be the same as the surface which contacts the chromatographic matrix. It is important to realize that a conformation that may not be favored in the mobile phase alone may be stabilized by interaction with the reversed-phase surface. The folding of peptides may cause their hydrophobicity to be lower, because of inaccessibility to the bonded phase, or higher, due to the exposure of large hydrophobic patches. The hydrophobic and hydrophilic residues attempt to coalesce on the surface of larger peptides and proteins into respective regions, forming an amphipathic structure. An 18-mer of leucine (L) and lysine (K) was synthesized as a model peptide known to form an amphipathic α-helix, where all lysines are on one side and leucines on the other (49). The axial and lateral projections are shown in Fig. 7.14. This kind of hydrophobic area or patch interacts strongly with a chromatographic surface, producing hydrophobic interaction which is greater than the theoretical models.

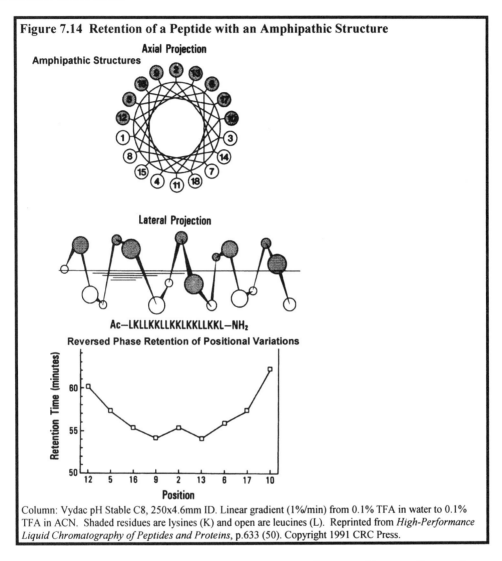

Figure 7.14 Retention of a Peptide with an Amphipathic Structure

Column: Vydac pH Stable C8, 250x4.6mm ID. Linear gradient (1%/min) from 0.1% TFA in water to 0.1% TFA in ACN. Shaded residues are lysines (K) and open are leucines (L). Reprinted from *High-Performance Liquid Chromatography of Peptides and Proteins*, p.633 (50). Copyright 1991 CRC Press.

The notion that the higher order structures common to large peptides contribute to the formation of their complex hydrophobic surfaces was tested in the following manner. When the lysine residues were substituted with leucine singly, in each of the nine positions, retention time by RPC was used to indicate positional effects due to a break in the helix. The graph in Fig. 7.14 shows the effect on retention of the leucine residue as it was "walked" along the hydrophobic helix pictured above. Insertion of leucine near either end of the peptide had little effect on retention, presumably because the remainder of the peptide could still form a helix. Insertion points deeper into the lysine region of the peptide reduced retention substantially. These data suggest that the positional effects of the single amino acid substitutions described previously may have been the result of their influence on peptide secondary structure.

4. Linear Solvent Strength Model

The effect of the concentration of organic modifier on the capacity factor (k) in isocratic elution varies tremendously between small molecules and peptides or proteins, as seen in Fig. 7.15a.

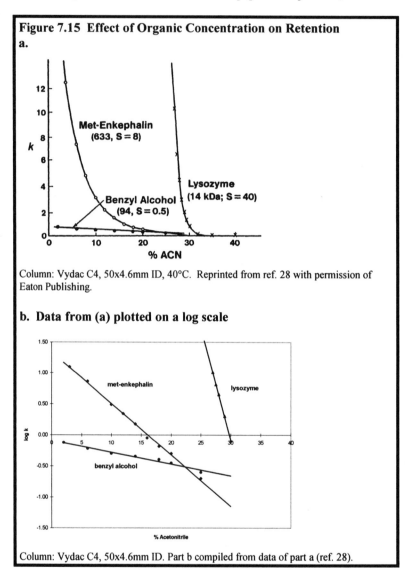

Figure 7.15 Effect of Organic Concentration on Retention

a.

Column: Vydac C4, 50x4.6mm ID, 40°C. Reprinted from ref. 28 with permission of Eaton Publishing.

b. Data from (a) plotted on a log scale

Column: Vydac C4, 50x4.6mm ID. Part b compiled from data of part a (ref. 28).

Benzyl alcohol, a small molecule, exhibited a fairly linear relationship between k and the content of organic in the mobile phase. This relationship was radically different for lysozyme and met-enkephalin, macromolecules which have multiple sites of interaction. It can be seen that met-enkephalin, a small peptide (633 Da), experienced major changes in retention when ACN was varied from 2 - 10%, but thereafter k was only minimally changed. Lysozyme, a small protein (14,000 Da), did not elute until more than 25% acetonitrile was used; subsequently, very small increments of acetonitrile produced large changes in k. When these k data are plotted on a log scale, as in Fig. 7.15b, the slopes of the lines can be used to describe the retention (51):

$$\log k = \log k_{w} - S\Phi \qquad\qquad\qquad \text{Eq. 7.1}$$

Where k_w is the capacity factor (k) in water as the eluent, S is a constant for the given system, and Φ is the volume fraction of organic solvent. This equation was also discussed in Chapter 2 (Eq. 2.33). An estimation of S can be made by (51):

$$S = 0.48 M^{0.44}$$ Eq. 7.2

where M is the molecular weight. This model indicates that the slopes of the plots in Fig. 7.15b are related to the molecular weight of the solutes. Dr. Lloyd Snyder and his associates developed these equations and relationships as the linear solvent strength model (LSS) to predict the chromatographic behavior of peptides and proteins (51, 52). The slope is generally in the range of 100 for proteins or other large molecules, 10 for peptides, and 1 or less for small organic molecules; in this example, the slope for lysozyme is about 40 and that for met-enkephalin is about 8 (28). It is hypothesized that these differences result from proteins having correspondingly greater hydrophobic surfaces which bind to reversed-phase ligands than do peptides or small molecules. The LSS concept also emphasizes why, in practice, isocratic elution is impractical or impossible for the separation of many peptides and proteins. The concentration of organic modifier which causes desorption of a weakly bound protein will likely be a concentration that leaves a more hydrophobic protein bound indefinitely. Conversely, a concentration of organic modifier that desorbs a more strongly bound protein will result in no separation between it and more weakly bound solutes, but rather simultaneous desorption. Additionally, a very slight variation in mobile phase results in a very large difference in retention for macromolecules.

The LSS model has been tested with various peptides and proteins ranging in size from 165 to 230,000 D (53-57). As shown in Table 7.11, a good agreement of slope (S) with protein molecular weight was found for both porous and nonporous packings.

Table 7.11 S Values For Select Proteins					
PROTEIN	M (kD)	S (theoretical)	S (experimental)		
			Porous	**Nonporous**	
			ref. 57	ref. 55	ref. 56
Cytochrome C	12.4	30.3	40	30.2	30
Catalase	232.0	110.0		44.0	
Lysozyme	14.4	32.4	40	34.5	
BSA	66.5	63.5		42.7	
Phe	0.2	4.5		5.3	
PhePhe	0.3	6.0		5.9	
PhePhePhe	0.5	7.1		6.1	
Leu-enkephalin	0.6	8.0	11		

Data from ref. 55-57, as indicated.

Data sets like these which verify the LSS model show that proteins which deviate significantly from their predicted retention may be composed of subunits, like catalase, or they may have unusual amounts of hydrophilic or hydrophobic residues compared to "average" proteins of similar molecular weight (54). Initial studies with RPC on nonporous supports showed no correlation for S values (54); however, later studies produced S values similar to the theoretical (55, 56).

One would predict that for the case of lysozyme shown in Fig. 7.15, the use of concentrations of acetonitrile above 32% would create retention similar to that obtained with 32%; however, this is not always the case. Increased retention is sometimes observed at higher concentrations of organic solvent (27, 59). As shown in Fig. 7.16, four peptides of 5 - 15 amino acid residues run isocratically at increasing concentrations of organic modifier varied in retention, reaching a minimum during a window of organic concentration.

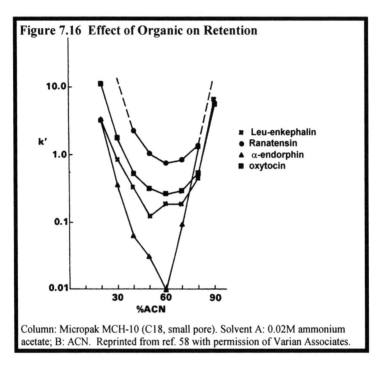

Figure 7.16 Effect of Organic on Retention

* Leu-enkephalin
* Ranatensin
▲ α-endorphin
■ oxytocin

Column: Micropak MCH-10 (C18, small pore). Solvent A: 0.02M ammonium acetate; B: ACN. Reprinted from ref. 58 with permission of Varian Associates.

In each case, a minimum retention was reached on the reversed-phase matrix, followed by an increase in peptide retention at higher concentrations of organic solvent. This surprising phenomenon has been well documented with many peptides and proteins and is related to the multi-modal or non-hydrophobic characteristics of the column (28, 59). The diagram shown in Fig. 7.17 of the effects which occur during silica-based RPC explains the separation mechanisms:

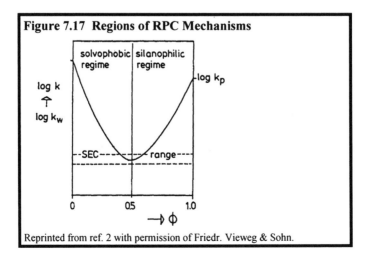

Figure 7.17 Regions of RPC Mechanisms

Reprinted from ref. 2 with permission of Friedr. Vieweg & Sohn.

Three regions of separation can be defined. As Φ increases from 0 to 0.5 (or a value defined by the minimum k of the system), k decreases. Standard RPC is occurring under these conditions. As solvent strength increases beyond that which causes the minimum retention or k, normal phase chromatography becomes dominant. Separations that occur at solvent concentrations at the minimum k are primarily governed by size exclusion principles because the mobile phase and the stationary phase are of similar polarity. The normal phase behavior on some columns makes it possible to run with a reversed gradient (beginning with organic) for certain applications (28). This has found special utility when samples containing SDS as an impurity are analyzed. These multimodal effects are attributable to matrix and bonded phase properties; supports with small pores (8 - 12nm), such as the column used in Fig. 7.16, are more likely to exhibit effects than those with large pores ($\geq$ 30nm) (28). Similar results have been observed for polymeric supports (4).

Although the exact mechanism of RPC is not totally defined for peptides, several pragmatic conclusions emerge from the experimental findings. The maximum organic concentration used for reversed-phase elution of peptides and proteins should be no greater than that which provides the minimum retention or k (often 50% for ACN and 65% for methanol). Using greater concentrations may increase the retention from multimodal forces and slow the elution. Similarly, cleaning a column of adsorbed protein should be done with concentrations of about 50% organic rather than 100%, or, preferably by using repeated gradients. Washing a column with pure organic solvent would only be appropriate for stripping accumulated lipids or small molecules, which bind by a purely hydrophobic mechanism.

C. Protein Behavior

1. Chromatographic Behavior

The utility of RPC for peptide analysis is unquestionable; this technique has been invaluable in all avenues of peptide investigation. In the case of proteins, however, RPC is not universally useful due to solute lability. Many proteins are denatured by acids or organic solvents, the usual constituents of the mobile phase, and others are denatured by the alkyl chain of the bonded phase (53). Sometimes denaturation is reversible, but not always.

Common and, unfortunately, frequent deviations from ideal chromatographic behavior of proteins during RPC are listed in Table 7.12.

Table 7.12 Nonideal Behavior of Proteins in RPC
• Broad peaks
• Asymmetrical peaks
• Multiple peaks from a pure compound
• Low recovery
• "Ghosting"

These problems, including "ghosting" (where peaks are observed without sample injection in blank runs following an analysis) are familiar to anyone who has analyzed proteins by RPC. One model which has been developed to explain these common difficulties is based on the assumption that a pure protein can exist in both a native, folded conformation and an uncoiled or randomly coiled, denatured form. The model by Barry Karger and associates assumes that there are dynamic transitions between these two states, as a protein partitions from the mobile phase to the stationary phase, as diagrammed in Fig. 7.18 (60).

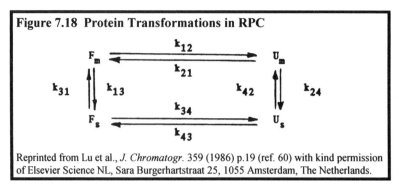

Figure 7.18 Protein Transformations in RPC

Reprinted from Lu et al., *J. Chromatogr.* 359 (1986) p.19 (ref. 60) with kind permission of Elsevier Science NL, Sara Burgerhartstraat 25, 1055 Amsterdam, The Netherlands.

The folded, (F), native state of the protein can exist in either phase, the mobile (F_m) or the stationary phase (F_s), as can the denatured, unfolded protein (U). The four possible states of the protein have corresponding equilibrium constants. The model permits only stepwise transitions - the protein can either change its structure or its phase with each transition, but not both simultaneously; therefore, there are no diagonal transitions in the diagram.

This theory was tested in two ways using purified ribonuclease as a model protein (36, 60). Ribonuclease was run at three different flowrates while keeping the gradient rate constant, as seen in Fig. 7.19.

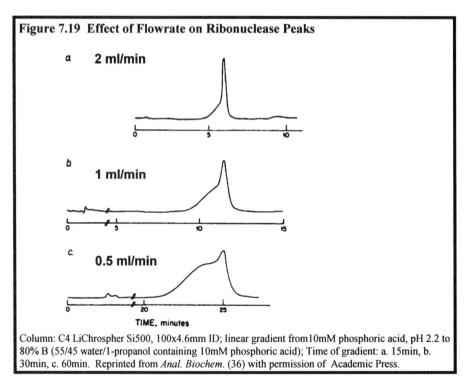

Figure 7.19 Effect of Flowrate on Ribonuclease Peaks

a **2 ml/min**

b **1 ml/min**

c. **0.5 ml/min**

TIME, minutes

Column: C4 LiChrospher Si500, 100x4.6mm ID; linear gradient from 10mM phosphoric acid, pH 2.2 to 80% B (55/45 water/1-propanol containing 10mM phosphoric acid); Time of gradient: a. 15min, b. 30min, c. 60min. Reprinted from *Anal. Biochem.* (36) with permission of Academic Press.

At the fastest flowrate, a relatively sharp peak was obtained with only a small leading front. Using absorbance and circular dichromism to measure changes caused by denaturation, it was determined that the main peak was denatured, more hydrophobic, and therefore later-eluting than the native protein. As the flowrate was slowed, more early-eluting material was seen because the slower flowrates allowed more time for the desorbed ribonuclease to renature as it passed through the column. In the parlance of the model of Fig. 7.18, slower flow permitted a more complete k_{42}, k_{21} transition than was afforded at higher flowrates.

In a second experiment, the effect of temperature on ribonuclease conformation was examined (36). As shown in Fig. 7.20, the peak sharpened and the front disappeared as the temperature was increased. In this case, the higher temperatures favored denaturation, whereas operating the separation at lower temperatures ($20°$ - $25°C$) maintained the native, more hydrophilic state.

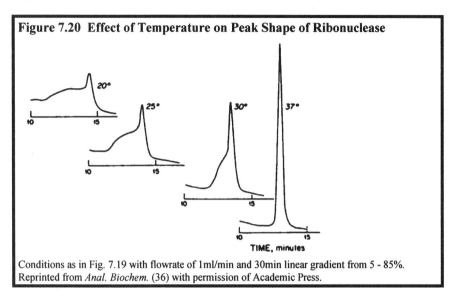

Figure 7.20 Effect of Temperature on Peak Shape of Ribonuclease

Conditions as in Fig. 7.19 with flowrate of 1ml/min and 30min linear gradient from 5 - 85%. Reprinted from *Anal. Biochem.* (36) with permission of Academic Press.

2. Protein Structure

The role of protein structure in chromatographic behavior was qualitatively discussed in an article by Fred Regnier in *Science* (61). The basic conclusions are listed in Table 7.13:

Table 7.13 Roles of Protein Structure in Chromatographic Behavior

- The weak chemical forces that govern protein conformation and surface recognition (ionic, nonpolar, van der Waals, hydrogen bonding) are the same as those involved in chromatographic interactions.
- It is not possible for all the amino acids in a protein to simultaneously contact the stationary phase surface.
- Only residues located at the protein surface have an impact on chromatographic behavior and only a fraction of those are involved in stationary phase interactions.
- Heterogeneous distribution of residues on the protein surface allows some portions of the surface to dominate chromatographic behavior; interactive regions may not be the same for different chromatographic modes.
- Structural changes which alter the protein surface can change chromatographic behavior if they occur within the contact area or alter the structure of the contact area.
- Interaction with the stationary or mobile phase can alter protein structure.

These ideas provide plausible explanations for the nonideal chromatographic behavior of proteins. It is clear from studies such as the one described above for ribonuclease that proteins can exist in more than one state (36). This idea has been amply confirmed by studies of enzyme transition states in enzyme-substrate interactions and the binding of antibodies to antigens. If a protein undergoes a conformational change during chromatography, it is clear that the transition(s) takes time. If the transition time is on the same order as the migration time, peak broadening would be expected to occur. If the transition time is rapid, then an "ideal" single, narrow peak should result. A very slow transition might produce multiple peaks. Such conformational transitions may account for protein loss, poor recovery, and ghosting. If a protein undergoes a transition while deep within a pore, new hydrophobic surfaces could be exposed that bind via multiple sites to the stationary phase. If these interactions stabilize an otherwise fleeting and unstable conformation, they could cause it to remain "stuck" in a pore throughout an entire run. A portion may refold and elute in a subsequent run giving rise to the phenomenon of ghosting.

D. Oligonucleotides

Oligonucleotides are sometimes difficult to differentiate by HPLC or other techniques because of their similar compositions. Separations by anion-exchange chromatography are fundamentally by size, and it is usually impossible to isolate variants of the same length which differ by only a base pair. Ion-pair RPC at elevated temperatures has proven effective in separating variants of even large DNA fragments. The technique, called denaturing high performance liquid chromatography (DHPLC), employs nonporous beads composed of PSDVB with C-18 chains (23, 62). Elevated temperatures, often 50°C, produce excellent resolution and also inhibit formation of secondary structures as illustrated in Fig. 7.21 for a complex mixture of restriction fragments. DHPLC can even distinguish between heteronucleotides of the same chain length which only differ in the terminal base, as seen in Fig. 7.22. Those with the adenine (A) and thymine (T) bases eluted unresolved and later than those with cytosine (C) or guanine (G).

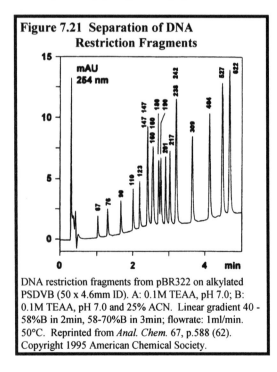

Figure 7.21 Separation of DNA Restriction Fragments

DNA restriction fragments from pBR322 on alkylated PSDVB (50 x 4.6mm ID). A: 0.1M TEAA, pH 7.0; B: 0.1M TEAA, pH 7.0 and 25% ACN. Linear gradient 40 - 58%B in 2min, 58-70%B in 3min; flowrate: 1ml/min. 50°C. Reprinted from *Anal. Chem.* 67, p.588 (62). Copyright 1995 American Chemical Society.

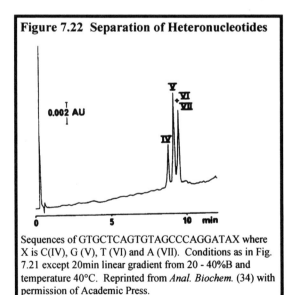

Figure 7.22 Separation of Heteronucleotides

Sequences of GTGCTCAGTGTAGCCCAGGATAX where X is C(IV), G (V), T (VI) and A (VII). Conditions as in Fig. 7.21 except 20min linear gradient from 20 - 40%B and temperature 40°C. Reprinted from *Anal. Biochem.* (34) with permission of Academic Press.

VI. START-UP PROCEDURES

It is difficult to detail an absolute scheme for running small molecules by RPC because of the frequent mixed-mode interactions which may occur. It is advisable to follow the column manufacturer's recommendations. An excellent start-up procedure can be found in ref. 63.

RECOMMENDATIONS FOR RPC START-UP

Small Molecules
1. Choose a C-18 support with 8 - 12nm pores, or a base-deactivated column, if cationic solutes will be run.
2. Use a column of 4.6mm ID and 15 - 25cm long for standard isocratic analyses. These columns have adequate efficiency and do not require special hardware.
3. Use a flowrate of 1 ml/min for a 4.6mm ID column and condition the column with at least ten column volumes of mobile phase. If the column was shipped in a solvent incompatible with the mobile phase, first wash with ten column volumes of distilled water or a bridge solvent like isopropanol.
4. Use the mobile phase recommended by the manufacturer and run 1μl of the sample shown in the test chromatogram to verify that the system and column are operating efficiently. Alternatively, use hydrocarbons, such as those recommended by Tanaka (41), to test the column. Inject 1μl of the test mixture made up in the mobile phase (approx. 1 mg/ml). Make sure that the plates are similar to those achieved by the manufacturer. If they are not, try the measures suggested below in troubleshooting.
5. Clean the injector. Inject 1 - 5μl of standard compounds which will be run as analytes, dissolved in the mobile phase.
6. Adjust the mobile phase organic, pH, ion-pair agent, etc. to achieve a capacity factor (k) of 2 - 7 for the peaks (56). Ionic solutes usually require a buffered mobile phase for optimum performance and reproducibility.
7. If k values vary by tenfold or more, a gradient may be necessary.
8. After the analyses are finished, clean the column in a stronger solvent to wash off highly bound materials.
9. Store in a solvent which will inhibit bacterial growth, usually organic without ion-pair agents. Storage in 0.1% TFA or another acid may promote hydrolysis of the bonded phase. Follow the column manufacturer's recommendations.

Peptides
1. Choose a reversed-phase column with at least 30nm pores.
2. Use a column of 4.6mm ID and 5 - 25cm long for standard analyses. These columns do not require special hardware.
3. Test as in steps #3 and #4 above.
4. Condition the column with a 30min gradient from 0.1% TFA in HPLC-grade water to 0.1% TFA in ACN. Then condition with the aqueous mobile phase for ten column volumes.
5. Run a peptide standard mixture like that from Alberta Peptide Institute. Use their conditions or a linear gradient of 1%B/min. Peaks should be sharp without tailing. If they are not, follow troubleshooting recommendations.
6. Recondition the column and run the sample under the same conditions. It is best to dissolve the sample in the initial mobile phase and inject 1 - 5μl.
7. After completing the analyses, clean the column, if necessary, and store as in #9 above.

VII. TROUBLESHOOTING

TROUBLESHOOTING REVERSED-PHASE CHROMATOGRAPHY

	Symptom	Cause	Remedy
1	Poor plates on new column (small molecule)	sample too concentrated volume too large old or contaminated extracolumn dead volume	dilute inject less make new sample minimize extra-column volume
2	Poor plates on used column (small molecule)	see #1 above column fouling column void	wash per manufacturer's recommendations organics will remove hydrophobic molecules 0.1% TFA will remove some ionic molecules replace column; use guard column in future; topping off may be temporary cure
3	Poor peak shape for peptides	sample specific peptide problem mobile phase column not conditioned ion-pair concentration too low column	as in #1 above run peptide standards to check column performance run 10 column volumes (or more, if not adequate) prepare fresh solvents see #2 above
4	peaks too early	gradient too stong peptides too hydrophilic	slow gradient rate use weaker organic solvent use more hydrophobic ion-pair agent
5	peaks too late	gradient too long or too slow peptides too hydrophobic column too hydrophobic	increase gradient rate begin with some organic in mobile phase use stronger organic solvent use more hydrophilic ion-pair agent use column with shorter ligand chain

VIII. REFERENCES

1. G.S. Howard and A.J.P. Martin, *Biochem. J.* 46 (1950) 532.
2. K.K. Unger, R. Janzen, and G. Jilge, *Chromatographia* 24 (1987) 144.
3. U. Esser and K.K. Unger, "Reversed-Phase Packings for the Separation of Peptides and Proteins by Means of Gradient Elution HPLC" in *High-Performance Liquid Chromatography of Peptides and Proteins* (ed. C.T. Mant and R.S. Hodges), CRC Press, Boca Raton (1991) 273.
4. K.A. Tweeten and T.N. Tweeten, *J. Chromatogr.* 359 (1986) 111.
5. H. Engelhardt and M. Jungheim, *Chromatographia* 29 (1990)59.
6. H.H. Freiser, M.P. Nowlan, and D.L. Gooding, *J. Liq. Chromatogr.* 12 (1989) 827.
7. M.J. Wirth, LC-GC 12 (1994) 656.
8. R.J. Steffeck, S.L. Woo, R.J. Weigand, and J.M. Anderson, *LC-GC* 13 (1995) 720.
9. M. Kawakatsu, H. Kotaniguchi, H. Freiser, and K.M. Gooding, *J. Liq. Chromatogr.* 18(4) (1995) 633.
10. J. Frenz, W.S. Hancock, W.J. Henzel, and Cs. Horvath, "RPC in Analytical Biotechnology of Proteins" in *HPLC of Biological Macromolecules* (ed. K.M. Gooding and F.E. Regnier), Marcel Dekker, New York (1990) 145.
11. J.J. Kirkland, *Current Issues in HPLC Technology (LC-GC)*, May 1997, S46.
12. C.T. Mant and R.S. Hodges, "Requirements for Peptide Standards to Monitor Column Performance and the Effect of Column Dimensions, Organic Modifiers, and Temperature in Reversed-Phase Chromatography" in *High-Performance Liquid Chromatography of Peptides and Proteins* (ed. C.T. Mant and R.S. Hodges), CRC Press, Boca Raton (1991) 289.
13. K.D. Nugent, "Commercially Available Columns and Packings for Reversed-Phase HPLC of Peptides and Proteins" in *High-Performance Liquid Chromatography of Peptides and Proteins* (ed. C.T. Mant and R.S. Hodges), CRC Press, Boca Raton (1991) 279.
14. C.T. Mant and R.S. Hodges, "HPLC of Peptides" in *HPLC of Biological Macromolecules* (ed. K.M. Gooding and F.E. Regnier), Marcel Dekker, New York (1990) 301.
15. H.H. Freiser, M.P. Nowlan, and D.L. Gooding, unpublished work.
16. C.T. Mant and R.S. Hodges, "The Effects of Anionic Ion-pairing Reagents on Peptide Retention in Reversed-Phase Chromatography" in *High-Performance Liquid Chromatography of Peptides and Proteins* (ed. C.T. Mant and R.S. Hodges), CRC Press, Boca Raton (1991) 327.
17. C.T. Mant and R.S. Hodges, 'Optimization and Prediction of Peptide Retention Behavior in Reversed-Phase chromatography" in *HPLC of Proteins, Peptides and Polynucleotides* (ed. M.T.W. Hearn), VCH Publishers, New York (1991) 277.
18. Cs. Horvath, W. Melander and I. Molnar, *J. Chromatogr.* 125 (1976) 129.
19. C.T. Mant, N.E. Zhou, and R.S. Hodges, "Amino Acids and Peptides" in *Chromatography, 5th Edition* (ed. E. Heftmann), Elsevier Science Publishers, Amsterdam (1992) B75.
20. N.E. Hoffmann and J.C. Liao, *Anal. Chem.* 49 (1977) 2231.
21. H.P.J. Bennett, "Manipulation of pH and Ion-pairing Reagents to Maximize the Performance of Reversed-Phase Columns" in *High-Performance Liquid Chromatography of Peptides and Proteins* (ed. C.T. Mant and R.S. Hodges), CRC Press, Boca Raton (1991) 319.
22. R. Bischoff and L.W. McLaughlin, "Resolution of Oligonucleotides and Transfer RNAs by HPLC" in *HPLC of Biological Macromolecules: Methods and Applications* (K.M. Gooding and F.E. Regnier, eds.), Marcel Dekker, New York, 1990.
23. C.G. Huber, P.J. Oefner, and G.K. Bonn, *Anal. Biochem.* 212 (1993) 351.
24. J.E. Rivier, *J. Liq. Chromatogr.* 1 (1978) 343.
25. D. Guo, C.T. Mant, and R.S. Hodges, *J. Chromatogr.* 386 (1987) 205.
26. D. Guo, C.T. Mant, A.K. Taneja, J.M.R. Parker, and R.S. Hodges, *J. Chromatogr.* 359 (1986) 499.

27. R. Feldhoff, *Tech. Protein Chem. II* (1991) 55.

28. R.J. Simpson and R.L. Moritz, "Chromatography of Proteins at High Organic Concentrations: an Inverse-gradient RPC method for Preparing Samples for Microsequence Analysis" in *High-Performance Liquid Chromatography of Peptides and Proteins* (ed. C.T. Mant and R.S. Hodges), CRC Press, Boca Raton (1991) 399.

29. M.W. Dong, J.R. Gant, and B.R. Larsen, *BioChromatography* 4 (1989) 19.

30. K.M. Gooding and F.E. Regnier, "Size Exclusion Chromatography" in *HPLC of Biological Macromolecules: Methods and Applications* (K.M. Gooding and F.E. Regnier, eds.), Marcel Dekker, New York, 1990.

31. D. Guo, C.T. Mant, A.K. Taneja, and R.S. Hodges, *J. Chromatogr.* 359 (1986) 519.

32. T.W.L. Burke, C.T. Mant, and R.S. Hodges, "The Effect of Varying Flowrate, Gradient-rate and Detection Wavelength on Peptide Elution Profiles in Reversed-Phase Chromatography" in *High-Performance Liquid Chromatography of Peptides and Proteins* (ed. C.T. Mant and R.S. Hodges), CRC Press, Boca Raton (1991) 307.

33. K. Kalghati and Cs. Horvath, "Micropellicular Sorbents for Rapid Reversed-Phase Chromatography of Proteins and Peptides" in *Analytical Biotechnology, Capillary Electrophoresis and Chromatography* (ed. Cs. Horvath and J.G. Nikelly), American Chemical Society (1990) 162.

34. K.K. Unger, G. Jilgs, J.N. Kinkel, and M.T.W. Hearn, *J. Chromatogr.* 359 (1986) 61.

35. M.N. Schmuck, K.M. Gooding, and D.L. Gooding, *J. Liq. Chromatogr.* 7 (1984) 2863.

36. S. Cohen, K. Benedek, Y. Tapuhi, J.C. Ford, and B.L. Karger, *Anal. Biochem.* 144 (1985) 275.

37. C.A. Hoeger, R. Galyean, R.A. McClintock, and J.E. Rivier, "Practical Aspects of Preparative RPC of Synthetic Peptides" in *High-Performance Liquid Chromatography of Peptides and Proteins* (ed. C.T. Mant and R.S. Hodges), CRC Press, Boca Raton (1991) 753.

38. J.R. Grun and R. Reinhardt, "Discontinuous RPC: a Separation Method for Complex Protein Mixtures Applicable in Both Analytical and Large Scale HPLC" in *High-Performance Liquid Chromatography of Peptides and Proteins* (ed. C.T. Mant and R.S. Hodges), CRC Press, Boca Raton (1991) 409.

39. T.W.L. Burke, J.A. Black, C.T. Mant, and R.S. Hodges, "Preparative RP Shallow Gradient Approach to the Purification of Closely-related Peptide Analogs on Analytical Instrumentation" in *High-Performance Liquid Chromatography of Peptides and Proteins* (ed. C.T. Mant and R.S. Hodges), CRC Press, Boca Raton (1991) 783.

40. F.D. Antia and Cs. Horvath, "Displacement Chromatography of Peptides and Proteins" in *High-Performance Liquid Chromatography of Peptides and Proteins* (ed. C.T. Mant and R.S. Hodges), CRC Press, Boca Raton (1991) 809.

41. K. Kimata, K. Iwaguchi, S. Onishi, K. Jinno, R. Eksteen, K. Hosoya, M. Araki, and N. Tanaka, *J. Chromatogr. Sci.* 27 (1989) 721.

42. N. Tanaka, presented at *HPLC 95*, Innsbruck.

43. L.C. Sander and S.A. Wise, *LC-GC* 8 (1990) 378.

44. M.A. Stadalius, J.S. Berus, and L.R. Snyder, *LC-GC* 6 (1988) 494.

45. J.L. Meek, *Proc. Natl. Acad. Sci.* (USA) 77 (1980) 1632.

46. I. Molnar and Cs. Horvath, *J. Chromatogr.* 142 (1977) 623.

47. J.L. Meek and Z.L. Rossetti, *J. Chromatogr.* 211 (1981) 15.

48. C.T. Mant, T.W.L. Burke, J.A. Black, and R.S. Hodges, *J. Chromatogr.* 458 (1988) 193.

49. R.A. Houghten and S.T. DeGraw, *J. Chromatogr.* 386 (1987) 223.

50. J.M. Ostresh, K. Buttner, and R.A. Houghten, "RPC: the Effect of Induced Conformations on Peptide Retention" in *High-Performance Liquid Chromatography of Peptides and Proteins* (ed. C.T. Mant and R.S. Hodges), CRC Press, Boca Raton (1991) 633.

51. M.A. Stadalius, H.S. Gold, and L.R. Snyder, *J. Chromatogr.* 296 (1984) 31.

52. M.A. Stadalius and L.R. Snyder, "HPLC Separations of Large Molecules: a General Model" in *HPLC - Advances and Perspectives, Vol. 4* (ed. Cs. Horvath), Academic Press, New York (1986) 195.

53. R. Janzen, K.K. Unger, H. Giesche, J.N. Kinkel, and M.T.W. Hearn, *J. Chromatogr.* 397 (1987) 81.
54. K.K. Unger, G. Jilge, J.N. Kinkel, and M.T.W. Hearn, *J. Chromatogr.* 359 (1986) 61.
55. G. Jilge, R. Janzen, H. Giesche, K.K. Unger, J.N. Kinkel, and M.T.W. Hearn, *J. Chromatogr.* 397 (1987) 71.
56. D.C. Lommen and L.R. Snyder, *LC-GC* 11 (1993) 222.
57. M.A. Stadalius, H.S. Gold, and L.R. Snyder, *J. Chromatogr.* 327 (1985) 27.
58. C.T. Wehr and L. Correia, *LC at Work* LC-121, Varian Associates.
59. M.I. Aguilar and M.T.W. Hearn, "Reversed-Phase and Hydrophobic-Interaction Chromatography of Proteins" in *HPLC of Proteins, Peptides and Polynucleotides* (ed. M.T.W. Hearn), VCH Publishers, New York (1991) 247.
60. X.M. Lu, K. Benedek, and B.L. Karger, *J. Chromatogr.* 359 (1986) 19.
61. F.E. Regnier, *Science* 238 (1987) 319.
62. C.G. Huber, P.J. Oefner, and G.K. Bonn, *Anal. Chem.* 67 (1995) 578.
63. J.J. Kirkland, *LC-GC* 14 (1996) 486.

CHAPTER 8
HYDROPHOBIC INTERACTION CHROMATOGRAPHY

I. INTRODUCTION

Hydrophobic Interaction Chromatography (HIC) is a mode of separation in which molecules in a high salt environment interact hydrophobically with a nonpolar surface. Such interactions have been observed as unwanted secondary effects during separation of proteins by size exclusion chromatography (1,2) and by affinity chromatography using supports synthesized with alkyl spacer arms (3). The discovery that this hydrophobic mechanism provided the basis for an effective means of protein resolution prompted the development of HIC as a distinct method (4-6).

HIC has been predominantly used to analyze proteins, nucleic acids and other biological macromolecules when maintenance of the three-dimensional structure is a primary concern. The main applications of HIC have been in the area of protein purification because the recovery is frequently quantitative in terms of mass and biological activity. Its discrimination by hydrophobicity makes HIC complementary to size exclusion and ion-exchange chromatography. The paucity of applications in the literature is due more to a lack of understanding of the potential of HIC than to its lack of utility.

II. MECHANISM

In HIC, a high salt environment causes association of hydrophobic patches on the surface of a protein with the nonpolar ligands of the bonded phase. Elution is generally effected by an "inverse" gradient to lower salt concentration. This is considered "inverse" because it is the opposite of gradients used for ion-exchange chromatography. Effective salts for HIC are those which are "antichaotropic", that is, they promote the ordering of water molecules at interfaces (7-9). HIC is a mild method, yielding high recoveries of biological activity. Because interaction is only with the surface of the protein, the number of amino acids involved in the chromatography is relatively small, and changes in surface structure can cause differential binding, and hence, separation.

Reversed phase and hydrophobic interaction chromatography are both based on interactions between hydrophobic moieties, but the operational aspects of the techniques render selectivities totally different. The selectivities of the two methods are contrasted in Fig. 8.1A and B for a mixture of three proteins.

Figure 8.1 HIC vs. RPC for a Protein Mixture

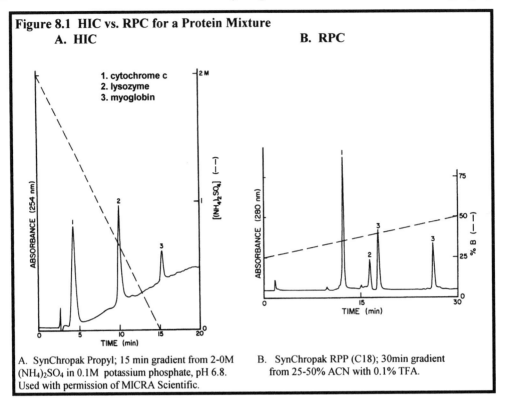

A. HIC   B. RPC

1. cytochrome c
2. lysozyme
3. myoglobin

A. SynChropak Propyl; 15 min gradient from 2-0M (NH$_4$)$_2$SO$_4$ in 0.1M potassium phosphate, pH 6.8. Used with permission of MICRA Scientific.

B. SynChropak RPP (C18); 30min gradient from 25-50% ACN with 0.1% TFA.

It can be seen that the selectivity, and even the number of peaks, varies between the two modes. All the proteins are retained on the C-18 reversed phase column but cytochrome c has no retention on the C-3 hydrophobic interaction column. A primary reason for the vast difference is the mobile phase environment for each method. The organic solvents and generally acidic conditions used in RPC cause denaturation of most proteins and even splitting into subunits, whereas the high salt concentrations at neutral pH used in HIC result in stabilization of globular or three-dimensional structures for biological macromolecules (7, 8). Additionally, the bonded phase of HIC supports consists of a hydrophilic matrix into which hydrophobic chains are inserted, generally in low density. This can be contrasted with the long chain organosilane chemistry used in RPC. Fig. 8.2 illustrates these physical dimensions of the two methods.

Figure 8.2 RPC vs. HIC

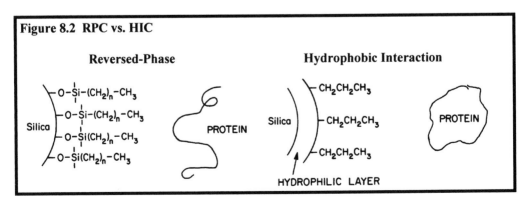

Reversed-Phase   Hydrophobic Interaction

The hydrophobic amino acid residues of globular proteins are generally folded inside the structure or located in a few patches on the surface. As a protein is denatured, the buried amino acids are exposed, yielding more sites for hydrophobic binding. The hydrophobic interaction system thus encounters only surface amino acids - far fewer hydrophobic residues than the reversed phase.

III. SUPPORTS

Bonded phases for HIC consist of a hydrophilic polymeric layer into which hydrophobic ligands are inserted (10-15). The hydrophilic layer totally covers the silica or polymer matrix, providing a wettable and noninteractive surface which is neutral to the protein, frequently being suitable for SEC. In HIC, even short ligands cause substantial binding and there is a definite relationship between ligand chain length and retention (12, 16), contrary to the minimal effect of chain length observed in RPC. Fig. 8.3 illustrates the effect of ligand arm on retention for several proteins (17).

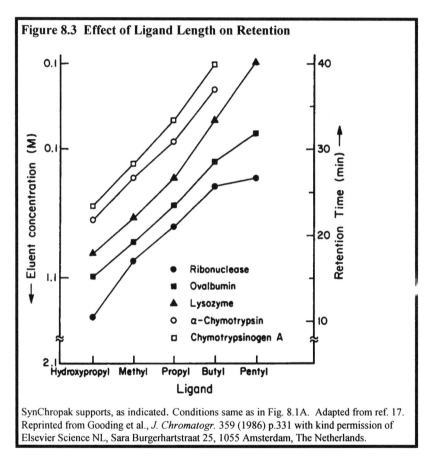

Figure 8.3 Effect of Ligand Length on Retention

SynChropak supports, as indicated. Conditions same as in Fig. 8.1A. Adapted from ref. 17. Reprinted from Gooding et al., *J. Chromatogr.* 359 (1986) p.331 with kind permission of Elsevier Science NL, Sara Burgerhartstraat 25, 1055 Amsterdam, The Netherlands.

The ligand chains are postulated either to interact with hydrophobic surface patches on proteins or to be inserted into their hydrophobic pockets (7); it is the latter interaction which is strengthened by and related to chain length. The strength of the binding causes some proteins to bind irreversibly if the ligand is too long; therefore, most ligands are either aromatic or 1 - 3 carbon alkyl chains. Short chains generally bind, and also release, most proteins under HIC conditions.

One variation of HIC uses weak ion-exchange supports with HIC conditions (7, 18). Separations are achieved when antichaotropic salts and inverse gradients are used on these columns because the slightly hydrophobic crosslinking agents of the ion-exchange bonded phases serve as HIC ligands. The use of such columns is a good option when weak ion-exchange supports are the only available columns to test the feasibility of HIC. Otherwise, it is best to use columns designed for HIC because their characteristics have been defined and tested and their ionic properties are minimized.

Because HIC supports are designed for macromolecules, they either possess pore diameters of at least 30nm to allow penetration (10-14) or are nonporous (19, 20). Both silica and polymer matrices are used because the hydrophilic polymeric coating minimizes or eliminates most matrix-based effects. Table 8.1 lists selected supports for HIC and their characteristics.

Table 8.1 Selected HIC Supports

Product	Chemistry	Pore (nm	Size (µm)	Support	Manufacturer
Alkyl Superose	neopentyl		13	agarose	Amersham Pharmacia Biotech
Bio-Gel MP7 HIC	methyl	80	7	PSDVB	Bio-Rad
LC-HINT	glycerylpropyl		5	silica	Supelco
Phenyl Superose	phenyl		13	agarose	Pharmacia
PolyEthyl A	ethyl	30	5	silica	PolyLC
PolyPropyl A	propyl	30	5	silica	PolyLC
POROS PE	phenyl ether	100	20	PSDVB	Perseptive Biosystems
POROS BU	butyl	100	20	PSDVB	Perseptive Biosystems
POROS ET	ether	100	20	PSDVB	Perseptive Biosystems
Resource	ether, phenyl		15	PSDVB	Amersham Pharmacia Biotech
Spherogel CAA-HIC	methyl	30	5	silica	Beckman
SynChropak Propyl	propyl	30	6	silica	MICRA
TSK Ether-5PW	ether	100	10	methacrylate copolymer	TosoHaas
TSK Phenyl-5PW	phenyl	100	10	methacrylate copolymer	TosoHaas
TSK Butyl-NPR	butyl	nonporous	2.5	methacrylate copolymer	TosoHaas

The absolute retention and selectivity of an HIC support is not only based on the ligand, but also on the specific composition of the bonded phase, as illustrated in Fig. 8.4 where seven proteins are resolved differently on three hydrophobic interaction columns, two of which have propyl functional groups (21).

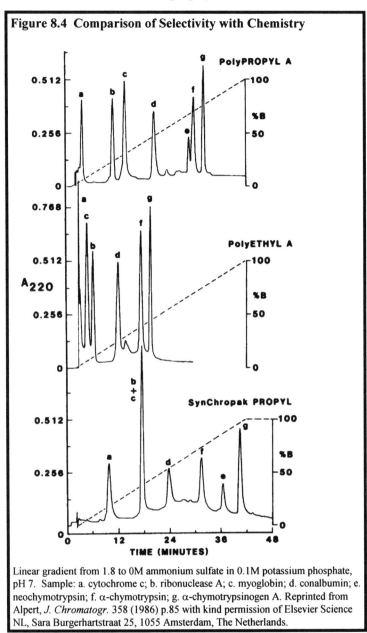

Figure 8.4 Comparison of Selectivity with Chemistry

Linear gradient from 1.8 to 0M ammonium sulfate in 0.1M potassium phosphate, pH 7. Sample: a. cytochrome c; b. ribonuclease A; c. myoglobin; d. conalbumin; e. neochymotrypsin; f. α-chymotrypsin; g. α-chymotrypsinogen A. Reprinted from Alpert, *J. Chromatogr.* 358 (1986) p.85 with kind permission of Elsevier Science NL, Sara Burgerhartstraat 25, 1055 Amsterdam, The Netherlands.

In these examples, retention of all the proteins except α-chymotrypsin was higher on the SynChropak Propyl column. Ribonuclease and myoglobin were separated on PolyPropyl A, but neochymotrypsin and chymotrypsin were totally resolved on the SynChropak Propyl.

IV. OPERATION

A. Mobile Phase

In HIC, the concept of weak and strong solvents can be confusing. The weak solvent, or the one which promotes binding, is that containing high salt concentration. The strong solvent, or one which causes elution, is that with low salt concentration.

1. Salt

The most important variable in HIC retention, other than the ligand chain, is the composition of the salt used to promote binding. The effectiveness is based on the molal surface tension increment, which is parallel to the Hofmeister salting-out series for precipitation of proteins. The strength of HIC binding for some commonly used salts is:

$$K_3\text{citrate} > Na_2SO_4 > (NH_4)_2SO_4 > Na_2HPO_4 > NaCl.$$

Fig. 8.5 A and B show the effects of different salts on the retention and resolution of several proteins (22).

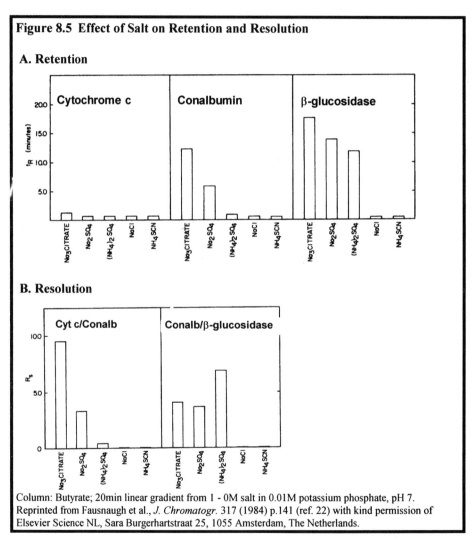

Figure 8.5 Effect of Salt on Retention and Resolution

A. Retention

B. Resolution

Column: Butyrate; 20min linear gradient from 1 - 0M salt in 0.01M potassium phosphate, pH 7.
Reprinted from Fausnaugh et al., *J. Chromatogr.* 317 (1984) p.141 (ref. 22) with kind permission of Elsevier Science NL, Sara Burgerhartstraat 25, 1055 Amsterdam, The Netherlands.

Although sodium citrate and sodium sulfate cause stronger retention, ammonium sulfate is the most popular choice for HIC. Besides being effective for retention, it is highly soluble, stabilizing for enzymes and resistant to microbial growth (7). Ammonium sulfate is available in high purity because of its use for salt fractionation. Sodium sulfate is less soluble and may precipitate under conditions of high concentration. The initial concentration of salt must be at a level high enough to cause binding of the proteins to the bonded phase or retention will vary with the salt concentration (23). In HIC, the concentration of antichaotropic salt is proportional to log k (24). Fig. 8.6 shows this relationship for conalbumin in four different salts. The exact relationship varies for each salt, as well as for the specific protein (24). Maximum binding is usually achieved with 2M ammonium sulfate for most proteins.

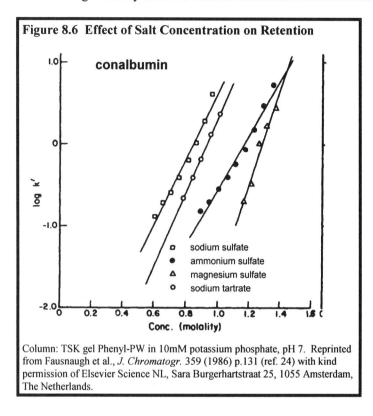

Figure 8.6 Effect of Salt Concentration on Retention

conalbumin

□ sodium sulfate
● ammonium sulfate
△ magnesium sulfate
○ sodium tartrate

Column: TSK gel Phenyl-PW in 10mM potassium phosphate, pH 7. Reprinted from Fausnaugh et al., *J. Chromatogr.* 359 (1986) p.131 (ref. 24) with kind permission of Elsevier Science NL, Sara Burgerhartstraat 25, 1055 Amsterdam, The Netherlands.

2. pH

In HIC, the mobile phase should be buffered to provide control of ionization because amino acids which are not ionized are more hydrophobic than those which are charged. Guidelines for buffer selection in specific pH regions can be found in Chapter 9. The effect of pH on hydrophobicity produces some variation of retention with pH (13-23); however, it is not directly related to the pI because only surface amino acids interact with the ligands. In a study of the effect of pH on retention by HIC for a series of lysozymes from different bird species, those containing histidine residues in the contact region exhibited deviation with pH, as seen in Fig. 8.7 (24). Lysozymes from ring necked pheasants have such histidine residues, whereas lysozymes from hen egg whites have no histidines in the contact area.

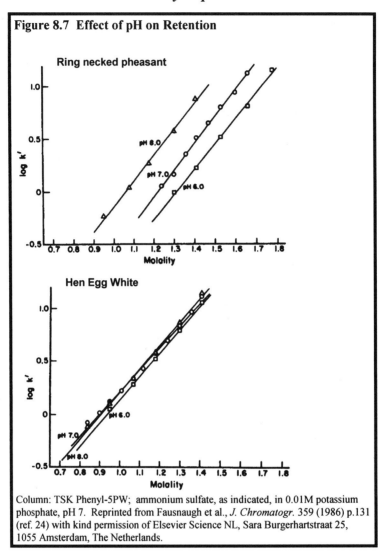

Figure 8.7 Effect of pH on Retention

Column: TSK Phenyl-5PW; ammonium sulfate, as indicated, in 0.01M potassium phosphate, pH 7. Reprinted from Fausnaugh et al., *J. Chromatogr.* 359 (1986) p.131 (ref. 24) with kind permission of Elsevier Science NL, Sara Burgerhartstraat 25, 1055 Amsterdam, The Netherlands.

3. Additives

Because HIC is based on surface tension phenomena, changing those characteristics by the addition of surfactants affects retention. When Wetlaufer et al. examined the effects of surfactants on retention of proteins (25-26), they found that the addition of CHAPS ((3-[(3-cholamidopropyl) dimethylammonio]-1-propane sulfonate) to the mobile phase resulted in shortened retention, improvement of peak shape, and a change in peak order for enolase and bovine pancreatic trypsin inhibitor (BPTI), as seen in Fig. 8.8 (27).

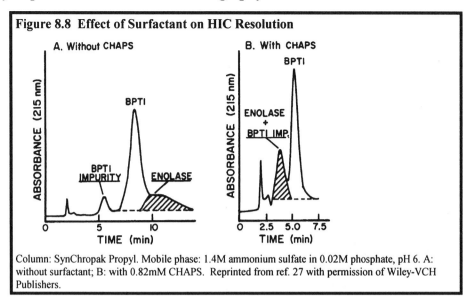

Figure 8.8 Effect of Surfactant on HIC Resolution

Column: SynChropak Propyl. Mobile phase: 1.4M ammonium sulfate in 0.02M phosphate, pH 6. A: without surfactant; B: with 0.82mM CHAPS. Reprinted from ref. 27 with permission of Wiley-VCH Publishers.

These effects are dependent on the concentration of the surfactant, as seen for trypsin inhibitor, lysozyme and ribonuclease A in Fig. 8.9 (25). The value k'/k_o' is the normalized retention where k_o' is that in the absence of CHAPS.

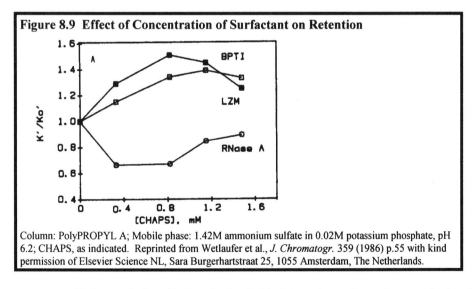

Figure 8.9 Effect of Concentration of Surfactant on Retention

Column: PolyPROPYL A; Mobile phase: 1.42M ammonium sulfate in 0.02M potassium phosphate, pH 6.2; CHAPS, as indicated. Reprinted from Wetlaufer et al., *J. Chromatogr.* 359 (1986) p.55 with kind permission of Elsevier Science NL, Sara Burgerhartstraat 25, 1055 Amsterdam, The Netherlands.

Surfactants can usually be washed easily from hydrophobic interaction columns because the bonded phases are neither highly hydrophobic nor ionic.

The hydrophobic basis of HIC means that alcohols may reduce interaction with supports; however, disruption of protein conformation may also occur (23, 28). Because of the high salt concentrations used in HIC, organic solvents should only be added after compatibility with the mobile phase has been tested to insure that precipitation will not take place. Generally, no more than 10% organic is added.

Other additives which increase the stability of a given protein can often be included in the mobile phase for HIC without adversely effecting the separation. In one example of such a mobile phase modification, peak shapes and recoveries of estrogen receptors were improved when the stabilizing agent, sodium molybdate, was added to the mobile phase (29).

B. Flowrate and Gradient

Nearly all HIC separations are performed in the gradient mode because proteins bind with multipoint interactions. The effects of flowrate and gradient on retention in HIC follow the linear solvent strength model, as discussed in Chapter 7 (30). Thus there is some effect of flowrate and gradient rate on resolution. In Fig. 8.10 it can be seen that the resolution decreased as the flowrate was increased. There was less variation when a 60min gradient was used (23). Because the gradient time was held constant, the gradient rate was also changing in this study.

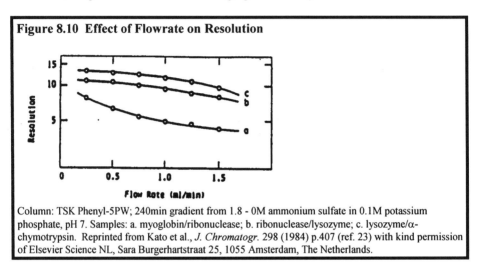

Figure 8.10 Effect of Flowrate on Resolution

Column: TSK Phenyl-5PW; 240min gradient from 1.8 - 0M ammonium sulfate in 0.1M potassium phosphate, pH 7. Samples: a. myoglobin/ribonuclease; b. ribonuclease/lysozyme; c. lysozyme/α-chymotrypsin. Reprinted from Kato et al., *J. Chromatogr.* 298 (1984) p.407 (ref. 23) with kind permission of Elsevier Science NL, Sara Burgerhartstraat 25, 1055 Amsterdam, The Netherlands.

The time of the gradient (t_G) is another determinant in improving resolution (8, 31), as shown in Fig. 8.11. Longer gradients provide increased resolution.

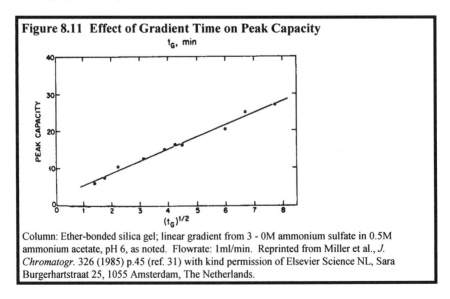

Figure 8.11 Effect of Gradient Time on Peak Capacity

Column: Ether-bonded silica gel; linear gradient from 3 - 0M ammonium sulfate in 0.5M ammonium acetate, pH 6, as noted. Flowrate: 1ml/min. Reprinted from Miller et al., *J. Chromatogr.* 326 (1985) p.45 (ref. 31) with kind permission of Elsevier Science NL, Sara Burgerhartstraat 25, 1055 Amsterdam, The Netherlands.

In this case, the resolving power is presented in terms of peak capacity (PC), which is the approximate number of peaks which can be resolved under the gradient conditions:

$$PC = \frac{t_G}{4\sigma_t}$$ Eq. 8.1

where σ_t is the time based standard deviation of the width of the peak (31). Generally, a 20 - 60 min gradient from 2M - 0M ammonium sulfate in 0.02M buffer at neutral pH, with a moderate flowrate (1ml/min for 4.6mm ID), will provide a satisfactory starting point for an HIC analysis.

C. Temperature

HIC is different than the other modes of HPLC because it is an entropy-driven process, characterized by increased retention with increased temperature (7, 28). This is a major benefit when subambient temperatures must be used to preserve the structure and biological activity of labile proteins, such as estrogen receptors (29). Retention is decreased somewhat rather than increased substantially as temperatures are lowered. Fig. 8.12 shows the effect of temperature on retention for several proteins (32).

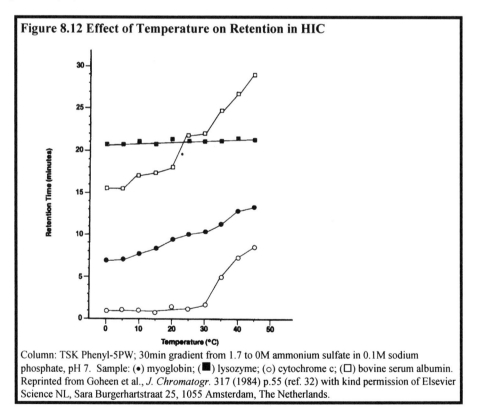

Figure 8.12 Effect of Temperature on Retention in HIC

Column: TSK Phenyl-5PW; 30min gradient from 1.7 to 0M ammonium sulfate in 0.1M sodium phosphate, pH 7. Sample: (•) myoglobin; (■) lysozyme; (o) cytochrome c; (□) bovine serum albumin. Reprinted from Goheen et al., *J. Chromatogr.* 317 (1984) p.55 (ref. 32) with kind permission of Elsevier Science NL, Sara Burgerhartstraat 25, 1055 Amsterdam, The Netherlands.

The retention of lysozyme was relatively unchanged throughout this temperature range, whereas albumin exhibited two peaks which changed in proportion with temperature. Some of the increase in retention with elevated temperatures, in this or other studies, can be attributed to protein unfolding and the increased exposure of hydrophobic residues, especially when peak broadening also occurs (7).

D. Loading

Loading capacities for proteins on HIC columns are quite high and similar to those for ion-exchange chromatography because proteins retain their globular forms during the procedure (28, 17). High loading does not generally appear to cause denaturation or other alterations which may reduce recoveries of biological activity. Table 8.2 lists some dynamic and absolute loading capacities for different HIC supports. As was discussed in Chapter 5, loading is related to the relative sizes of the pore diameter and the solute, with 30nm giving maximum capacity for many proteins.

Table 8.2 Loading Capacities for HIC		
HIC Support	**Dynamic Loading**	**Absolute Loading**
TSK Phenyl 5-PW	4 mg/ml	10 - 30mg/ml
SynChroprep Propyl	9 mg/ml	48 mg/ml
Phenyl Superose HR	10 mg/ml	
From ref. 25 and product catalogs.		

V. APPLICATIONS

A. Proteins

The major use of HIC is in protein analysis. An excellent application of HIC is to monitor changes caused by solvents, temperature, and contact time on the conformations of proteins. In one study, the spectroscopic characteristics of peaks were observed during HIC with various conditions to verify that conformational changes due to partial unfolding of the proteins caused longer retention (33). Although HIC is a relatively gentle chromatographic technique, the surface of the support can disrupt the quaternary structure of some proteins. This was demonstrated when tumor necrosis factor (TNF), a trimeric protein, was dissociated into monomers during elution (34). The extent of dissociation was related to operational factors such as temperature and flowrate. The unfolding was reversible, with the trimer reforming after elution, either in solution or during subsequent SEC.

The initial mobile phase conditions of high salt make the use of HIC directly after salting-out steps very convenient. The chromatography in this case both separates the components and reduces the salt in the sample. Samples purified by this means can subsequently be applied to an ion-exchange column (15) after dilution, if necessary. Similarly, samples which have been previously purified by IEC can be applied to a hydrophobic interaction column to be further purified after addition of salt to the sample.

HIC has proven effective for separating antibodies from albumin (15, 35-36). Fig. 8.13 illustrates the fractionation of a crude monoclonal antibody (mAb) formulation by HIC, before and after purification by IEC on a Mono Q column (36).

Figure 8.13 Analysis of Monoclonal Antibodies by HIC

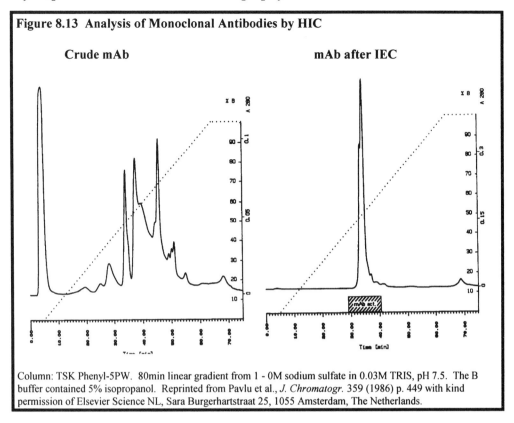

Column: TSK Phenyl-5PW. 80min linear gradient from 1 - 0M sodium sulfate in 0.03M TRIS, pH 7.5. The B buffer contained 5% isopropanol. Reprinted from Pavlu et al., *J. Chromatogr.* 359 (1986) p. 449 with kind permission of Elsevier Science NL, Sara Burgerhartstraat 25, 1055 Amsterdam, The Netherlands.

The recovery of antibody from the HIC column was about 75%. It can be seen that excellent peak shapes were obtained for the pure protein, as well as good selectivity for the components in the crude mixture.

B. Peptides

The best HPLC method for peptide analysis is RPC because of its excellent power of resolution and selectivity. HIC offers a different selectivity than RPC for those peptides which possess three-dimensional conformations under high salt conditions. When the separation of several peptide mixtures by HIC and RPC were compared, peaks were generally narrower on RPC, but some peptides, such as those from snake venom, could only be resolved by HIC (37).

In a study of calcitonin variants, it was seen that peptides with certain amino acid substitutions could not be resolved by RPC, but were separated by HIC (38). A comparison of the retention of three variants with that of native calcitonin is seen in Fig. 8.14.

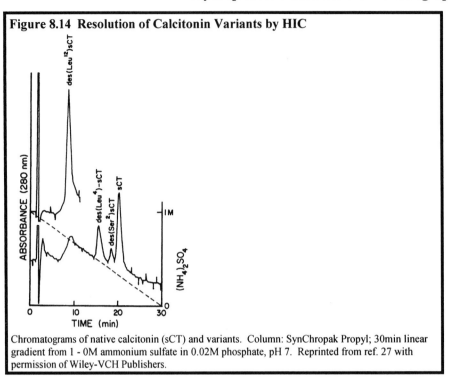

Figure 8.14 Resolution of Calcitonin Variants by HIC

Chromatograms of native calcitonin (sCT) and variants. Column: SynChropak Propyl; 30min linear gradient from 1 - 0M ammonium sulfate in 0.02M phosphate, pH 7. Reprinted from ref. 27 with permission of Wiley-VCH Publishers.

The main utility of HIC for peptide separations seems to lie in applications for extremely hydrophilic or hydrophobic peptides, or those with three-dimensional structures which are stable in high salt.

C. t-RNA

Another application of HIC for biological macromolecules has been the separation of nucleic acids. The tertiary structure of t-RNA has made analysis under the gentle conditions of HIC very feasible (39-40). Fig. 8.15 shows an example of the purification of t-RNA molecules specific for different amino acids on a 100nm polyol HIC column (40).

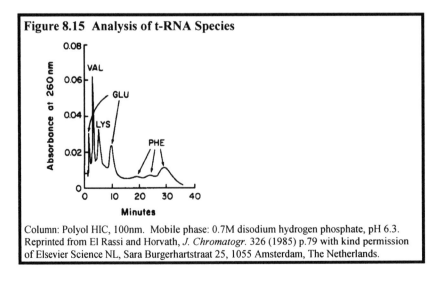

Figure 8.15 Analysis of t-RNA Species

Column: Polyol HIC, 100nm. Mobile phase: 0.7M disodium hydrogen phosphate, pH 6.3. Reprinted from El Rassi and Horvath, *J. Chromatogr.* 326 (1985) p.79 with kind permission of Elsevier Science NL, Sara Burgerhartstraat 25, 1055 Amsterdam, The Netherlands.

Separation of t-RNA molecules has also been accomplished successfully by using HIC conditions on supports with alkylamino ligands, which are functionally similar to those traditionally used to separate nucleic acids (39).

VI. START-UP PROCEDURES

RECOMMENDATIONS FOR HIC START-UP

1. Choose an HIC column from Table 8.1. If capacity is important, choose the smallest pore diameter which still includes the protein of interest.

2. Use a column of 4.6mm ID and 10 - 25cm long if standard analyses will be run. These columns do not require special hardware.

3. Use the mobile phase recommended by the manufacturer and run the sample shown in the test chromatogram to make sure that the system and column are operating efficiently. The mobile phase should contain at least 0.02M buffer at an appropriate pH.

4. Use a flowrate of 1 ml/min for 4.6mm ID columns and condition with at least ten column volumes of mobile phase. If the column was shipped in a solvent incompatible with the mobile phase, first wash with ten column volumes of distilled water or a miscible solvent.

5. Inject 1µl of the test mixture sample in the mobile phase (~ 1 mg/ml). Make sure that the plates are similar to those achieved by the manufacturer. If they are not, try the measures suggested below in Troubleshooting.

6. To analyze proteins or biological macromolecules, use mobile phase conditions which enhance the biological stability of the sample. Solvent A can be 2M ammonium sulfate in 0.02 - 0.1M buffer at an appropriate pH (*see* Table 9.7). Solvent B should be the buffer without ammonium sulfate. Adjust the pH after addition of salt. Ammonium sulfate should be high quality, suitable for HPLC.

7. Run the sample isocratically in solvent B. If it does not elute immediately, change the pH and/or add up to 10% organic, and try again. If there is still no elution, the ligand chain may be too long.

8. Condition the column with 10 column volumes of solvent B, followed by 10 column volumes of solvent A.

9. Run a mixture of standard proteins similar to those in the sample, dissolved in solvent A. Run a 30 min linear gradient from 0 - 100% B at 1ml/min (4.6mm ID). Peaks should be sharp without tailing. If they are not, follow the troubleshooting recommendations.

10. Recondition the column and run the sample under the same conditions. It is best to dissolve the sample in the initial mobile phase and inject 1 - 5µl. The sample must contain high salt concentration.

11. After the analyses, wash the column with water and store in a solvent which will inhibit bacterial growth, such as 5% isopropanol in water. Follow the manufacturer's recommendations.

12. Wash the system thoroughly with water.

VII. TROUBLESHOOTING

The most common operational error in HIC is switching the weak and strong mobile phases because they intuitively seem to be the opposite of those used in many other modes of chromatography.

TROUBLESHOOTING HYDROPHOBIC INTERACTION CHROMATOGRAPHY

	Symptom		Cause	Remedy
1	Poor plates on new column (small molecule*)	a.	sample	
			has less salt than mobile phase	add salt to sample
			too concentrated	dilute
			volume too large	inject less
			old or contaminated	make new sample
		b.	dead volume	minimize extra-column volume
2	Poor plates on used column (small molecule*)	a.	see #1 above	
		b.	column fouling	wash per manufacturer's recommendations
				0.1% TFA will remove many ionic molecules
				10% organic in Solvent B may remove hydrophobic molecules
		c.	column void	replace column; use guard column in future
				topping off may be temporary cure
3	Poor peak shape for proteins	a.	sample	as in #1 above
			specific protein problem	run standard proteins to compare with original
		b.	mobile phase	
			column not conditioned	run 10 column volumes (or more, if not adequate)
			bad	prepare fresh solvents
		c.	column	see #2 above
4	peaks too early	a.	solvents A and B switched	put high salt as A
		b.	salt	use stronger salt (higher in Hofmeister series)
		c.	gradient too stong	slow gradient rate
		d.	pH	adjust pH
		e.	ligand chain too short	use longer ligand
5	peaks too late	a.	gradient too slow	increase gradient rate
		b.	salt	use weaker salt (lower in Hofmeister series)
		c.	hydrophobic interaction	use 5 - 10% alcohol in mobile phase
		d.	ligand chain too long	use shorter ligand
6	peaks did not elute		conditions too strong	see #5 above and #7 in Start-Up

* Theoretical plates are always measured with small molecules even though they are not the standard analytes.

VIII. REFERENCES

1. J. Porath, *Biochim. Biophys. Acta* 39 (1960) 193.
2. B. Gelotte, *J. Chromatogr.* 3 (1960) 330.
3. P. Cuatrecasas and C.B. Anfinsen, *Ann. Rev. Biochem.* 40 (1971) 259.
4. B.H.J. Hofstee, *Biochem. Biophys. Res. Comm.* 53 (1973) 1137.
5. J. Porath, L. Sundberg, N. Fornstedt, and I. Olsson, *Nature* 245 (1973) 465.
6. S. Hjerten, *J. Chromatogr.* 87 (1973) 325.
7. R.E. Shansky, S.-L. Wu, A. Figueroa, and B.L. Karger, "Hydrophobic Interaction Chromatography of Proteins" in *HPLC of Biological Macromolecules* (ed. K.M. Gooding and F.E. Regnier) Marcel Dekker, New York (1990) 95.
8. M.I. Aguilar and M.T.W. Hearn, "Reversed-Phase and Hydrophobic-Interaction Chromatography of Proteins" in *HPLC of Proteins, Peptides and Polynucleotides* (ed. M.T.W. Hearn), VCH Publishers, New York (1991) 247.
9. W.R. Melander, D. Corradini, and Cs. Horvath, *J. Chromatogr.* 317 (1984) 67.

10. Y. Kato, T. Kitamura, and T. Hashimoto, *J. Chromatogr.* 266 (1983) 49.
11. J.L. Fausnaugh, E. Pfannkoch, S. Gupta, and F.E. Regnier, *Anal. Biochem.* 137 (1984) 464.
12. D.L. Gooding, M.N. Schmuck, and K.M. Gooding, *J. Chromatogr.* 296 (1984) 107.
13. N.T. Miller, B. Feibush, and B.L. Karger, *J. Chromatogr.* 316 (1984) 519.
14. J.-P. Chang, Z. El Rassi, and Cs. Horvath, *J. Chromatogr.* 319 (1985) 396.
15. *Hydrophobic Interaction Chromatography*, Pharmacia (1993).
16. M.N. Schmuck, M.P. Nowlan, and K.M. Gooding, *J. Chromatogr.* 371 (1986) 55.
17. D.L. Gooding, M.N. Schmuck, M.P. Nowlan, and K.M. Gooding, *J. Chromatogr.* 359 (1986) 331.
18. L.A. Kennedy, W. Kopaciewicz, and F.E. Regnier, *J. Chromatogr.* 359 (1986) 73.
19. R. Janzen, K.K. Unger, H. Giesche, J.N. Kinkel, and M.T.W. Hearn, *J. Chromatogr.* 397 (1987) 91.
20. Y. Kato, T. Kitamura, S. Nakatani, and T. Hashimoto, *J. Chromatogr.* 483 (1989) 401.
21. A.J. Alpert, *J. Chromatogr.* 359 (1986) 85.
22. J.L. Fausnaugh, L.A. Kennedy, and F.E. Regnier, *J. Chromatogr.* 317 (1984) 141.
23. Y. Kato, T. Kitamura, and T. Hashimoto, *J. Chromatogr.* 298 (1984) 407.
24. J.L. Fausnaugh and F.E. Regnier, *J. Chromatogr.* 359 (1986) 131.
25. D.B. Wetlaufer and M.R. Koenigbauer, *J. Chromatogr.* 359 (1986) 55.
26. J.J. Buckley and D.B. Wetlaufer, *J. Chromatogr.* 464 (1988) 61.
27. K.M. Gooding and M.N. Schmuck, "Comparative Performance of Silica-Based Adsorbents for Ion-Exchange and Hydrophobic-Interaction Chromatography" in *HPLC of Proteins, Peptides and Polynucleotides* (ed. M.T.W. Hearn), VCH Publishers, New York (1991) 177.
28. R.H. Ingraham, "Hydrophobic Interaction Chromatography of Proteins" in *High-Performance Liquid Chromatography of Peptides and Proteins* (ed. C.T. Mant and R.S. Hodges), CRC Press, Boca Raton (1991) 425.
29. S.M. Hyder, J. Dong, P. Folk, and J.L. Wittliff, "High-Performance Hydrophobic Interaction Chromatography of a Labile Regulatory Protein" in *High-Performance Liquid Chromatography of Peptides and Proteins* (ed. C.T. Mant and R.S. Hodges), CRC Press, Boca Raton (1991) 451.
30. L.R. Snyder, "Gradient Elution Separation of Large Biomolecules" in *HPLC of Biological Macromolecules* (ed. K.M. Gooding and F.E. Regnier) Marcel Dekker, New York (1990) 95.
31. N.T. Miller and B.L. Karger, *J. Chromatogr.* 326 (1985) 45.
32. S.C. Goheen and S.C. Engelhorn, *J. Chromatogr.* 317 (1984) 55.
33. S.-L. Wu and B.L. Karger, "On-line Conformational Monitoring of Proteins in HPLC" in *High-Performance Liquid Chromatography of Peptides and Proteins* (ed. C.T. Mant and R.S. Hodges), CRC Press, Boca Raton (1991) 613.
34. M.G. Kunitani, R.L. Cunico, and S.J. Staats, *J. Chromatogr.* 443 (1988) 205.
35. S.C. Goheen and R.S. Matson, *J. Chromatogr.* 326 (1985) 235.
36. B. Pavlu, U. Johansson, C. Nyhlen, and A. Wichman, *J. Chromatogr.* 359 (1986) 449.
37. A.J. Alpert, *J. Chromatogr.* 444 (1988) 269.
38. M.L. Heinitz, E. Flanigin, R.C. Orlowski, and F.E. Regnier, *J. Chromatogr.* 443 (1988) 229.
39. R. Bischoff and L.W. McLaughlin, "Resolution of Oligonucleotides and Transfer RNAs by HPLC" in *HPLC of Biological Macromolecules* (ed. K.M. Gooding and F.E. Regnier) Marcel Dekker, New York (1990) 641.
40. Z. El Rassi and Cs. Horvath, *J. Chromatogr.* 326 (1985) 79.

CHAPTER 9
ION-EXCHANGE CHROMATOGRAPHY

I. INTRODUCTION

Ion-exchange chromatography (IEC) has been the primary method for protein analysis and purification, due to its excellent resolution under generally nondenaturing conditions. Proteins are usually recovered quantitatively, retaining their biological activity. IEC represents a fundamental approach to the separation of molecules by their charge and is complementary to separation methods based on differences in molecular size or hydrophobicity. As a rule, a purification process should use alternating principles of separation (orthogonal steps) so that different mechanisms are implemented. Given the high charge densities on many IEC supports (up to 0.5 mmol/ml of packing) and the feasible use of high flowrates, IEC is usually a part of modern protein purification processes at early stages when large volumes of fluid need to be processed. Methods with high loading capacity are used earlier than those with inherently lower capacity, especially for the analysis of large dilute samples. Because of its selectivity based on charge, IEC is also widely used to separate amino acids and carbohydrates whose small size and chemical similarity make them recalcitrant to other methods of separation. Nucleic acids, with their negatively-charged phosphate groups, are also amenable to separation by IEC.

II. MECHANISM

A. General

In ion-exchange chromatography, molecules bind by the reversible interaction of electrostatic charges located on the outer surface of the solute molecule with dense clusters of groups with an opposite charge on an ion-exchanger. To maintain electrical neutrality, the charges on both the molecule(s) of interest and the matrix are associated with ions of opposite charge, termed counterions, which are either provided by pre-equilibration with the mobile phase or during manufacturing. Because a solute must displace the counterions on the matrix to attach to it, the technique is termed "ion-exchange". If the support carries a positive charge, it is an anion-exchange packing (AEX); if it carries a negative charge, it is a cation-exchange support (CEX). Generally, the molecule of interest will have a charge that is opposite (positive or negative) of that on the support and the same as the competitively-displaced counterions. A diagram of the ion-exchange process for a protein is shown in Fig. 9.1.

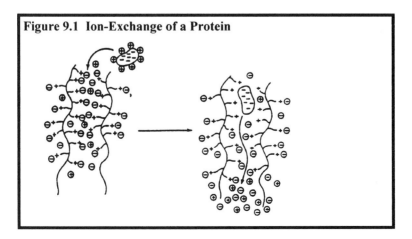

Figure 9.1 Ion-Exchange of a Protein

When a negatively-charged protein, associated with positively-charged ions, is applied to an anion-exchanger, the protein displaces the negative counterions associated with the bonded phase when it binds. The ions originally associated with both the protein and the support elute from the column.

For small molecules, interaction with an ion-exchanger usually consists of partitioning or readily reversible binding, leading to separation of molecules by their charge. The differential binding of charged molecules is diagrammed in the idealized ion-exchange separation shown in Fig. 9.2.

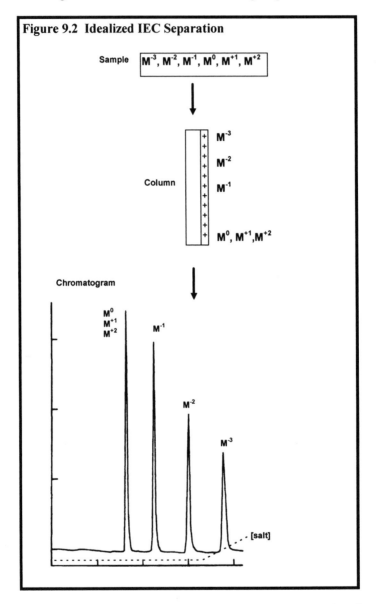

Figure 9.2 Idealized IEC Separation

Six classes of molecules are depicted at an acidic pH: three anions (M^{-1}, M^{-2} and M^{-3}), a neutral molecule (M^0), and two cations (M^{+1} and M^{+2}). When these molecules are applied to an anion-exchanger isocratically at low ionic strength, the cations and neutral molecule elute unresolved in a single fraction at the deadvolume. M^{-1} and M^{-2} interact with the matrix and are separated under the isocratic conditions; M^{-3} binds tightly to the matrix and does not elute under the low ionic strength conditions; however, it is desorbed by increasing the salt concentration. Eventually, even tightly bound molecules will "leach" off a column if enough initial (weak) solvent is run, but peaks may be excessively broad. For macromolecules such as proteins, interaction is generally so strong that gradients must be used to obtain distinct and narrow peaks.

B. Amino Acid Ionization

Because ionic interactions are at the heart of IEC, it is essential that the effects of pH on the ionization of molecules be clearly understood. In single (free) amino acids, the amino and carboxyl groups are fully charged in the normal physiological range of proteins (pH 6 - 8) and the amino acids are thus zwitterions. In stronger acids and bases, the amino and carboxyl groups titrate, being 50% ionized at their respective pK_a, as shown in Fig. 9.3 for an amino acid with an uncharged side chain. For example, at pH 9.6, half of the amino groups will possess a charge and half will not.

Figure 9.3 The Effect of pH on the Charge of Amino Acids

$$pK_{a1} = 2.3 \qquad pK_{a2} = 9.6$$

$$H_3{}^+NCHRCO_2H \longrightarrow H_3{}^+NCHRCO_2{}^- \longrightarrow H_2NCHRCO_2{}^-$$

pH 1	pH 6	pH 11
charge = +1	net charge = 0	charge = -1

R is uncharged in these examples.

Every amino acid is distinguished by the chemistry of its side group (R). When the side chain is ionizable (25% of the amino acids), its charge will be affected by changes in pH. Because several of the charged amino acids are among the most naturally abundant, virtually all proteins have positively and negatively charged groups which make them potentially amenable to separation by IEC. Based on their net charge at pH 7, amino acids with ionic side chains can be divided into two groups, basic and acidic (Table 9.1).

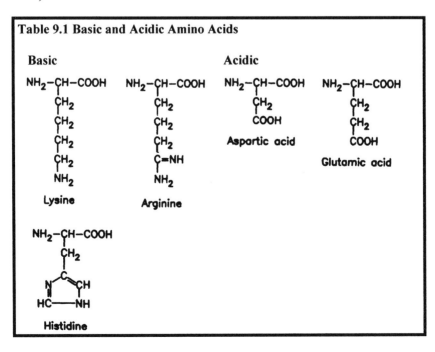

Table 9.1 Basic and Acidic Amino Acids

A summary of ionization data for the R groups of charged amino acids is presented in Table 9.2 (1).

Table 9.2 p*K*a Values for Charged Amino Acids					
Amino Acid	Functional Group (R)	pK_{a1}	pK_{a2}	pK_{aR}	Average Occurrence
Neutral	--	2.3	9.7	--	
aspartic acid	β-carboxyl	2.1	9.8	3.9	5.5%
glutamic acid	γ-carboxyl	2.2	9.7	4.3	6.2%
lysine	ε-amino	2.2	8.9	10.5	7.0%
arginine	guanidinium	2.2	9.0	12.5	4.7%
histidine	imidazole	1.8	9.2	6.0	2.1%
From ref. 1.					

The functional carboxyl group of aspartic acid has a pK_{aR} value of 3.9 which means that in buffer environments of pH greater than 3.9, the ionized form (COO⁻) is more prevalent than the protonated form (COOH). A useful guideline is that at the p*K* of a molecule, 50% of the groups are charged. At one pH unit from the p*K* in the direction of ionization, 90% of the groups are charged. At two pH units from the p*K*, 99% of the R groups are charged. The pI is the pH where an amino acid or protein has a net charge of zero; that is, the sum of the charges of all the groups equals zero. When an amino acid is part of a protein, its p*K* may be influenced somewhat by its immediate chemical environment and it may vary slightly from its value as a free amino acid.

The functional amino groups of lysine and arginine are positively charged, except at very high pH, because their pK_{aR} values are 10.5 and 12.5, respectively. Histidine, which contains an imidazole group, is unique among the amino acids in that its pK_{aR} is in the physiological pH range. It is frequently located at the center of biochemical reaction sites (e.g. enzyme binding sites) where its sensitivity to pH plays a key role in binding.

C. Protein Ionization

To appreciate certain key aspects of the charged nature of proteins and their behavior during IEC, it is important to review their primary structure (amino acid sequence) as shown in Fig. 9.4.

Figure 9.4 Peptide Structure
$(NH_3)^+$-$CHR_1CONHCHR_2CONHCHR_3CONHCHR_4CO$~$NHCHR_n(COO)^-$

All amino acids in a peptide or protein are sandwiched between the amino and carboxy terminal amino acids. Because amino and carboxyl groups of individual amino acids link in peptide (or amide) bonding, only those groups on the terminal amino acids and the R group side chains are free to participate in ionic interactions. The isoelectric point (pI) of a protein is the pH at which the net charge or the sum of its charges is zero. Proteins with more basic amino acids than acidic have a pI greater than 7 and are called basic proteins. Similarly, acidic proteins have a greater number of acidic amino acids and a pI less than 7. The plot of retention time vs. pH on both anion-exchange and cation-exchange columns yields a retention map which depicts pH-dependent chromatographic behavior (2-4). It would seem that a protein should have no ion-exchange binding at its pI and bind by anion- or cation-exchange as the pH varies from the pI, as in the idealized protein retention map in Fig. 9.5A.

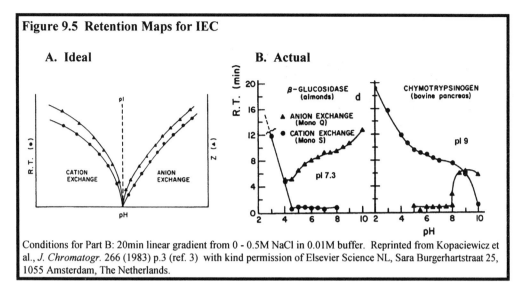

Figure 9.5 Retention Maps for IEC

Conditions for Part B: 20min linear gradient from 0 - 0.5M NaCl in 0.01M buffer. Reprinted from Kopaciewicz et al., *J. Chromatogr.* 266 (1983) p.3 (ref. 3) with kind permission of Elsevier Science NL, Sara Burgerhartstraat 25, 1055 Amsterdam, The Netherlands.

Such idealized retention is rarely encountered. Charged amino acids are often grouped on the exterior surface of a folded three-dimensional protein, forming charge clusters which are sometimes responsible for high ionic interactions (2). A protein frequently binds at its pI, when its overall charge is neutral, because of charged groups on the surface. In actuality, both the charge of individual amino acids on the surface of a protein and their density cause the binding in IEC; therefore, retention maps of individual proteins often vary from the ideal, as seen in Fig. 9.5B. It can be seen that chymotrypsinogen has little binding to an anion-exchange support except at its pI. β-glucosidase, on the other hand, does not bind to a cation-exchange support until the pH is significantly lower than its pI. These exceptions should not obscure an important guideline for ion-exchange: choosing a pH away from the pI imparts an overall charge to the molecule, and increases the likelihood of binding to an ion-exchanger. Theoretically, for most large, multi-chained proteins, either cation or anion-exchange is possible, but frequently one mode is more useful, as was seen in Fig. 9.5B.

Because IEC in nondenaturing solvents essentially maintains the tertiary structures of proteins, only surface amino acids are exposed to the ion-exchanger and involved in the binding process. This is one reason why IEC is so sensitive to slight variations in amino acid composition that it can provide total resolution of proteins which differ by only a single amino acid substitution, such as sickle cell hemoglobin (Hb S) and normal hemoglobin (Hb A) (5). The quantity of amino acids on the surface is a small percentage of the total amino acids in a protein, and a substitution of one of them, or the disruption in tertiary structure caused by an internal modification, can cause a major change in chromatographic properties (6).

III. SUPPORTS

There are several major variables which distinguish high performance ion-exchange packings and determine their utility for specific classes of solutes and for analytical or preparative applications. Those variables are listed in Table 9.3.

Table 9.3 High Performance Ion-Exchange Characteristics
1. Chemistry of the functional group 2. Composition of the support matrix 3. Pore diameter 4. Charge density and related nominal capacity

An understanding of the importance of these parameters will enable optimal column selection for any specific application.

A. Functional Groups

The focus thus far has been on the ionization of the solutes or the amino acid constituents of the proteins undergoing separation; however, the functional groups of the ion-exchanger are also susceptible to the effects of pH. This is the basis for distinguishing two types of ion-exchangers - strong and weak. These designations do not refer to the strength of binding or to the capacity of the gel, but simply to the pK of the ionizable ligand group, similar to the designations for acids and bases. The capacities of typical strong and weak ion-exchange groups as a function of pH are shown in Fig. 9.6.

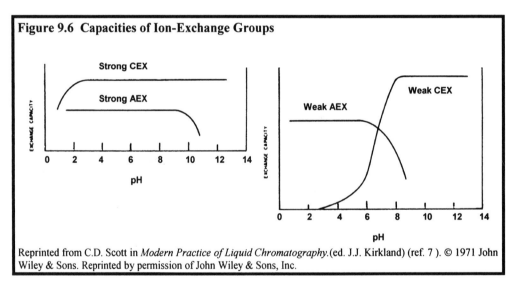

Figure 9.6 Capacities of Ion-Exchange Groups

Reprinted from C.D. Scott in *Modern Practice of Liquid Chromatography.*(ed. J.J. Kirkland) (ref. 7). © 1971 John Wiley & Sons. Reprinted by permission of John Wiley & Sons, Inc.

The strong ion-exchangers retain their charge over a wide range of pH with binding capacity dropping off at the extremes. For example, quaternary ammonium (Q) resins are strong anion-exchangers which are effective throughout the normal physiological range of proteins. Similarly, sulfonyl groups are strong cation-exchangers that remain negatively charged until acidic pH levels of approximately 3 are used. Strong ion-exchangers can be thought of as possessing a permanent positive or negative charge.

Weak ion-exchangers only have maximum capacity in the pH range where they maintain charge - pH less than 6 for the AEX support in Fig. 9.6 and pH greater than 8 for the CEX. The diminished capacities near neutral pH result in less predictable separations, if operation in this range is necessary for protein stability. In these cases, the use of a strong ion-exchanger would allow the pH of the mobile phase to be manipulated to protonate or deprotonate the R groups without changing the ionic properties of the packing. For example, the binding of positively charged amino acids in CEX would be enhanced at pH less than 4 (protonating the carboxyl group), but would require full charge on the matrix.

Clearly, careful consideration of titration curves such as these is an essential aspect of designing appropriate conditions for a separation. A complete description of the charged group of an ion-exchanger is necessary to understand its pH characteristics, which are dependent on the exact chemical composition of the bonded phase and the matrix. Convenient descriptions such as, "strong", "S", "stable weak ion-exchange", etc. do not sufficiently describe the ionic characteristics of the packing.

The two most commonly used ion-exchange groups for protein analysis on carbohydrate gel supports have been diethylaminoethyl (DEAE) and carboxymethyl (CM). Other functional groups which have seen popularity for high performance IEC include polyethyleneimine (PEI) and quaternary ammonium (Q) for anion-exchange and sulfonyl (S) for cation-exchange. The structures of strong and weak exchange groups are provided in Table 9.4.

Table 9.4 Strong and Weak Ion-Exchange Ligands

Anion-Exchange (AEX)	**Cation-Exchange (CEX)**
Weak	**Weak**
DEAE (diethylaminoethyl)	CM (carboxymethyl)
$-O-CH_2-CH_2-N^+H(CH_2CH_3)_2$	$-O-CH_2-COO^-$
PEI (polyethyleneimine)	
$(-NHCH_2CH_2)_n-N(CH_2CH_2-)_{n'}$	
$\quad\quad\quad\mid$	
$\quad\quad CH_2CH_2NH_2$	
Strong	**Strong**
Q (quaternary ammonium)	S (sulfonate)
$-CHOH-CH_2-N^+(CH_3)_3$	$-CH_2-CH_2-CH_2SO_3^-$

B. Matrix

Ion-exchange supports were initially based on derivatized cellulose and agarose. When higher performance was required, the smaller particle size and concomitantly higher pressures demanded less compressible supports like silica and cross-linked polymers (2, 8). Most supports for high performance IEC are based on silica or rigid polymers, whereas low pressure methods employ carbohydrate gel matrices (4).

1. Silica

Silica-based ion-exchangers are composed of 5 - 10μm silica bonded with a charged ligand group. Operating pH is limited to pH 2 - 8 due to the silica backbone. Although some small pore silica-based ion-exchangers are synthesized with silane bonding, large pore supports (≥ 30nm) designed for protein analysis have polymeric layers containing ionic functional groups (2, 8-10). These are very stable layers and even protect the silica matrix from erosion. Tentacle IEC bonded phases, as discussed in Chapter 5, are an alternate design which are either based on silica or a polymer (11). Silica columns have several advantages, as enumerated previously:

1. High mechanical stability due to the coupling of the charged groups onto rigid silica.
2. Minimal shrinkage or swelling so that changes in counter ions can be made without repacking the column.
3. Stability to organic modifiers (with the restriction of salt solubility).

4. High capacity (2 - 10meq/g packing).
5. Good mass transfer.
6. Large variety of available particle and pore sizes.

2. Polymers

Polymeric matrices are also widely available for IEC (12-14). Polystyrene crosslinked with divinylbenzene (PSDVB) is one such polymer, typically available with pore diameters exceeding 100nm. The repetitive structure of polystyrene permits reproducible coupling of both strong and weak ion-exchange groups, and the crosslinking adds the rigidity required for high pressure applications. This durable support was originally developed for specialized water treatment aboard submarines and served as one of the initial core products of Bio-Rad, Inc. The structure of PSDVB ion-exchangers is shown in Fig. 9.7 where "X" is the ionic functional group.

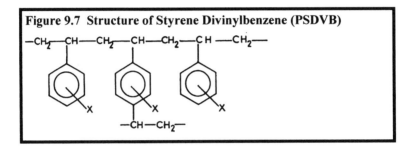

Figure 9.7 Structure of Styrene Divinylbenzene (PSDVB)

Methacrylate copolymers are also used as matrices in IEC (12-13). These are more hydrophilic than PSDVB and can be porous or nonporous.

3. Pellicular

A third group of ion-exchange supports are pellicular, consisting of a solid inert core made of PSDVB agglomerated with 350nm functionalized latex (15). The design of the beads is shown in Fig. 9.8. The quaternary amine groups are closely and uniformly bound on the microbeads, improving flow and reducing nonspecific retention. These pellicular supports are primarily used for carbohydrate analysis.

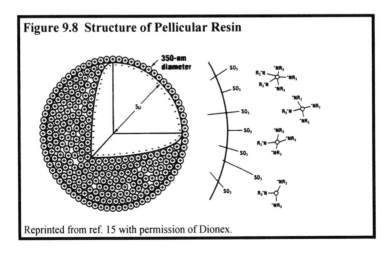

Figure 9.8 Structure of Pellicular Resin

Reprinted from ref. 15 with permission of Dionex.

4. Comparison

Characteristics of silica, polymeric and pellicular packings are compared in Table 9.5.

Table 9.5 Characteristics of IEC Matrices

Property	Silica	Polymeric Porous	Polymeric Nonporous	Pellicular
Typical d_p	5 - 10μm	10 - 20μm	2.5 - 7μm	10μm
Typical ion-exchange capacity	110[c]	30 -120[c]	5[c]	3μeq/ml[b]
rigidity to pressure	very good	good	very good	good
pressure drop	high	moderate	very high	moderate
efficiency	very good	very good	excellent	very good
pH range	2 - 7 or 9[a]	0 - 12	2 - 12	0 - 14
regeneration rates	moderate	moderate	fast	fast

[a]depends on bonded phase [b]suitable for carbohydrates [c]mg BSA/ml

C. Pore Diameter

As the size of the pore diameter decreases, there is a tremendous increase in surface area, as discussed in Chapter 5. Matrices with the smallest pores exhibit the highest ion-exchange capacities for small, totally included solutes, as shown in Table 9.6.

Table 9.6 Ion-Exchange Capacity Relative to Pore Diameter

Pore (nm)	Surface Area (m²/g)	Ion-Exchange Capacity picric acid μmol/g	Ion-Exchange Capacity ovalbumin mg/g	Ion-Exchange Capacity BSA mg/g
10	250	1415[a]	59[a]	64[b]
30	100	656[a]	98[a]	130[b]
50	50	308[a]	76[a]	59[b]
100	20	129[a]	26[a]	57[b]

PEI bonded phase. [a] from ref. 16. [b] Printed with permission from *High-Performance Liquid Chromatography of Peptides and Proteins* (9). Copyright 1991 CRC Press.

The ion-exchange capacities of picric acid correlate with surface area; however, those of the proteins do not, because they are partially excluded by size from portions of the smaller pores and are effectively prevented from reaching all the reactive exchange sites. It can be seen that the 30nm pore had the maximum capacity for ovalbumin and bovine serum albumin (BSA) because they could permeate and bind to the optimum available surface area.

D. Comparison of Selected Ion-Exchange Columns

Examples of some microparticulate packings for IEC and their characteristics are presented in Table 9.7.

Table 9.7 Selected HPIEC Supports

Product	Chemistry	Pore (nm)	Size (µm)	Support	Manufacturer
Weak AEX					
Bio-Gel MA7P	PEI	nonporous	7	PSDVB	Bio-Rad
LiChrospher DEAE	DEAE	100	20-40	silica	E. Merck
Nucleogen DEAE	DEAE	6, 50, 400	7	silica	Macherey Nagel
Nucleosil PEI	PEI	400	7	silica	Macherey Nagel
POROS PI	PEI	100	20	PSDVB	Perseptive Biosystems
SynChropak AX	PEI	10, 30, 100	5-7	silica	MICRA
TSK DEAE-5PW	DEAE	100	10	methacrylate copolymer	TosoHaas
TSK DEAE-3SW	DEAE	25	10	silica	TosoHaas
TSK DEAE-2SW	DEAE	12	5	silica	TosoHaas
TSK DEAE-NPR	DEAE	nonporous	2.5	methacrylate copolymer	TosoHaas
Vydac 301VHP	DEAE	90	5	PSDVB	Separations Group
Strong AEX					
Bio-Gel MA7Q	Q	nonporous	7	PSDVB	Bio-Rad
CarboPak PA10	Q	pellicular	10	PSDVB	Dionex
Mono Q	Q	>100	10	PSDVB	Amersham Pharmacia
Nucleogel SAX	Q	100, 400	8	PSDVB	Macherey Nagel
POROS Q	Q	100	10	PSDVB	Perseptive Biosytems
SynChropak Q	Q	30	6	silica	MICRA
TSK Q-5PW	Q	100	10	methacrylate copolymer	TosoHaas
Vydac 300VHP	Q	90	5	PSDVB	Separations Group
Zorbax SAX	Q	7	5	silica	Hewlett Packard
Weak CEX					
Bio-Gel MA7C	CM	nonporous	7	PSDVB	Bio-Rad
LiChrospher COO⁻	CM	100	20-40	silica	E. Merck
PolyCat A	Asp	30	5	silica	PolyLC
POROS CM	CM	100	20	PSDVB	Perseptive Biosystems
SynChropak CM	CM	10, 30	5-6	silica	MICRA
TSK CM-5PW	CM	100	10	methacrylate copolymer	TosoHaas
TSK CM-3SW	CM	25	10	silica	TosoHaas
Strong CEX					
Aminex HPX-87C	S		9	PSDVB	Bio-Rad
Bio-Gel MA7S	S	nonporous	7	PSDVB	Bio-Rad
LiChrospher SO₃⁻	S	100	20-40	silica	E. Merck
Mono S	S	>100	10	PSDVB	Amersham Pharmacia
Nucleogel SCX	S	100, 400	8	PSDVB	Macherey Nagel
PolySulfoethyl Aspartamide	S	30	5	silica	PolyLC
POROS S	S	100	10	PSDVB	Perseptive Biosystems
SynChropak S	S	30	6	silica	MICRA
TSK S-5PW	S	100	10	methacrylate copolymer	TosoHaas
TSK SP-NPR	S	nonporous	2.5	methacrylate copolymer	TosoHaas
Vydac 400VHP	S	90	5	PSDVB	Separations Group
Zorbax SCX	S	30	5	silica	Hewlett Packard

E. Hydroxyapatite Chromatography

Hydroxyapatite chromatography is a technique which separates molecules, particularly proteins and nucleic acids, by their differential adsorption to a crystalline matrix. Hydroxyapatite is actually $Ca_{10}(PO_4)_6(OH)_2$ which has positive sites based on calcium ions and negative sites based on phosphates (17). This mode is distinct from IEC, generally resulting in less differentiation than anion-exchange. The binding is based on adsorption to the charged groups but the separation is by conformational differences, such as the presence of a carbohydrate side chain (18). Operation is generally in sodium phosphate containing calcium chloride. Binding is promoted with low ionic strength phosphate and elution is usually achieved with a phosphate gradient.

IV. OPERATION

A. Mobile Phase

1. pH

The previous discussion of IEC has emphasized the importance of pH, which dictates the charge of both the solutes and the ion-exchanger, and therefore, their affinity for one another. Adjustment of pH is one way to change selectivity in IEC or to release bound molecules. The critical nature of pH in the process necessitates its exact control; therefore, any mobile phase used for IEC must contain an effective buffer (0.02 - 0.1M) within its optimum pH range. A series of common buffers which cover the range of pH used in IEC is listed in Table 9.8.

Table 9.8 Buffers Used for IEC		pK_a	pH Range
Buffer			
Phosphate			
pK_1		2.1	1.5-2.7
pK_2		7.2	6.6-7.8
pK_3		12.3	11.7-12.9
Citrate			
pK_1		3.1	2.5-3.7
pK_2		4.7	4.1-5.3
pK_3		5.4	4.8-6.0
Formate*		3.8	3.2-4.4
Acetate*		4.8	4.2-5.4
MES	2-[N-morpholino]ethanesulfonic acid	6.1	5.5-6.7
Bis-Tris	bis(2-hydroxyethyl)iminotris-(hydroxymethyl)methane	6.5	5.8-7.2
PIPES	piperazine-N,N'-bis(2-ethanesulfonic acid)	6.8	6.1-7.5
BES	N,N'-bis(2-hydroxyethyl)-2-aminoethanesulfonic acid	7.1	6.4-7.8
MOPS	3-(N-morpholino)propanesulfonic acid	7.2	6.5-7.9
HEPES	N-(2-hydroxyethyl)piperazine-N'-ethanesulfonic acid	7.5	6.8-8.2
Tris	tris(hydroxymethyl)aminomethane	8.3	7.7-8.9
Ammonia*		9.2	8.6-9.8
Borate		9.2	8.6-9.8
Diethylamine*		10.5	9.9-11.1

*Volatile buffers for concentration or LC-MS. Reprinted with permission from *High-Performance Liquid Chromatography of Peptides and Proteins* (9). Copyright 1991 CRC Press.

A secondary reason for buffering the mobile phase is seen when the mechanism of ion-exchange shown in Fig. 9.1 is viewed from the molecular level. If an acidic protein is bound to an anion-exchanger and the adsorbent and sample are equilibrated in a Tris-Cl buffer, the counterions associated with the protein are $Tris^+$ and those with the adsorbent are Cl^-. As the protein binds to the column, chloride ions are displaced from the ion-exchanger and Tris ions from the protein - producing Tris-Cl. If a concentrated sample of protein is applied, the Tris-Cl could actually change the effective pH if the mobile phase were not buffered, as well as increase the ionic strength ahead of the adsorbed protein. This effect would increase with protein concentration. If no buffer were used, the generation of salt and increased ionic strength could decrease the binding capacity of the matrix and lead to irreproducible separations whenever high protein concentrations were applied to a column. These problems are eliminated, or at least minimized, by including a buffer of at least 20mM in the mobile phase, at a pH where it exerts its greatest buffering capacity (within 0.3 pH units of its pK).

2. Salt

For proteins which are strongly bound by an ion-exchanger, elution with a gradient to increasing concentrations of salt is the most common and readily controlled method to achieve displacement. Generally, gradients go from 0 - 1M salt in buffer at a suitable pH. The salt counterions competitively displace solute ions from the charged sites on the stationary phase. Smaller, more highly charged ions are most effective at this displacement. Specifically, the strength of displacement for cations is:

$$Mg^{++} > Ca^{++} > NH_4^+ > Na^+ > K^+$$

and for anions, it is:

$$SO_4^{-2} > HPO_4^{-2} > Cl^- > CH_3COO^-$$

The strength of the ions for displacement is not necessarily related to optimum selectivity or resolution. Selectivity is dictated by the effect of the salt on the protein and the bonded phase (3, 17-18). Fig. 9.9 illustrates the effect of salt on the resolution of four standard proteins by a weak anion-exchange column.

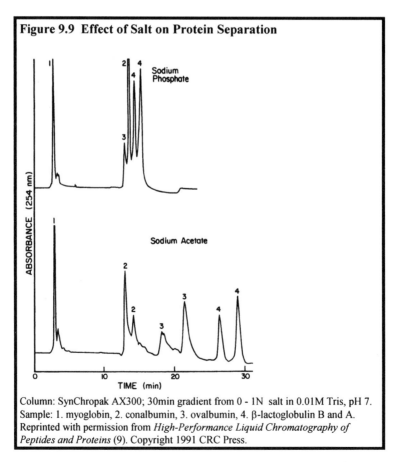

Figure 9.9 Effect of Salt on Protein Separation

Column: SynChropak AX300; 30min gradient from 0 - 1N salt in 0.01M Tris, pH 7.
Sample: 1. myoglobin, 2. conalbumin, 3. ovalbumin, 4. β-lactoglobulin B and A.
Reprinted with permission from *High-Performance Liquid Chromatography of Peptides and Proteins* (9). Copyright 1991 CRC Press.

The gradients are formed with 1N salt so that ionic concentrations are equal. Not only does the substitution of acetate for phosphate produce longer retention, it also resolves some of the proteins into their constituents. These salt effects are observed for both anions and cations in AEX and CEX, as shown in Fig. 9.10, implying that the selectivity occurs because of ionic interactions with the functional groups of both the support and the solute.

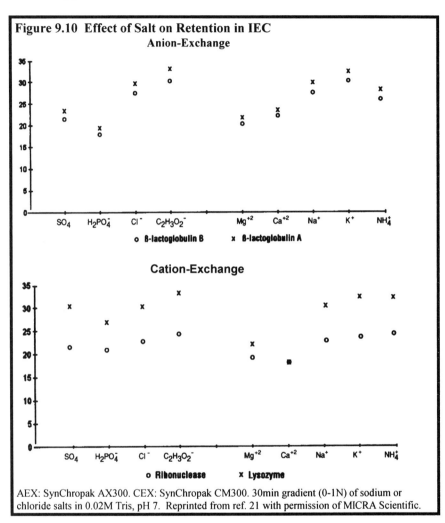

Figure 9.10 Effect of Salt on Retention in IEC

Anion-Exchange

o B-lactoglobulin B   x B-lactoglobulin A

Cation-Exchange

o Ribonuclease   x Lysozyme

AEX: SynChropak AX300. CEX: SynChropak CM300. 30min gradient (0-1N) of sodium or chloride salts in 0.02M Tris, pH 7. Reprinted from ref. 21 with permission of MICRA Scientific.

This ability to change selectivity so dramatically by varying the salt significantly broadens the utility of IEC as a tool for protein analysis. The only restrictions on choice of salt are those involving protein stability.

3. Surfactants and Organic Solvents

Secondary separation mechanisms have all the usual caveats, primarily because the chromatography is no longer predictable. Deviations from an "idealized, pure IEC" mode of separation can be caused by size exclusion or matrix hydrophobicity, as diagrammed in Fig. 9.11 where c is the concentration of the eluent. Generally, hydrophobic interactions are only observed under conditions of high salt. For proteins, hydrophobic binding is not generally significant under the normal elution conditions for IEC using salts like sodium chloride or sodium acetate.

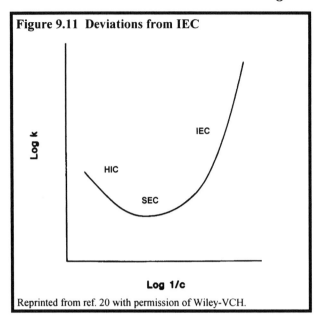

Figure 9.11 Deviations from IEC

Reprinted from ref. 20 with permission of Wiley-VCH.

Hydrophobicity is due to the matrix or crosslinking agents which were employed in the synthesis of the bonded phase. Any hydrophobic interactions are fundamentally undesirable and can be minimized by the addition of 1 - 10% of an organic solvent, such as methanol, ethanol or acetonitrile, to the running buffer. The solubility of the salt in the organic mobile phase should always be verified to avoid precipitation. Nonionic detergents may also reduce hydrophobic interactions with a column.

Although salt conditions should generally be chosen so that only the ion-exchange mechanism is operational, secondary interactions can sometimes be exploited during refinement of a well understood separation. Hydrophobic interaction has been promoted on weak ion-exchange supports by using high concentrations of chaotropic salts like those used in hydrophobic interaction chromatography (22).

Nonionic detergents can be added to ion-exchange mobile phases to aid in the solubilization of membrane or other insoluble proteins (23-24). Such detergents are easy to equilibrate and remove from the columns, contrary to the deleterious effects of many detergents on reversed phase columns. Ionic detergents should be avoided because they may bind very strongly to the column or solutes.

B. Flowrate and Gradient

Small molecules can often be effectively separated isocratically by IEC using the guidelines for capacity factor, plates, and selectivity given in Chapter 2 for maximizing performance. Isocratic IEC of proteins, and most biological macromolecules, is not usually feasible, yielding no resolution and extreme tailing due to multipoint interactions.

In IEC, molecules elute at a specific salt concentration, generally without binding from secondary effects, as the gradient proceeds to higher levels. The effect of gradient conditions on elution ($k*$) can be described by (25):

$$k* = 0.87\, t_\mathrm{G}\ \frac{F}{V_\mathrm{M}}\left(\log\frac{C_2}{C_1}\right) Z \qquad\qquad \text{Eq. 9.1}$$

where C_1 and C_2 are the total salt concentrations (salt plus buffer) at the beginning and the end of the gradient, respectively; Z is the effective charge on the solute molecule; F is the flowrate; V_M is the total volume; and t_G is the gradient time (25). The Z number will vary with solute and pH. An ion-exchange protocol for proteins using a 20 - 30min linear gradient from 0 - 1M salt in buffer will usually yield a separation which can be later optimized, if necessary. Flowrates of 1ml/min for a 4.6mm ID column are satisfactory.

C. Loading

One of the main benefits of IEC is its high loading capacity. Not only are both the dynamic and absolute capacities high, but high loads do not usually have deleterious effects on proteins. Table 9.9 lists dynamic loading capacities for several common column configurations of a weak anion-exchanger with a 30nm pore.

Table 9.9 Dynamic Loading Capacities of Columns	
Column Dimensions	**Dynamic Loading (mg BSA)**
250 x 2.1mm ID	5
250 x 4.6mm ID	25
250 x 10mm ID	120
250 x 21.2mm ID	475
Column: SynChropak AX300; 30min linear gradient of 0 - 1M sodium acetate in 0.02M Tris, pH8. All linear velocities were the same.	

Dynamic loading is the amount which can be applied before peaks broaden significantly during chromatography (26). The data on the 30nm pore diameter support shown in Table 9.9 reflects maximal loading of BSA, as was previously noted in Table 9.6.

V. APPLICATIONS

Markers for deadtime (t_o) or unretained peaks in IEC are best chosen from small molecules which are neutral or have the same charge as the support. Nucleotides can be used for cation-exchangers and nucleosides for anion-exchangers, but any marker should be tested in the particular column-mobile phase system to confirm noninteraction. Mobile phases for deadtime measurements should always be buffered and contain salt so that the pH is controlled and ion-exclusion is avoided.

A. Amino Acids

Given their small size and well understood charge characteristics, the separation of amino acids provides a good example of IEC. IEC is the basis of amino acid analyzers because of its resolution capabilities (27). The separation of seventeen pure amino acids, using a citrate buffer on a sulfonic acid strong cation-exchange polystyrene based column, is seen in Fig. 9.12.

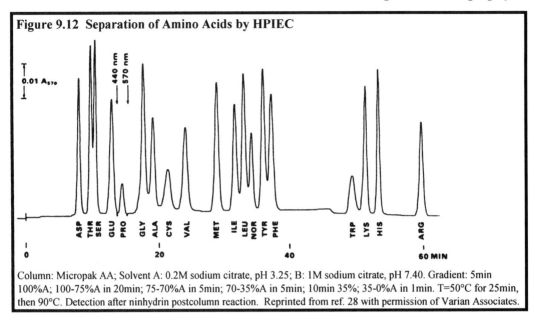

Figure 9.12 Separation of Amino Acids by HPIEC

Column: Micropak AA; Solvent A: 0.2M sodium citrate, pH 3.25; B: 1M sodium citrate, pH 7.40. Gradient: 5min 100%A; 100-75%A in 20min; 75-70%A in 5min; 70-35%A in 5min; 10min 35%; 35-0%A in 1min. T=50°C for 25min, then 90°C. Detection after ninhydrin postcolumn reaction. Reprinted from ref. 28 with permission of Varian Associates.

The sample was eluted with concurrent gradients of sodium chloride concentration and pH from 3.25 to 7.4. As expected, the acidic and polar amino acids (aspartic acid, threonine, serine, and glutamic acid) eluted first, at low pH, into fairly well resolved peaks. The basic amino acids eluted near the end of the run, following an increase in temperature from 50° to 90°C. Due to its hydrophobic attraction to the polystyrene matrix, tryptophan eluted late under these conditions.

B. Peptides

Separation of peptides by IEC has gained increasing popularity in recent years, primarily because it can be used in a complementary manner with RPC due to its different mechanism. Although both anion- and cation-exchange have been used successfully for peptide analysis, strong cation-exchange at low pH (~4) has been most popular (10). Synthetic peptides have been developed by Hodges and colleagues to monitor the ion-exchange properties of cation-exchange packings in terms of overall charge, charge distribution, and hydrophobicity (29). These model peptides contain eleven amino acids, in the range of peptides generated by proteolytic digestion. Table 9.10 lists the sequences of the four peptide components (C1 - C4) of a cation-exchange peptide standard, which have net charges of +1 to +4.

Table 9.10 Structure of Cation-Exchange Peptide Standards

Peptide	Sequence	Net Charge
C1	Ac-Gly-Gly-Gly-Leu-Gly-Gly-Ala-Gly-Gly-Leu-Lys-amide	+1
C2	Ac-Lys-Tyr-Gly-Leu-Gly-Gly-Ala-Gly-Gly-Leu-Lys-amide	+2
C3	Ac-Gly-Gly-Ala-Leu-Lys-Ala-Leu-Lys-Gly-Leu-Lys-amide	+3
C4	Ac-Lys-Tyr-Ala-Leu-Lys-Ala-Leu-Lys-Gly-Leu-Lys-amide	+4

From ref. 29 and 30.

When these 11-mers were analyzed by strong cation-exchange chromatography at pH 3 and pH 6.5 using simple linear salt gradients for elution, differential retention was observed, as shown in Fig. 9.13.

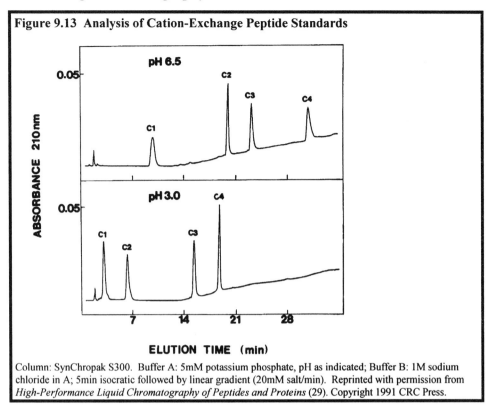

Figure 9.13 Analysis of Cation-Exchange Peptide Standards

Column: SynChropak S300. Buffer A: 5mM potassium phosphate, pH as indicated; Buffer B: 1M sodium chloride in A; 5min isocratic followed by linear gradient (20mM salt/min). Reprinted with permission from *High-Performance Liquid Chromatography of Peptides and Proteins* (29). Copyright 1991 CRC Press.

At pH 6.5, C1, the singly charged peptide, eluted during the isocratic wash before the gradient began, whereas C2 - C4 were well fractionated by the salt gradient. At pH 3.0, both C1 and C2 eluted isocratically, and C3 and C4 eluted during the gradient. The shorter retention times for all the peptides at pH 3 reflect the reduced charge and capacity of this bonded phase at pH 3, compared to pH 6.5.

When these same peptides were applied to two other strong cation-exchangers using buffer conditions identical to those at pH 6.5 shown in Fig. 9.13, the peptides exhibited apparent hydrophobic interactions (29). Addition of acetonitrile to the mobile phase at a level of 10% decreased retention and resulted in excellent peak shape in one case, and improved in the other. This comparison of three supports which all have sulfonyl functionalities illustrates the importance of the chemical composition of the bonded phase and/or the matrix on the selectivity. An additional difference was that one of the supports showed no diminished capacity at pH 3. This again emphasizes the importance of understanding the characteristics of each specific column packing.

Nonionic interactions with ion-exchange columns have been exploited to achieve additional selectivity in a technique called cation-exchange hydrophilic interaction chromatography (CEC-HILIC) (30). Hydrophilic interaction chromatography (HILIC) employs a gradient from organic to aqueous solvents to separate solutes by hydrophilicity, or reverse order of hydrophobicity (31). When a mixture of peptides which varied in hydrophobicity and charge was run with several concentrations of acetonitrile, different resolution and selectivity were obtained. At 90% acetonitrile, HILIC was the primary mechanism; whereas at 20%, ion-exchange was dominant. The use of an acetonitrile gradient, in addition to the salt gradient, yielded the best resolution and peak shapes.

C. Proteins

One of the most common uses of IEC in biochemistry is the analysis and purification of proteins (9, 32-33). The excellent selectivity, generally nondenaturing conditions, and compatibility with nonionic detergents (23) make it a powerful and versatile separation technique. High recoveries are obtained, even for very labile proteins like estrogen receptors (34), allowing purification of many proteins for further characterization. There have been studies which suggest that recoveries may decrease with residence time on the column for some proteins like fibrinogen (35) but in such cases, faster analysis or adjustment of conditions, like pH, may improve recovery. The excellent resolution potential of IEC for proteins can be illustrated by the analysis of human hemoglobin variants, both glycosylated and chain variations, which are routinely quantitated for clinical evaluations by IEC (5), as seen in Fig. 9.14. Hemoglobin A_0 and Hb S differ by only one amino acid in two of their four subunits. Hb A_{1c} is glycosylated Hb A_0, and therefore elutes earlier due to its blocked amines.

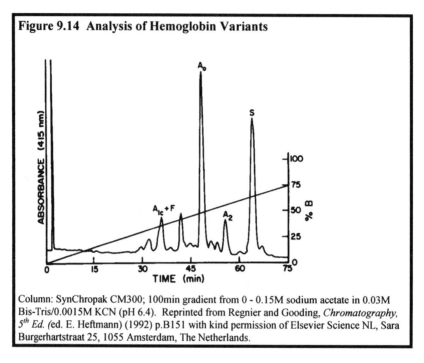

Figure 9.14 Analysis of Hemoglobin Variants

Column: SynChropak CM300; 100min gradient from 0 - 0.15M sodium acetate in 0.03M Bis-Tris/0.0015M KCN (pH 6.4). Reprinted from Regnier and Gooding, *Chromatography, 5th Ed. (*ed. E. Heftmann) (1992) p.B151 with kind permission of Elsevier Science NL, Sara Burgerhartstraat 25, 1055 Amsterdam, The Netherlands.

Chromatofocusing is a variation of IEC which uses multiple polybuffers to form a pH gradient along a weak anion-exchanger (36). Chromatofocusing methods can resolve proteins which differ by only a small increment in pI; however, this technique has not found wide acceptance due to the high cost of the mobile phase and the contamination of proteins by the polybuffers interfering in downstream processing.

D. DNA Restriction Fragments

As mentioned previously, IEC is an effective method for analysis of polynucleotides because of their high charge densities. Nucleic acids are so highly charged, however, that elution often requires very high salt concentrations. Individual nucleotides can be effectively resolved on small pore anion-exchangers by both their phosphate content and their bases (21). Small polynucleotides have also been separated successfully by anion-exchange on supports with 6 - 30nm pores (37).

Because polynucleotides have linear or helical structures, their radii of gyration are quite large, especially compared to globular proteins. For this reason, nucleic acid fragments and large polynucleotides have been most successfully separated on large pore (400nm) (37) or nonporous supports (13, 38). Separation is usually by size because the charge is related directly to the number of base pairs; further resolution by base composition is generally not possible. Fig. 9.15 illustrates the resolution of components of a nucleic acid digest on a 2.5μm nonporous DEAE support. This support also effectively separates PCR products from their primers (39).

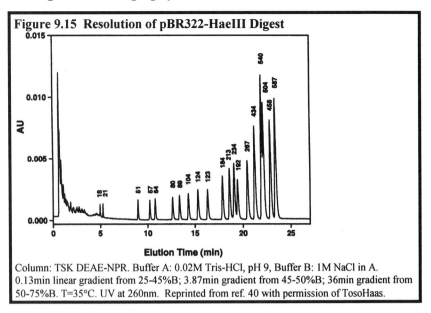

Figure 9.15 Resolution of pBR322-HaeIII Digest

Column: TSK DEAE-NPR. Buffer A: 0.02M Tris-HCl, pH 9, Buffer B: 1M NaCl in A.
0.13min linear gradient from 25-45%B; 3.87min gradient from 45-50%B; 36min gradient from
50-75%B. T=35°C. UV at 260nm. Reprinted from ref. 40 with permission of TosoHaas.

E. Carbohydrates

Sugars are a difficult class of molecules to analyze because of their chemical similarities and
low detectability; however, two different ion-exchange methods have proven effective. When sugars
are subjected to high pH in dilute base, they form anions which can be can be separated by pellicular
anion-exchange supports (41). By using pulsed amperometric detection (PAD), the carbohydrates can
be detected at subnanomole levels (15, 42). Fig. 9.16 illustrates a separation of three sialic acids, N-
acetylneuraminic acid (Neu5Ac), N-glycolylneuraminic acid (Neu5Gc), and 3-deoxy-d-glycero-d-
galacto-2-nonulosonic acid (KDN), on a quaternized pellicular support.

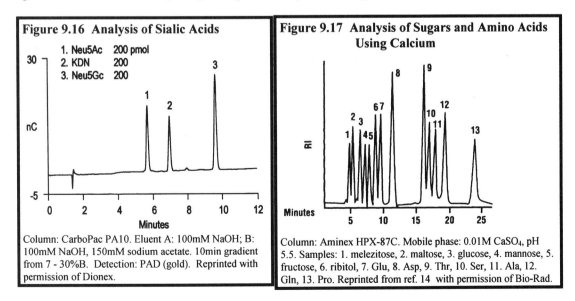

Figure 9.16 Analysis of Sialic Acids

Column: CarboPac PA10. Eluent A: 100mM NaOH; B:
100mM NaOH, 150mM sodium acetate. 10min gradient
from 7 - 30%B. Detection: PAD (gold). Reprinted with
permission of Dionex.

**Figure 9.17 Analysis of Sugars and Amino Acids
Using Calcium**

Column: Aminex HPX-87C. Mobile phase: 0.01M $CaSO_4$, pH
5.5. Samples: 1. melezitose, 2. maltose, 3. glucose, 4. mannose, 5.
fructose, 6. ribitol, 7. Glu, 8. Asp, 9. Thr, 10. Ser, 11. Ala, 12.
Gln, 13. Pro. Reprinted from ref. 14 with permission of Bio-Rad.

An alternative method of carbohydrate analysis employs strong cation-exchange supports
which have been pre-equilibrated with a specific ion, most notably, calcium (14, 43). The technique is

referred to as ion-moderated partitioning (IMP). Excellent resolution is achieved for mixtures of sugars, as well as amino acids, as seen in Fig. 9.17.

VI. START-UP

RECOMMENDATIONS FOR IEC START-UP

Small Molecules
1. Select a small pore AEX column for acids or other anionic molecules or a CEX column for amines or other cationic species.
2. Use a column of 4.6mm ID and 15 - 25cm long if standard isocratic analyses will be run.
3. Use the mobile phase recommended by the manufacturer and run the sample shown in the test chromatogram to make sure that the system and column are operating efficiently. The mobile phase must contain at least 0.02M buffer at a pH within 0.3 units of its pK_a.
4. Use a flowrate of 1 ml/min for a 4.6mm ID column and condition with at least ten column volumes of mobile phase. If the column was shipped in a solvent incompatible with the mobile phase, wash with ten column volumes of distilled water first.
5. Inject 1µl of the test mixture made up in the mobile phase (approx. 1 mg/ml). Verify that the plates are similar to those achieved by the manufacturer. If they are not, try the measures suggested below in troubleshooting.
6. Clean the injector. Inject 1 - 5µl of standard solutions of the analytes, dissolved in mobile phase.
7. Adjust the mobile phase to achieve a capacity factor (k) of 2 - 7 for the peaks. Adjust the salt concentration, pH, or type of salt.
8. If k values vary by tenfold or more, a gradient may be necessary.
9. After the analyses are finished, wash the column with higher salt and/or with a different pH to remove highly bound materials. Wash with water and then store in a solvent which will inhibit bacterial growth, e.g. 10% isopropanol in water. Storage in 100% organic after using high salt may result in precipitation. Follow the column manufacturer's recommendations.

Proteins
1. Choose an IEC column with pores $\geq$ 30nm. If loading capacity is important, choose the smallest pore diameter which allows permeation of the protein of interest. Base the choice of functional group on pl or, preferably, on the retention map. If pI > 7, use CEX; if pI < 7, use AEX.
2. Use a column of 4.6mm ID and 5 - 25cm long if standard analyses will be run.
3. Test as in steps #3 and #4 above.
4. Use mobile phase conditions which enhance the biological stability of the sample. Choose a pH that is 1 - 2 units from the pI. Solvent A should be 0.02 - 0.1M buffer which has good buffering capacity at that pH (see Table 9.8). Solvent B should be 1M sodium acetate or another salt in solvent A. Adjust the pH after addition of the salt.
5. Condition the column with 10 column volumes of Solvent B, followed by 10 column volumes of Solvent A.
6. Run a mixture of standard proteins, the same or similar to those in the sample, dissolved in Solvent A. Run a 30min linear gradient from 0 - 100% B at 1ml/min (4.6mm ID). Peaks should be sharp without tailing. If they are not, follow troubleshooting recommendations.
7. Recondition the column and run the sample under the same conditions. It is best to dissolve the sample in the initial mobile phase and inject 1 - 5µl.
8. After the analyses, clean the column, if necessary, and store as in #9 above.

VII. TROUBLESHOOTING

TROUBLESHOOTING ION-EXCHANGE

Symptom	Cause	Remedy
1 Poor plates on new column (small molecule)	sample	
	too concentrated	dilute
	volume too large	inject less
	old or contaminated	make new sample
	dead volume	minimize extra-column volume
2 Poor plates on used column (small molecule)	see #1 above	
	column fouling	wash per manufacturer's recommendations
		0.1% TFA will remove many ionic molecules
		10% organic in Solvent A may remove hydrophobic molecules
	column void	replace column; use guard column in future
		topping off may be temporary cure
3 Poor peak shape for proteins	sample	as in #1 above
	specific protein problem	run standard proteins to compare with original
	mobile phase	
	column not conditioned	run 10 column volumes (or more, if not adequate)
	bad	prepare fresh solvents
	column	see #2 above
4 peaks too early	gradient too stong	slow gradient rate
	salt	use weaker salt
	pH	adjust pH (AEX, lower; CEX, higher)
5 peaks too late	gradient too slow	increase gradient rate
	salt	use stronger salt
	hydrophobic interaction	use 10% ACN in mobile phase

VIII. REFERENCES

1. R.E. Creighton, *Proteins, Structures and Molecular Principles*, Freeman, New York (1983) Chap. 1 - 2.
2. F.E. Regnier and R.M. Chicz, "Ion-Exchange Chromatography" in *HPLC of Biological Macromolecules: Methods and Applications* (K.M. Gooding and F.E. Regnier, eds.), Marcel Dekker, New York (1990) 77.
3. W. Kopaciewicz, M.A. Rounds, J. Fausnaugh, and F.E. Regnier, *J. Chromatogr.* 266 (1983) 3.
4. *Ion-Exchange Chromatography, Principles and Methods,* Pharmacia Biotech, Sweden.
5. J.B. Wilson, "Separation of Human Hemoglobin Variants by HPLC" in *HPLC of Biological Macromolecules: Methods and Applications* (K.M. Gooding and F.E. Regnier, eds.), Marcel Dekker, New York (1990) 457.
6. F.E. Regnier, *Science* 238 (1987) 319.
7. C.D. Scott in *Modern Practice of Liquid Chromatography* (ed. J.J. Kirkland), Wiley-Interscience, New York (1971).
8. S.H. Chang, K.M. Gooding, and F.E. Regnier, *J. Chromatogr.* 125 (1976) 103.
9. M.P. Nowlan and K.M. Gooding, "HPIEC of Proteins" in *High-Performance Liquid Chromatography of Peptides and Proteins* (ed. C.T. Mant and R.S. Hodges), CRC Press, Boca Raton (1991) 203.
10. A.A. Alpert, "Ion-Exchange High-Performance Liquid Chromatography of Peptides" in *High-Performance Liquid Chromatography of Peptides and Proteins* (ed. C.T. Mant and R.S. Hodges), CRC Press, Boca Raton (1991) 187.
11. W. Muller, *J. Chromatogr.* 510 (1990) 133.

12. O. Mikes and J. Coupek, "Organic Supports" in *HPLC of Biological Macromolecules: Methods and Applications* (K.M. Gooding and F.E. Regnier, eds.), Marcel Dekker, New York (1990) 25.

13. TosoHaas Catalog, 1996.

14. *HPLC Columns; Methods and Applications*, Bio-Rad, 1993.

15. *Analysis of Carbohydrates by HPAE-PAD*, Technical Note 20, Dionex 1993.

16. G. Vanecek and F.E. Regnier, *Anal. Biochem.* 109 (1980) 345.

17. B. Pavlu, "HPLC Purification of Monoclonal Antibodies" in *HPLC of Proteins, Peptides and Polynucleotides* (ed. M.T.W. Hearn), VCH Publishers, New York (1991) 599.

18. N. Takahashi and F.W. Putnam, "Glycoproteins" in *HPLC of Biological Macromolecules: Methods and Applications* (K.M. Gooding and F.E. Regnier, eds.), Marcel Dekker, New York (1990) 571.

19. K.M. Gooding and M.N. Schmuck, *J. Chromatogr.* 296 (1984) 321.

20. M.I. Aguilar, A.N. Hodder, and M.T.W. Hearn, "HPIEC of Proteins" in *HPLC of Proteins, Peptides and Polynucleotides* (ed. M.T.W. Hearn), VCH Publishers, New York (1991) 199.

21. *SynChronotes* 8(1) (1993), SynChrom, Inc.

22. M.L. Heinitz, L. Kennedy, W. Kopaciewicz, and F.E. Regnier, *J. Chromatogr.* 443 (1988) 173.

23. G.W. Welling and S. Welling-Wester, "Anion-Exchange HPLC of Viral Proteins" in *High-Performance Liquid Chromatography of Peptides and Proteins* (ed. C.T. Mant and R.S. Hodges), CRC Press, Boca Raton (1991) 223.

24. G. W. Welling, R. van der Zee, and S. Welling-Wester, "HPLC of Membrane Proteins" in *HPLC of Biological Macromolecules: Methods and Applications* (K.M. Gooding and F.E. Regnier, eds.), Marcel Dekker, New York (1990) 333.

25. L.R. Snyder, "Gradient Elution Separation of Large Biomolecules" in *HPLC of Biological Macromolecules: Methods and Applications* (K.M. Gooding and F.E. Regnier, eds.), Marcel Dekker, New York (1990) 231.

26. M.N. Schmuck, K.M. Gooding, and D.L. Gooding, *J. Liq. Chromatogr.* 7(14) (1984) 2863.

27. C. Lazure, J.A. Rochemont, N.G. Seidah, and M. Chretien, " Amino Acids in Protein Sequence Analysis" in *HPLC of Biological Macromolecules: Methods and Applications* (K.M. Gooding and F.E. Regnier, eds.), Marcel Dekker, New York (1990) 263.

28. "Amino Acid Analysis with Ninhydrin Postcolumn Derivatization," *LC at Work*, Varian Associates.

29. C.T. Mant and R.S. Hodges, "The Use of Peptide Standards for Monitoring Ideal and Non-ideal Behavior in Cation-Exchange Chromatography" in *High-Performance Liquid Chromatography of Peptides and Proteins* (ed. C.T. Mant and R.S. Hodges), CRC Press, Boca Raton (1991) 171.

30. B.-Y. Zhu, C.T. Mant, and R.S. Hodges, *J. Chromatogr.* 594 (1992) 75.

31. A.J. Alpert, *J. Chromatogr.* 499 (1990) 177.

32. F.E. Regnier and K.M. Gooding, "Proteins" in *Chromatography, 5th Edition* (ed. E. Heftmann), Elsevier Science Publishers, Amsterdam (1992) B151.

33. F.B. Rudolph, D.P. Wiesenborn, J. Greenhut, and M.L. Harrison, "Preparative Enzyme Purification by HPLC" in *HPLC of Biological Macromolecules: Methods and Applications* (K.M. Gooding and F.E. Regnier, eds.), Marcel Dekker, New York (1990) 333.

34. R.D. Wiehle and J.L. Wittliff, "Assessment of Steroid Receptor Polymorphism by HPIEC" in *High-Performance Liquid Chromatography of Peptides and Proteins* (ed. C.T. Mant and R.S. Hodges), CRC Press, Boca Raton (1991) 255.

35. S.C. Goheen and J.L. Hilsenbeck, *J. Chromatogr.*, submitted (ISPPP '97).

36. T.W. Hutchens, R.D. Wiehle, N.A. Shahabi, and J.L. Wittliff, *J. Chromatogr.* 266 (1983) 115.

37. M. Colpan and D. Reisner, *J. Chromatogr.* 296 (1984) 339.

38. T. Hashimoto, *J. Chromatogr.* 544 (1991) 257.

39. J.M. Wages, Jr. And E.D. Katz, "The Application of HPLC for Nucleic Acid Analysis" in *High Performance Liquid Chromatography: Principles and Methods in Biotechnology* (ed. E.D. Katz), John Wiley & Sons, New York (1996) 351.

40. S. Nakatani, T. Tsuda, Y. Yamasaki, H. Moriyama, H. Watanabe, and Y. Kato, *Technical Report 78*, Toso Haas, 1995.

41. R.R. Townsend, "High-pH Anion Exchange Chromatography of Recombinant Glycoprotein Glycans " in *High Performance Liquid Chromatography: Principles and Methods in Biotechnology* (ed. E.D. Katz), John Wiley & Sons, New York (1996) 381.

42. *Glycoprotein Oligosaccharide Analysis Using High Performance Anion-Exchange Chromatography*, Technical Note 42, Dionex, 1997.

43. S.C. Churms, "Carbohydrates" in *Chromatography, 5th Edition* (ed. E. Heftmann), Elsevier Science Publishers, Amsterdam (1992) B229.

CHAPTER 10
AFFINITY CHROMATOGRAPHY AND RELATED TECHNIQUES

David Burke, Matrix Pharmaceutical, Inc.

I. AFFINITY CHROMATOGRAPHY

A. Mechanism

Affinity chromatography encompasses a family of techniques in which an immobilized ligand is used to capture a specific molecule from solution based on highly biospecific binding (1-4). In an affinity separation, resolution depends primarily upon the selectivity of the immobilized compound rather than on differential migration through the chromatographic bed; therefore, it often uses larger particle diameter supports than normally used for HPLC. Adsorption is essentially an all or nothing event, and desorption is usually achieved by simply switching solvents or adding a displacer, rather than by using a gradient. For analytical applications, it is desirable to have a well-packed column so that peak shapes and peak areas are consistent; however, resolution may not be significantly improved by increasing column length or efficiency.

Ligands for affinity chromatography include many classes of compounds, such as proteins, peptides, carbohydrates, hormones, lipids, and nucleic acids. Table 10.1 lists some examples of ligand-analyte pairs that have been used in affinity applications.

Table 10.1 Common Ligand-Analyte Pairs Used in Affinity Separations

Immobilized Compound (Ligand)	Compound to be Purified (Analyte)	Means of Elution (Elution Buffer)
Antibody	Antigen	Low pH, chaotrope
Protein A (Protein G)	IgG antibody	Low pH
Enzyme inhibitors or co-factors	Enzyme	Change pH, add soluble inhibitor or co-factor
Benzamidine	Serine proteases	Low pH
Lectin	Glycoprotein	Soluble sugar
Dye (Cibacron® Blue)	Nucleotide binding protein	High salt
Hormone	Receptor*	Chaotropes, soluble hormone
Heparin	Heparin-binding proteins	High salt
Polymyxin	Endotoxin removal	Base
Boronate (5)	Vicinal cis-diols (proteins, carbohydrates, nucleic acids)	Low pH, competing diol, such as sorbitol

*A protein with one or more specific high affinity binding sites for a particular ligand.

The basic steps of affinity chromatography are illustrated in Fig. 10.1.

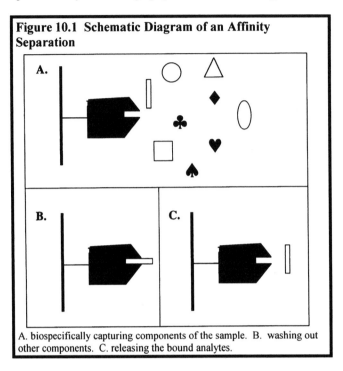

Figure 10.1 Schematic Diagram of an Affinity Separation

A. biospecifically capturing components of the sample. B. washing out other components. C. releasing the bound analytes.

In step A, a solution containing many different molecules is passed over an affinity column. Analytes with high affinity for the immobilized ligand are retained on the column while all other molecules elute in the column wash (step B). After washing to remove contaminants completely, the captured analyte is eluted by running the column with a buffer which is selected to promote dissociation of the ligand-analyte complex (step C). Elution buffers often cause changes in conformation or surface charge in either the analyte or the ligand, thus weakening interactions between them. Alternatively, the elution buffer may contain a soluble compound that competes for binding. Examples of elution buffers for specific applications were given in Table 10.1.

In order for an affinity separation to be effective, three criteria must be met:

1. The analyte must bind tightly to the ligand in the mobile phase with no binding of other sample components.
2. The bound analyte must be quantitatively released from the ligand in a second solvent, allowing elution of the analyte.
3. Both the analyte and the ligand must be stable under the conditions used in binding and elution; the immobilized ligand must also be stable during regeneration and storage.

B. Supports

1. General

The heart of an affinity separation is the column packing. In the usual configuration, this consists of an immobilized ligand connected via a linker (spacer arm) to a beaded, spherical chromatography support, as shown in Fig. 10.2.

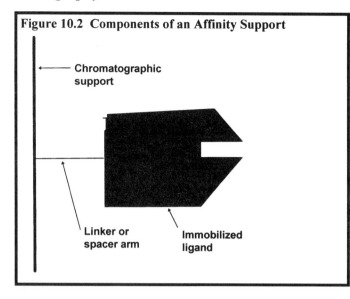

Figure 10.2 Components of an Affinity Support

As with other chromatographic supports, the matrix is selected to minimize unwanted interactions with substances other than the desired analyte. For most applications, the best support matrices are highly cross-linked, chemically inert, hydrophilic beads with no extraneous ionic or hydrophobic functionalities. Base materials include synthetic polymers, cross-linked polysaccharides, controlled pore glass, and polymer-coated silica. Pore diameters should be large enough to accommodate the desired analyte-ligand pair and selection of particle diameter should be based on column scale and desired flowrate.

The purpose of the linker is to make the immobilized ligand accessible to molecules in the mobile phase (4), as discussed in Chapter 5. These spacer arms are typically 6 - 20 carbons in length. Like the support matrix, the linker must be inert and chemically stable; however, it must also contain functional groups suitable for coupling the ligand.

Finally, the immobilized ligand must be highly pure, highly active, and well characterized with respect to stability. If contaminants in the ligand preparation become immobilized along with the ligand, these contaminants may adsorb unwanted compounds from the sample during operation of the column. Inactive ligands give rise to a support with little or no binding capacity, whereas unstable ligands result in a column that is only usable for a few cycles.

2. Immobilization Chemistry and Stereochemistry

As noted above, it is essential that the chemical linkage between the immobilized ligand and the support be stable. In the case of proteins, there are a number of potentially reactive side chains suitable for use in coupling, including carboxylic acids (aspartic acid and glutamic acid), amines (lysine), thiols (cysteine) and hydroxyls (serine and threonine). Coupling chemistries used for immobilization of proteins and other biomolecules are listed in Table 10.2. The undesirable linkages are those which are unstable or likely to leak. The best way to avoid leakage during operation is to strip the support with weak acid prior to use, thus removing weakly bound ligand.

Table 10.2. Activated Supports and Functional Groups

Functional Group on the Activated Support	Reacts with	Bond Formed
Tresyl (or tosyl)	1. Primary amine 2. Thiol 3. Hydroxyl	1. Secondary amine 2. Thioether 3. Ether
N-hydroxysuccinimide ester (NHS)	1. Primary amine 2. Thiol	1. Amide (stable) 2. Thioester (not very stable)
Carboxylic acid (plus soluble carbodiimide)	1. Primary amine 2. Thiol	1. Amide (stable) 2. Thioester (not very stable)
Epoxy	1. Primary amine 2. Thiol 3. Hydroxyl	1. Secondary amine 2. Thioether 3. Ether
Carbonyldiimidazole-activated	Primary amine	N-alkylcarbamate
Aldehyde	Primary amine	Secondary amine (Schiff base intermediate)
Iodoacetyl	Thiol	Thioether
Hydrazide	Aldehyde (oxidized sugar)	Hydrazone

In general, coupling reactions which result in formation of secondary amines, amides, ethers, and thioethers are highly preferred because these linkages are exceptionally stable. For coupling proteins to supports, tresyl, N-hydroxysuccinimide (NHS), and carbodiimide chemistries are particularly useful since they can be performed at neutral pH. Tosyl and epoxy couplings must be carried out at basic pH, which is suitable for immobilization of many chemically stable ligands but may damage some proteins.

It is very important that the immobilization process not obstruct ligand-analyte interactions by blocking interaction, as illustrated in Fig. 10.3. In this example, antibody molecules have been coupled to a support using a procedure which randomly links to amino acid residues distributed throughout the antibody structure. Although many of the antibodies are coupled in a way that leaves the binding sites unencumbered, some of the antibodies attach via amino acid residues at or near the actual binding sites, rendering these sites inactive.

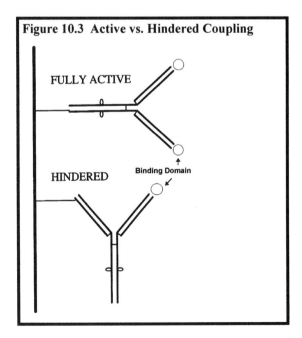

Figure 10.3 Active vs. Hindered Coupling

FULLY ACTIVE

HINDERED

Binding Domain

In some cases, it is possible to increase the binding capacity of a support by choosing a coupling chemistry which results in minimal hindered binding sites. Antibodies, for example, can be coupled to supports via carbohydrate chains that are located away from the binding sites. Antibodies, as diagrammed in Fig. 10.4, have two heavy (H) chains and two light (L) chains linked by disulfide bonds. The variable regions of the heavy (V_H) and light (V_L) chains are different from antibody to antibody and contain sequences that confer antigen binding specificity. The constant regions (C_H and C_L) have virtually the same sequences for all antibodies of the same subclass for each species. Most IgG heavy chains are glycosylated in the C_H region, and thus immobilization via the carbohydrate chains will leave the variable regions free to interact with antigens after immobilization.

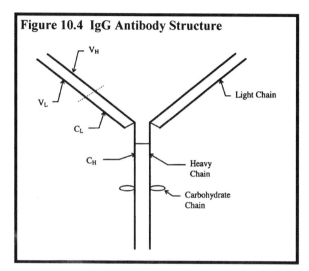

Figure 10.4 IgG Antibody Structure

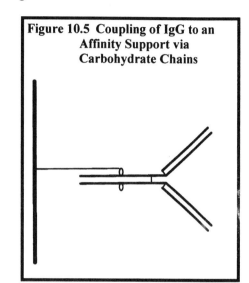

Figure 10.5 Coupling of IgG to an Affinity Support via Carbohydrate Chains

To attach the antibodies via their carbohydrate chains, as illustrated in Fig. 10.5, the oligosaccharides on the antibody heavy chains are oxidized with periodate and then coupled to a support containing hydrazide groups. With this method, few of the binding sites of the antibodies are blocked and resultant supports often have higher capacities than those which contain some hindered ligands. Some antibodies may be damaged by oxidation; therefore, before immobilization, they should be tested to determine whether the periodate reaction results in diminished binding of antigen.

3. Sources of Supports

Affinity supports with ligands such as those listed in Table 10.1 are obtained by synthesis using reactive coupling gels or by using commercially available group specific adsorbents. Coupling gels contain reactive groups to which ligands can be easily attached. An example of a coupling gel is Epoxy-activated Sepharose®. Group specific adsorbents are ready to use and are thus generally more expensive than coupling gels because they have already been modified with a ligand, such as Protein A. High performance affinity supports are based on polymers, highly crosslinked agarose, or silica, which are all stable to the pressures generated by HPLC procedures. As mentioned previously, high performance affinity supports yield minimal band spreading but do not generally affect resolution, due to the bind/release mechanism of the technique. Table 10.3 lists some vendors of affinity supports. The last two columns in the table describe special techniques of affinity chromatography which will be discussed later in this chapter. Manufacturers generally provide literature describing coupling procedures, operation, and handling of the supports and columns.

Table 10.3 Selected Vendors of Affinity Chromatography Supports

Company	Particle Diameter	Coupling Gels	Group Specific Gels	IMAC	Covalent Chromatography
Toso Haas	10μm	no	yes	yes	yes
	65μm	yes	yes	yes	yes
Amersham Pharmacia	34μm	yes	yes	yes	no
	90μm	numerous	numerous	yes	yes
Bio-Rad	Affigel	yes	yes	no	no
	Affiprep	yes	yes	no	no
E. Merck	25-40μm	no	no	yes	yes
Pierce Chemical	various	numerous	yes	yes	yes
Sigma	various	numerous	numerous	yes	yes
Supelco	20μm	no	numerous	yes	no
Perseptive Biosystems	20μm	yes	yes	yes	no

C. Operation of Affinity Supports Using Protein A as a Model

1. Introduction

One of the more common applications of affinity chromatography in biotechnology is the use of immobilized Protein A (or the functionally similar Protein G) to capture antibodies from biological fluids (3, 6-7). Protein A is a bacterial protein with high affinity binding sites for the constant regions of IgG heavy chains from a wide range of species. Chromatography using immobilized Protein A is widely used to purify IgG antibodies from complex solutions such as serum, plasma, ascitic fluids, and cell culture filtrates.

Protein A is also widely used analytically to measure antibody concentrations in complex solutions. At neutral or slightly basic pH, IgG antibodies bind to Protein A, while most other biomolecules do not. At low pH, antibodies dissociate from this complex. Thus, when a solution containing antibodies is passed over a Protein A column, the antibodies are bound and other solution components flow through the column. After washing the column, the antibodies are generally eluted with a low pH buffer and the IgG peak area is measured.

2. Quantitative Analytical Method Development

As noted above, a common application of Protein A chromatography is determination of the antibody (IgG) concentration of a complex solution. Like other quantitative HPLC assays, the technique must be selective, sensitive, linear, robust, and free of interfering substances. In order for the assay to be robust, the antibody of interest must both bind to the column and be released by the column in a quantitative way. The HPLC system, the column, the injection volume, and the detection system must be optimized to detect and quantify the IgG at the required sensitivity. For most applications, detection at 280 nm with a UV monitor is suitable.

It is essential that a purified antibody reference material be available to test for recovery and linearity of response. For the assay to be quantitative, this reference material must be highly pure, as verified by electrophoresis (SDS-PAGE) and/or HPLC techniques such as size exclusion or ion exchange chromatography. Only a small amount of the antibody is needed for evaluation of Protein A chromatography methods.

With a small amount of pure antibody, it is possible to evaluate binding to, and elution from an analytical column. For the technique to be useful, recovery of the antibody must be greater than 95%. Most IgG molecules will bind to a Protein A column equilibrated with Tris buffer (pH 8 - 9) containing moderate levels of salt (0.1 - 0.2M sodium chloride). It is best if the sample itself is also adjusted to pH 8 - 9 by addition of concentrated Tris buffer solution. If binding is not satisfactory, increasing either the pH or the salt concentration of both the sample and the equilibration buffer may improve interaction. After binding, most IgG antibodies can be desorbed from a Protein A column by using a glycine buffer at pH 3.

To test for recovery, 100 - 200μg of purified antibody solution should be loaded onto the equilibrated column. After washing with equilibration buffer, the antibody is eluted and collected into a clean, tared tube. The protein concentration is measured by absorbance at 280nm or by using the Bradford or bicinchoninic acid (BCA) assays. Peak volume is determined by weighing the fraction. The mass of antibody in the peak is calculated by:

$$Mass = (Concentration) \ X \ (Volume)$$

The recovery is the mass of antibody in the peak expressed as a percentage of the mass injected onto the column. Table 10.4 shows recovery data from Protein A chromatography obtained for six different antibodies. The data suggest that this technique was suitable for four of the six antibodies.

Table 10.4 Recovery Data for Six Antibodies

| | **Protein Content of Eluate** | | |
Antibody	Through Column	No Column Control	% Recovery
A	0.184 ± 0.006	0.178 ± 0.002	103
B	0.243 ± 0.011	0.240 ± 0.004	101
C	0.131 ± 0.003	0.129 ± 0.006	102
D	0.242 ± 0.004	0.238 ± 0.005	102
E	0.189 ± 0.003	0.221 ± 0.004	86
F	0.299 ± 0.009	0.322 ± 0.002	93

As will be described in Chapter 16 on method validation, it is necessary to determine the linear range of an assay. At the upper limit of the linear range, the curve flattens out as the column dynamic binding capacity is approached. At the lower end of the linear range, peak areas become small, and variability, expressed as percent of the coefficient of variation (CV), increases correspondingly. The lowest concentration giving an acceptable CV is called the limit of quantitation (LOQ). Fig. 10.6 shows the linear region of a binding curve for an analytical Protein A method. The slope and intercept can be determined by linear regression.

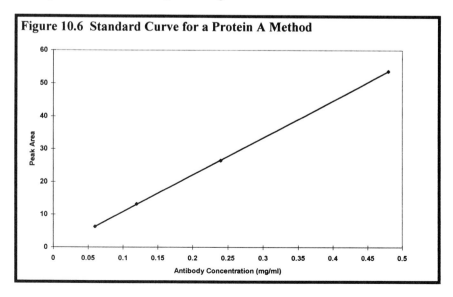

Figure 10.6 Standard Curve for a Protein A Method

For determination of the IgG concentration in unknown samples, the peak area is measured and compared to a standard curve. Sample preparation consists of adjusting the sample pH with concentrated Tris buffer to pH 8 or above. If the IgG peak area for the unknown is above the highest standard in the curve, the sample must be diluted and analyzed again. If the IgG concentration is below the LOQ, a larger sample volume must be injected. For samples with very low concentrations, it may be necessary to run a blank to subtract out baseline drift due to small absorbance differences in the HPLC solvents.

It is very important to demonstrate that an assay is detecting only the antibody of interest. This can be evaluated in several ways. The peak can be collected and analyzed by SDS-PAGE to determine whether contaminants co-elute with the antibody peak. If cell culture supernatants are being analyzed, samples from a similar cell line that does not produce IgG can be run to test whether other components bind to Protein A. Such supernatants can also be used for spiking experiments to demonstrate recovery from relevant test solutions, as shown in Fig. 10.7. Chromatogram A is a Protein A elution profile for a purified IgG (peak at 5.4min). Chromatogram B shows the same mass of antibody spiked into ascitic fluid from a non-IgG-producing cell line. Chromatogram C shows the same mass of antibody spiked into cell culture fluid from a non-IgG-producing cell line.

Figure 10.7 Spiking to Demonstrate Recovery

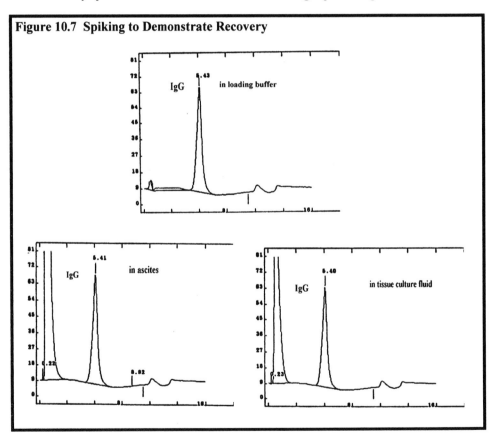

3. Affinity Purification using Protein A Supports

Purification of antibodies using Protein A is functionally similar to the analytical applications described above. Antibody recovery and purity are the principal objectives. An additional requirement is demonstration that the affinity-purified antibody is functionally active, which requires development of a sensitive binding assay. The principal drawback to large scale Protein A purification is the cost of the media; however, Protein A columns are very stable and can be used repeatedly without loss of capacity.

The great advantage of affinity purification with protein A is that highly pure antibody (>95%) can be obtained in a single step. Protein A separations have been used for purification of kilogram quantities of monoclonal antibodies. It removes virtually all protein contaminants, as well as unwanted DNA and lipids. For pharmaceutical applications, it is also a very good virus removal step. For these reasons, many scientists prefer to use Protein A columns relatively early in an antibody purification process, but because columns are expensive, care must be taken not to foul them. Fouling can be minimized by centrifugation and/or filtration of the samples to remove cells and particulates. Alternatively, an ion- exchange adsorptive step can be run upstream from the Protein A column to "clean up' the antibody solution prior to loading it onto the expensive affinity support.

One advantage that Protein A has over many other affinity supports is its remarkable stability. Several commercially available Protein A supports can be cleaned (sanitized) by short treatments with 0.1N sodium hydroxide (6). These supports can also be washed or eluted with acidic solutions down to pH 2 without loss of binding activity (6). Although some leakage of Protein A from the column can contaminate a purified antibody, a subsequent step, such as hydroxyapatite purification can effectively remove any Protein A-IgG complexes (8).

D. Separations using Lectins and Dyes

Plant proteins called lectins have high affinity sites for binding specific classes of carbohydrate chains; therefore, immobilized lectins can be used to purify glycoproteins and other glycosylated compounds. Because lectins bind carbohydrate chains and not specific proteins, this type of affinity chromatography is usually used as one step of a general purification strategy (9). Many different lectins have been used for affinity separations; each has a uniquely different specificity for the carbohydrate structures that it binds (9), as shown in Table 10.5.

Table 10.5 Selected Lectins Used for Affinity Separations

Immobilized Lectin	Sugar Specificity
Concanavalin A	α-linked mannose
Wheat germ agglutinin	N-acetylglucosamine dimer or trimer
Lens culinaris agglutinin	mannose
Ricinis communis agglutinin (RCA I)	terminal galactose
Helix pomatia lectin	N-acetyl-galactosamine
Arachis hypogaea lectin	galactose
Ulex europaeus lectin	fucose

The lectin Concanavalin A (Con A) binds to glycoproteins that contain mannose-rich carbohydrate structures, such as ovalbumin. After loading a glycoprotein solution onto a Con A column and washing to remove compounds lacking high-mannose structures, the retained glycoproteins can be eluted by applying a concentrated mannose solution. Because a complex solution may contain more than one glycoprotein with the same type of carbohydrate structure, lectin affinity chromatography does not always result in purification of a single component. Binding by lectin affinity is optimum at high ionic strength, up to 0.5M sodium chloride (9). Nonionic detergents, such as Triton X-100 and NP-40, are compatible with the method, but detergent/salt compatibility should always be considered. Equilibration may be slow, requiring up to several hours in a batch mode (9).

Proteins, such as serum albumin, which contain hydrophobic pockets for binding nucleotide cofactors, bind tightly to dyes like Cibacron® Blue. When serum is loaded onto a column containing immobilized Cibacron® Blue, serum albumin binds quantitatively, while other serum proteins are not retained. After washing the column with buffered saline, the purified albumin can be eluted using concentrated salt solutions (3M NaCl). It is possible to improve the specificity of this technique for certain proteins by structural modifications of the dyes (10).

E. Affinity Scale-Up

Although they are often used for preparative purification, affinity supports are generally most effective at only 30 - 50% of their loading capacity (11). During scale-up, it is very important to maintain the kinetics of binding the sample to the ligand (11). This necessitates the use of equal column lengths and linear flow velocities. Columns should be made wider but not longer, and flowrates should be adjusted to maintain the same linear velocity, as discussed in Chapter 2.

F. Advantages and Disadvantages of Affinity Chromatography

Affinity chromatography can be used for either analysis or purification, yielding very pure product in a single step. Because the columns are so selective, they can often be loaded, washed, and eluted at very high flow rates with little or no loss of product or decrease in purity. For analytical separations, large volumes of a crude sample can be loaded quickly, thus increasing the sensitivity of detection for very dilute analytes.

The main disadvantage of affinity supports is that they are often very expensive relative to other chromatographic media. Elution solvents may also be expensive and, if they are harsh, may cause a loss of biological activity. If the immobilized compound is unstable, the number of cycles of use may be limited.

G. Related Technology

1. Fusion Proteins

Genetic engineering technology has allowed molecular biologists to create fusion proteins containing either segments of two different proteins or proteins with a peptide "tail" of specific composition. Such proteins can be expressed at high levels and often have unique biochemical properties. In some cases, the fusion proteins are fully active, while in others, the proteins must be cleaved to liberate the desired segment. One advantage of this technology is that the engineered domains can be designed as a purification tool (12-15). For instance, if a protein is grafted onto an antibody heavy chain, the resulting complex may bind to Protein A, whereas the original protein would not do so. In another example, proteins which have been expressed with added sequences containing histidine residues can be purified in a single step using immobilized metal ion affinity supports because they chelate copper (14). In the future, it may become more commonplace to design recombinant proteins with functional groups incorporated solely for purification purposes.

Several parameters are important in the design of fusion proteins for purification (13):

1. The fused portion should allow specific adsorption to an affinity or ion-exchange support.
2. The fused portion should have minimal effect on the tertiary structure of the protein and no permanent effect on its biological activity.
3. The native protein must be able to be reconstituted through easy and specific removal of the fused portion.
4. A simple and accurate assay must be achieved.
5. The technology should be applicable to many different proteins.

Some of the successful strategies for protein design with purification as the primary consideration are listed in Table 10.6. Various chromatographic modes, including affinity, ion-exchange and immobilized metal affinity chromatography, which will be discussed later in the chapter, can be utilized. An affinity tail is attractive because it can be used to immobilize proteins or to recover them from a culture medium or lysate. It also permits final processing by site specific cleavage (12).

Table 10.6 Purification-Dictated Fusion Proteins		
Fused Peptide or Protein	Chromatographic Mode	References
poly(histidine)	IMAC	14, 16
poly(arginine)	CEX	13
β-galactosidase	affinity (substrate analog)	12
Protein A	affinity (IgG)	12
streptavidin	affinity (biotin)	12
glutathione S-transferase (GST)	covalent chromatography	17

2. Peptide Libraries

Peptides are excellent ligands for affinity chromatography because they are usually nontoxic, unlikely to cause an immune response, can be manufactured aseptically, and produce moderate interactions by affinity chromatography (16). Highly specific peptides for affinity interactions can be identified by using peptide libraries. For libraries produced by bacterial phages, the various phage particles are incubated with immobilized target protein. After the unbound particles are washed away, the retained phages are used for gene amplification, followed by sequencing. For example, when looking for a peptide that binds to a target protein or receptor, the phages which bind to the protein by their specific peptide must be isolated and identified.

For chemical synthesis peptide libraries, labeled target protein is applied to spots containing the individual peptides, and those which bind are identified (16). With either of these techniques, the highly specific peptide ligands can be attached to affinity matrices and used for purification.

3. Immobilized Enzymes

Besides being used as ligands on affinity chromatography supports, immobilized enzymes may also be implemented as catalysts for biochemical reactions. For example, immobilized pepsin and papain can cleave IgG molecules into Fab or F(ab')$_2$ subunits, as shown in Fig. 10.8. This method overcomes a problem that is encountered with soluble protease digestions, whereby the proteases digest and inactivate themselves, introducing unwanted fragments into the digestion mixture. With immobilized proteases, auto-digestion does not occur.

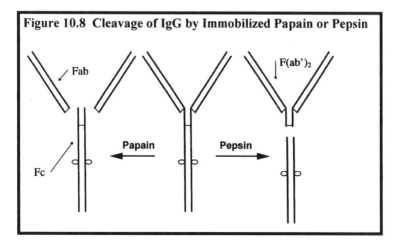

Figure 10.8 Cleavage of IgG by Immobilized Papain or Pepsin

One major advantage of immobilized enzyme techniques is that the enzymes do not need to be removed or inactivated at the end of the reaction. Instead, the immobilized enzymes remain on the columns or are simply removed from the products by filtration or centrifugation from batch processes, allowing further purification, as needed. Immobilized enzyme columns can often be washed and re-used numerous times.

II. IMMOBILIZED METAL ION AFFINITY CHROMATOGRAPHY (IMAC)

Immobilized metal ion affinity chromatography (IMAC) is a method which separates molecules by their differential affinities to metal ions which have been immobilized on a support (14, 18, 19). This technique was introduced by Helfferich for small molecules as ligand exchange chromatography (LEC) (20). Porath later showed its utility in separating proteins and nucleic acids, calling it metal chelate affinity chromatography (MCAC) (21). He later proposed the name "immobilized metal ion affinity chromatography" (IMAC). An alternative name, "metal interaction chromatography" (MIC), conforms to the standard nomenclature of chromatographic modes (18). IMAC has proven useful for the purification of serum proteins and interferons (11). Fusion proteins with histidine tails are also readily isolated by this technique.

A. Mechanism

In IMAC, as illustrated in Fig. 10.9, a metal ion (Me) is loaded onto a matrix containing a chelating group (B) linked with a spacer (A). Free coordination sites are occupied by ions from the mobile phase (L_l). A protein which can form a complex with the metal ion will bind, displacing the mobile phase ions. The protein is later eluted by displacing it with a molecule which has stronger binding or by changing the mobile phase pH, salt, or organic concentration. Metal ions can be removed from the support with a chelating agent like ethylenediaminetetraacetic acid (EDTA).

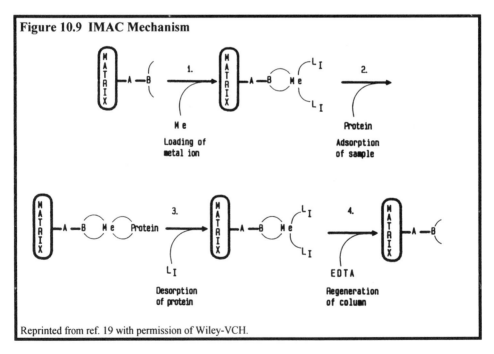

Figure 10.9 IMAC Mechanism

Reprinted from ref. 19 with permission of Wiley-VCH.

Binding in IMAC involves the interaction of electron donor groups on certain amino acids such as histidine, tryptophan, and cysteine, with the immobilized metal ion; therefore, IMAC is a general method for protein analysis. The side chains of histidine dominate protein binding to most chelated metals ions used in IMAC (14). Although cysteine has a strong affinity for metal ions, residues exposed on the surface of proteins are usually oxidized rapidly. Aromatic amino acids have a weaker, and possibly indirect, affinity to the metals (14). Consequently, retention in IMAC is often directly related to the number of surface histidines (14). The binding is distinct from that of metals in metalloproteins where they bind only at specific coordination sites (19). Protein conformation plays a major role in adsorption on IMAC supports because the interaction is with surface amino acids. Ionic interactions of the protein with the support ligand can be increased or diminished by adjusting the ionic strength and/or pH of the mobile phase (18).

B. Supports

Most IMAC supports are based on agarose (19), although some have been synthesized on silica (22). They require a "soft" or noninteractive layer into which are incorporated chelatogenic ligates (18). The most commonly used ligand for IMAC is iminodiacetic acid (IDA), which forms a tridentate bond with metal ions, as seen in Fig. 10.10. Another ligand is the pentadentate, tricarboxymethylenediamine (TED).

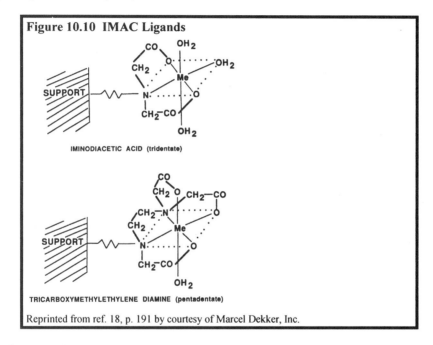

Figure 10.10 IMAC Ligands

IMINODIACETIC ACID (tridentate)

TRICARBOXYMETHYLETHYLENE DIAMINE (pentadentate)

Reprinted from ref. 18, p. 191 by courtesy of Marcel Dekker, Inc.

It is essential that the ligand in IMAC be at least tridentate to form a stable complex with the metal; yet it must also provide free coordination sites for interaction with the solutes. The metal ion-ligand complex varies in stability with the metal ions. For Superose 12 with IDA, the order of stability was: $Cu^{2+} > Ni^{2+} > Zn^{2+} \geq Co^{2+} > Fe^{2+} >> Ca^{2+}$ (19). Fig. 10-11 illustrates the postulated structure of Cu^{2+} immobilized on IDA.

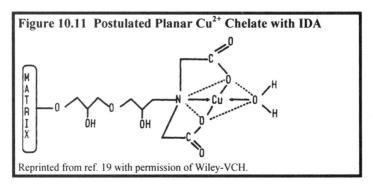

Figure 10.11 Postulated Planar Cu^{2+} Chelate with IDA

Reprinted from ref. 19 with permission of Wiley-VCH.

Some selected commercially available IMAC supports are listed in Table 10.7.

Table 10.7 Selected IMAC Supports

Product	Matrix	Ligand	Particle Diameter	Manufacturer
HiTrap® Chelating	agarose	IDA	34μm	Amersham Pharmacia
Chelating Sepharose®	agarose	IDA	22-30μm	Amersham Pharmacia
Fractogel® EMD Chelate 650(S)	polymer, tentacle	IDA	25-40μm	E. Merck
Immobilized Iminodiacetic Acid Gel	agarose	IDA	45-165μm	Pierce
Immobilized Iminodiacetic Gel	methacrylate	IDA	30-60μm	Pierce
TSKgel® Chelate 5PW	hydrophilic polymer	IDA	10μm	TosoHaas
Toyopearl® AF-Chelate 650M	methacrylate	IDA	65μm	TosoHaas

C. Operation

1. Metal Ions

The metal ion which is affixed to the ligand affords the selectivity in IMAC. Protein retention on different immobilized metals is related to the relative affinity of imidazole for them (14). The bindiing strength of ions for imidazole is $Cu^{2+} > Ni^{2+} > Zn^{2+} \sim Co^{2+}$ (14) and, in one study, the affinity for ribonuclease A, angiotensin I, and angiotensin II was $Cu^{2+} > Ni^{2+} > Co^{2+} > Zn^{2+}$ (23). The affinity of imidazole for Cu^{2+} is about fifteen times that for Ni^{2+}, which is about three times that for the other metals (14). Highest affinity, however, may not always reflect optimal separation conditions. Zn^{2+} and Cu^{2+} are most commonly used. Fe^{3+} has a group specificity for phosphoproteins and phosphoamino acids (11); therefore, it may be useful for isolating this type of protein from mixtures. If phosphoproteins are part of a pathway, they could be removed as a class for further study. In a field such as proteomics, general grouping of target molecules can be selected by using a specific metal ion.

The optimum pH and conditions for binding are specific for each metal-protein pair. For example, in a comparative study involving five metal ions, five proteins, and 2 - 3 pH levels, ovalbumin only bound on Cu^{2+}, Fe^{3+}, or Fe^{2+} when run at pH 6 (22). Ovalbumin only bound on Zn^{2+}, Ni^{2+}, or the bare IDA support when run at pH 5. The selectivities of immobilized Fe^{3+} and Fe^{2+} were different for each protein of the mixture and the elution order changed between different metal ions and pH levels.

2. Mobile Phase

Underlying IMAC is an ion-exchange mechanism because the bare ligands are, in fact, cation-exchangers. Due to this, the nature and concentration of salt ions affect the retention in IMAC. Tris, phosphate, and acetate are suitable buffers and should generally contain a high concentration of salt (0.5 - 1M NaCl) to minimize ionic interactions. Sometimes acidic proteins may be excluded at low ionic strength, and show more binding as the ionic strength is increased (18). Basic proteins may experience ionic attraction at low ionic strength which decreases as the concentration of salt is increased. Retention may also increase when high levels of antichaotropic salts, like ammonium sulfate, are used (18).

As mentioned previously, pH is an important factor in retention by IMAC. Besides the general effects of pH on ionic interactions and protein conformation, histidine has a pI of 6.0; therefore, adjustment of pH around neutrality will effect its ionization. Fig. 10.12 shows the separation of cytochromes c from three species; they vary in their surface histidine content (14). The inverse pH gradient effectively resolved them.

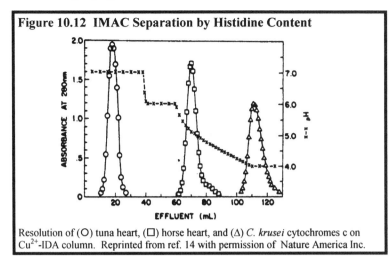

Figure 10.12 IMAC Separation by Histidine Content

Resolution of (O) tuna heart, (□) horse heart, and (Δ) *C. krusei* cytochromes c on Cu^{2+}-IDA column. Reprinted from ref. 14 with permission of Nature America Inc.

The affinity dimension of IMAC results in the feasibility of using a competing agent such as histidine, imidazole, histamine, glycine, or ammonia to effect elution via a step gradient. The agent displaces the protein from the immobilized metal ion. An even stronger displacer is EDTA, which also strips the metal ions off the support (24). Sequential step gradients of increasingly strong elution agents can be used to discriminate different fractions. For example, in a study by Porath, several factors in human serum were purified by steps of sodium acetate, imidazole, and EDTA, each step yielding a different fraction (24).

The effects of the addition of an organic solvent to the mobile phase on retention by IMAC is also specific to the protein and immobilized metal (18, 22). Fig. 10.13 shows that increasing concentrations of methanol decreased retention for lysozyme, but increased it for cytochrome c and β-lactoglobulin A. This may be due to disruptive effects of the organic solvent on the protein conformation and the resultant exposure of additional amino acids to the support.

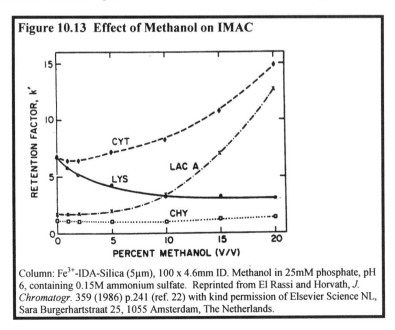

Figure 10.13 Effect of Methanol on IMAC

Column: Fe^{3+}-IDA-Silica (5μm), 100 x 4.6mm ID. Methanol in 25mM phosphate, pH 6, containing 0.15M ammonium sulfate. Reprinted from El Rassi and Horvath, *J. Chromatogr.* 359 (1986) p.241 (ref. 22) with kind permission of Elsevier Science NL, Sara Burgerhartstraat 25, 1055 Amsterdam, The Netherlands.

D. Design of Recombinant Proteins for IMAC Purification

Histidine, the predominant amino acid involved in IMAC, is relatively rare; therefore, its incorporation into a fusion peptide or handle usually results in significantly increased retention by IMAC. Poly(histidine) peptides are one type of handle which can be removed by carboxypeptidase A (25, 26). Other histidine-containing peptides have also been utilized as fusion peptides. Peptides of the structure, His-X$_3$-His, which is known to form an α-helix, has high affinity for bound metal ions (14). Because of its helical properties, retention of proteins containing this specific peptide may also yield information about protein folding.

E. General Guidelines

Some general guidelines for IMAC can be defined, as diagrammed in Table 10.8.

Table 10.8 Parameters of IMAC		Binding		
Parameter	Weaker		⟶	Stronger
Stationary Phase				
Metal Ion	Ca^{2+} Co^{2+}	Zn^{2+}	Ni^{2+}	Cu^{2+}
Chelating Ligand				
Type	TED			IDA
Amount Bound	Low			High
Mobile Phase				
pH	5	7		8
Buffer Ions	EDTA	ammonia		acetate
	citrate	Tris		phosphate
		ethanolamine		
Reprinted from ref. 19 with permission of Wiley-VCH.				

Changing parameters such as the buffer, pH, or metal ion can modify the binding, elution and thus, the selectivity of IMAC methods for various proteins. The start-up procedure described in Table 10.9 arbitrarily uses copper and a phosphate buffer but other variations could be substituted. Metal ions are loaded as their salts, for example, ZnCl$_2$ or CuSO$_4 \bullet$H$_2$0.

Table 10.9 Start-Up for IMAC

1. Select a column from those listed in Table 10.7.
2. Dialyze sample to remove components that might compete for binding (low molecular weight thiols or amines).
3. Load Cu^{2+} under weakly acidic to neutral conditions, using phosphate buffer (pH 7), containing 0.2M sodium chloride and 15mM copper nitrate.
4. Wash column with starting buffer (without copper nitrate) to remove excess Cu^{2+} and minimize bleed during the run.
5. Inject sample. Run with starting buffer until all the unbound compounds are eluted.
6. Use a step gradient to pH 3 to elute bound proteins (19). If nothing elutes, use a competitive ligand.
7. After use, strip the column with EDTA. Always strip columns with EDTA before changing to a different metal, especially one with weaker binding.
8. Recharge the column with metal ion before the next use.

F. Summary of IMAC

IMAC is a valuable tool for protein purification because of its high selectivity. Oxidation of proteins tends to occur more frequently with IMAC than other modes of chromatography, possibly due to the metal ions. Table 10.10 lists some of the useful features of the technique.

Table 10.10 Features of IMAC

- IMAC is a general technique for protein purification.
- Certain amino acids, primarily histidine, on the surface of the protein are required for adsorption.
- Protein conformation can affect binding.
- Nonspecific ion-exchange interactions can be modified by buffer and salt selection.
- Binding is influenced by pH, salt, and competitive molecules.
- Several elution techniques are feasible.

III. COVALENT CHROMATOGRAPHY

A. Mechanism

Covalent chromatography is a unique chromatographic method in that a stable covalent bond, which can be disrupted under mild conditions, is formed between the ligand on the support and the analyte (27, 28). Covalent chromatography is based on thiol-disulfide exchange, as illustrated in Fig. 10.14. A protein (P) with a free thiol binds to the ligand, concomitantly releasing the activation group. After the protein binding process, other sample components are washed through the column. The protein is released by using a thiol-containing molecule, like β-mercaptoethanol, in another thiol-disulfide exchange.

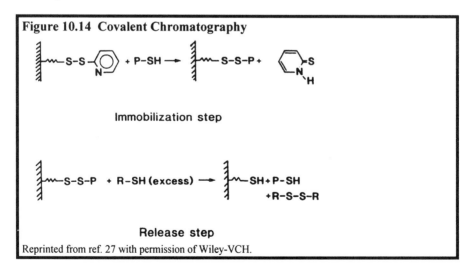

Figure 10.14 Covalent Chromatography

Immobilization step

Release step

Reprinted from ref. 27 with permission of Wiley-VCH.

B. Supports

Covalent chromatography does not require high performance supports because of its specificity and bind/release mechanism. Most commercially available supports are based on agarose which is easily derivatized and has good chromatographic properties. The gels are supplied either activated and thus, in the disulfide form with a pyridine derivative, or with thiol ligands which require activation, usually by reaction with 2,2'-dipyridyldisulfide (27). Table 10.11 lists some commercial column packings for covalent chromatography.

Table 10.11 Column Packings for Covalent Chromatography

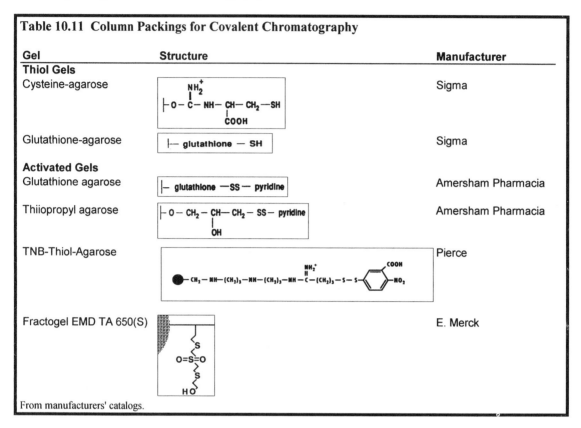

Gel	Structure	Manufacturer
Thiol Gels		
Cysteine-agarose		Sigma
Glutathione-agarose		Sigma
Activated Gels		
Glutathione agarose		Amersham Pharmacia
Thiiopropyl agarose		Amersham Pharmacia
TNB-Thiol-Agarose		Pierce
Fractogel EMD TA 650(S)		E. Merck

From manufacturers' catalogs.

C. Operation

The first step in covalent chromatography is to determine the column capacity for the protein under the binding conditions in the specific buffer. This is accomplished by binding dipyridyldisulfide, and then monitoring the release of 2-thiopyridine at 343nm by the elution buffer. Buffer composition can usually be suited to the sample. The common biological buffers, formate, acetate, phosphate, and Tris at pH 3 - 8, with or without denaturants like urea or guanidinium hydrochloride, can be used successfully. Buffers should not contain reducing agents like dithiothreitol (DTT) or cysteine which are used to effect elution (11).

Low molecular weight thiols such as glutathione should be removed from the sample by SEC or dialysis before application. The sample can be applied to the activated gel either in a batch or column mode and reacted for at least one hour. In a column, this is accomplished by stopping the flow. After that time, the support is washed with 0.1 - 0.3M sodium chloride to remove nonspecifically bound proteins or other sample components. This is easily carried out in a column, with 1 - 2 column volumes usually being adequate. Residual pyridyl groups can be selectively removed with an equimolar amount of low molecular weight thiol (DTT or β-mercaptoethanol), after which the protein is removed with an excess of the same agent. The excess is usually 10 - 25mM DTT or 25 - 50mM β-mercaptoethanol at pH 8. Generally, all protein is eluted at once, but it is possible to implement selective elution by successive steps of increasing strength (DTT > β-mercaptoethanol > cysteine). The eluted protein will contain excess thiol which can be subsequently removed by SEC. The gel can be reactivated by washing with DTT, followed by dipyridyldisulfide.

D. Applications

Covalent chromatography can be used to remove sulfhydryl-containing impurities from protein preparations. It has been successfully utilized to purify proteins and peptides, such as urease (29), papain (30), thiol peptides (31), and Band 3 proteins from human erythrocyte plasma membrane (32). In the latter example, shown in Fig. 10.15, the nonspecifically adsorbed and unbound molecules were removed by washing with Tris buffer at pH 7.2 containing sodium chloride, EDTA, and Triton X-100. After the washing, the bound material was removed with the same buffer containing 50mM cysteine. The Band 3 protein was purified to 95%, with 90% recovery.

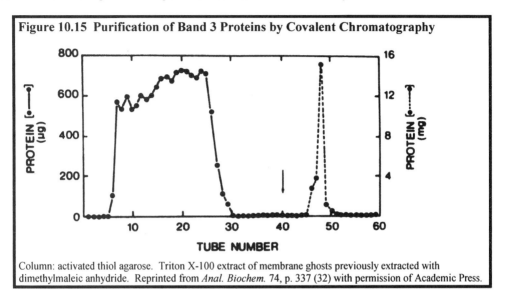

Figure 10.15 Purification of Band 3 Proteins by Covalent Chromatography

Column: activated thiol agarose. Triton X-100 extract of membrane ghosts previously extracted with dimethylmaleic anhydride. Reprinted from *Anal. Biochem.* 74, p. 337 (32) with permission of Academic Press.

IV. REFERENCES

1. S. Ostgrove, "Affinity Chromatography: General Methods" in *Meth. Enzymol.* 182 (M.P. Deutcher, ed.) (1990) 357.
2. J. Turkova, ed., *Affinity Chromatography,* Elsevier, New York (1978).
3. D. Josic, A. Becker, and W. Reutter, "Application of High-Performance Affinity Chromatographic Techniques" in *HPLC of Proteins, Peptides and Polynucleotides* (ed. M.T.W. Hearn), VCH Publishers, New York (1991) 469.
4. T.M. Phillips, "Affinity Chromatography" in *Chromatography, 5th Edition* (ed. E. Heftmann), Elsevier, Amsterdam (1992) A309.
5. J.R. Mazzeo and I.S. Krull, *BioChromatogr.* 4 (1989) 124.
6. P. Gagnon, *Purification Tools for Monoclonal Antibodies,* Validated Biosystems, Tucson, 1996.
7. L.R. Massom, C. Ulbright, P. Snodgrass, and H.W. Jarrett, *BioChromatogr.* 4 (1989) 144.
8. M. Mariani, F. Bonelli, L. Tarditi, R. Calogero, M. Camgna, E. Spranzi, E. Seccamani, G. Deleide, and G.A. Scassellati, *BioChromatogr.* 4 (1989) 149.
9. C. Gerard, *Meth. Enzymol.* 182 (ed. M.P. Deutscher) (1990) 529.
10. Y.D. Clonis, "Preparative Dye-Ligand Chromatography" in *HPLC of Proteins, Peptides and Polynucleotides* (ed. M.T.W. Hearn), VCH Publishers, New York (1991) 453.
11. S. Ostgrove and S. Weiss, *Meth. Enzymol.* 182 (ed. M.P. Deutscher) (1990) 371.
12. M. Uhlen, T. Moks, and L. Abrahmsen, *Biochem. Soc. Trans.* 16 (1988) 111.
13. H.M. Sassenfeld and S.J. Brewer, *Bio/Technology* 2 (1984) 76.
14. F.H. Arnold, *Bio/Technology* 9 (1991) 151.
15. F.A.O. Marston and D.L. Hartley, *Meth. Enzymol.* 182 (ed. M.P. Deutscher) (1990) 264.
16. P.Y. Huang and R.G. Carbonell, *Biotechnol. Bioeng.* 47 (1995) 288.

17. C.-C. Huang, D. Nguyen, R. Martinez, and C. Edwards, *Biochemistry* 31 (1992) 993.
18. Z. El Rassi and Cs. Horvath, "Metal Interaction Chromatography of Proteins" in *HPLC of Biological Macromolecules* (ed. K.M. Gooding and F.E. Regnier), Marcel Dekker, New York, 1990, p. 179.
19. L. Kagedahl, "Immobilized Metal Ion Affinity Chromatography" in *Protein Purification* (ed. J.C. Jansen and L. Ryden), VCH Publishers, New York (1989) 227.
20. F. Helfferich, *Nature* 189 (1961) 1001.
21. J. Porath, J. Carlsson, I. Olsson, and G. Belfrage, *Nature* 258 (1975) 598.
22. Z. El Rassi and Cs. Horvath, *J. Chromatogr.* 359 (1986) 241.
23. E. Sulkowski, K. Vastola, D. Oleszek, and W. von Muenchhausen, *Affinity Chromatography and Related Techniques* (ed. T.C.J. Gribnau, J. Visser, and R.F.J. Nevard), Elsevier, The Netherlands (1982) 313.
24. J. Porath, *J. Chromatogr.* 443 (1988) 3.
25. E. Hochuli, W. Bannwarth, H. Dobeli, R. Gentz and D. Stuber, *Bio/Technology* 6 (1988) 1321.
26. C. Ljungquist, A. Breitholtz, H. Brink-Nilsson, T. Moks, M. Uhlen, and B. Nilsson, *Eur. J. Biochem.* 186 (1989) 563.
27. L. Ryden and J.C. Jansen, "Covalent Chromatography " in *Protein Purification* (ed. J.C. Jansen and L. Ryden), VCH Publishers, New York (1989) 252.
28. *Affinity Chromatography, Principles and Methods*, Pharmacia Fine Chemicals, Sweden (1979) 35.
29. J. Carlsson, I. Olsson, R. Axen, and H. Drevin, *Acta Chem. Scand.* B30 (1976) 180.
30. K. Brocklehurst, J. Carlsson, M.P.J. Kierstan, and E.M. Crook, *Biochem. J.* 133 (1973) 573.
31. A. Svensson, J. Carlsson, and D. Eaker, *FEBS Lett.* 73 (1977) 171.
32. A. Kahlenberg and C. Walker, *Anal. Biochem.* 74 (1976) 337.

CHAPTER 11
COLUMN HANDLING AND MAINTENANCE

The high-performance liquid chromatography (HPLC) columns used in biomolecule separation, isolation, and characterization represent a considerable investment. With appropriate maintenance and troubleshooting procedures, HPLC columns should give satisfactory performance for a period from several months to over a year (1). Column handling procedures for chromatographic modes commonly used in biomolecule separations - size exclusion, ion-exchange and reversed-phase will be discussed in this chapter.

I. COLUMN INSTALLATION

A. General

Microparticulate HPLC columns consist of 1.5 - 10 μm packing materials based on silica or organic polymers (resins), packed at high pressure into stainless steel or polymeric tubing, as discussed in Chapter 5. Although most HPLC packings have the high mechanical strength needed to withstand the pressure drops of 150 - 5000psi encountered at typical flow rates, column beds will not tolerate undue mechanical shock. For this reason, columns should be handled carefully during installation, removal and storage to avoid shocks that might cause disruption of the packed bed.

B. Connection

The majority of commercially available HPLC columns are "universal"; that is, they may be installed on any HPLC pump or injector directly or with a simple adapter. When fitting a new column to an existing system, however, care must be taken to ensure that the connections do not introduce dead volume. Inlet and outlet nuts and ferrules must be mated appropriately with the end fittings, as discussed in Chapter 3. Frequently, polymeric fittings, with or without metal rings, are used for the inlet and outlet connectors to simplify compatibility and ensure minimal dead volume. Unions and connectors should have zero- or low dead volume and connecting tubing should be of low internal diameter (0.01 - 0.03" ID). Connections must be sufficiently tight to prevent leaks (approximately finger-tight plus a quarter-turn). If overtightening is required to prevent leakage, there is probably a mismatch between the end fitting and nut or ferrule that can be remedied only by replacement. In addition to leakage, fitting mismatch can result in band broadening due to excessive dead volume or unswept voids. Because there is no standardization of terminators and fittings among manufacturers, it is best that, in cases where columns from different sources are used on the same HPLC system, separate connector lines be fabricated for each kind of column, using fittings supplied or recommended by the column manufacturer.

The trend in HPLC continues to be towards the use of high efficiency columns packed with microparticulate supports ($\leq 5\mu m$). Although these columns are frequently 25 - 30cm long, 3 - 15cm lengths have become popular. These short columns, packed with materials smaller than 5μm, can be operated without excessive pressure at normal flow rates (1.0 ml/min) to achieve moderate efficiency and low solvent consumption, or at high flow rates (2 - 5 ml/min), to obtain very short analysis times. As the plate height or column length decreases, the extra-column contributions to band broadening become more significant. Use of high efficiency columns demands that extra-column dead volume be minimized by reducing the length of connecting tubing, using very small inner diameter tubing, and installing appropriate guard columns. With microbore or capillary HPLC columns ($\leq 2mm$ ID), extra-column effects are sufficiently severe that a conventional liquid chromatograph requires extensive modifications, including the use of low volume injectors and microbore transfer lines (0.005" ID), as well as detectors with micro flow cells ($< 5\mu l$) and rapid time constants ($< 100msec$).

C. Washing

When shipped from the vendor, most commercial HPLC columns contain an organic solvent, as designated in the column installation instructions. If this solvent is not compatible with the mobile phase, the column must be washed thoroughly (5 - 10 column volumes) with a mutually miscible solvent prior to use. For example, reversed-phase columns shipped in methanol or acetonitrile should be washed with water before introduction of buffers or salts. Columns shipped in hexane (e.g., nitrile or amino columns) must be washed with propanol or tetrahydrofuran before introduction of methanol, acetonitrile, or aqueous solvents. A new column should not be connected to the detector until at least one column volume of wash solvent or mobile phase solvent has passed through it to prevent residual packing solvent or silica fines from entering the detector.

D. Column Testing

A newly purchased HPLC column should always be tested upon receipt, using the manufacturer's suggested test compounds and separation conditions. Manufacturers generally subject columns to tests, as shown in Fig. 11.1, which verify minimal batch-to-batch variation in stationary phase characteristics and confirm that individual columns meet performance specifications. Test compounds are selected so that stationary phase parameters, such as bed efficiency, selectivity, free silanol content, and contamination by trace metals can be probed. Test compounds for column efficiency are usually small molecules selected for their stability, low toxicity, and availability (Fig. 11.1A). Column packings designed expressly for chromatography of proteins are usually tested as a batch by the manufacturer with a protein test mixture to check selectivity and/or recovery, as illustrated in Fig. 11.1B.

Figure 11.1 Column Vendor Test Chromatograms

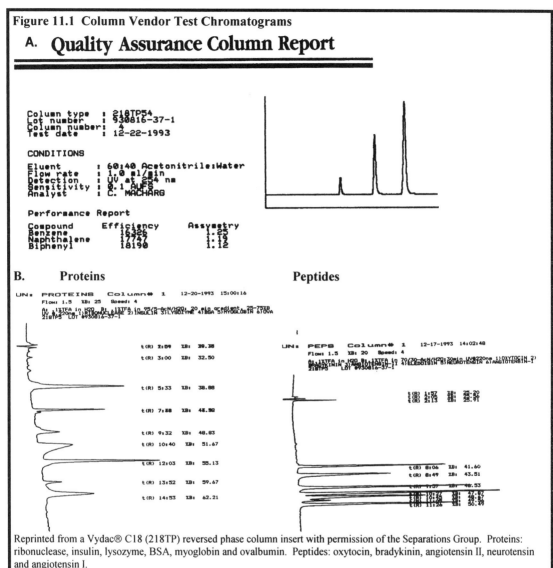

Reprinted from a Vydac® C18 (218TP) reversed phase column insert with permission of the Separations Group. Proteins: ribonuclease, insulin, lysozyme, BSA, myoglobin and ovalbumin. Peptides: oxytocin, bradykinin, angiotensin II, neurotensin and angiotensin I.

The user should evaluate every new column to verify that it meets the published specifications and has not been damaged in shipment; in some cases, failure to test a column within a given time period may void the manufacturer's warranty. The first test should be that for plate count and symmetry. Variances of up to 10% from the manufacturer's test results are not indicative of a defective column because they may reflect extra-column effects, flow accuracy, or small differences in solvent composition. In addition to confirming column integrity, testing a new column provides a baseline for monitoring column performance during use.

After verifying the plate count, it is usually advisable to test a new column with compounds chemically related to the user's samples. If these samples are peptides or proteins, a test sample composed of standard proteins or peptides will establish a performance baseline and help monitor column aging. Because interactions of polypeptides with HPLC stationary phases are often different than those of small molecules, this test mixture will probe column characteristics such as protein selectivity and recovery. A reversed-phase column intended for small molecules should be tested with a mixture containing hydrophobic, basic and acidic components to evaluate the characteristics for each class (2) (*see* Table 7.7) and to provide a baseline to monitor the effects of aging. The results of all initial tests should be recorded in a log for each individual column. Typical probes for column testing are listed in Table 11.1.

Table 11.1 Test Probes for HPLC Columns

- Manufacturer's test mixture for efficiency and tailing
- Small molecules relevant for user
- Manufacturer's protein/peptide test mixture
- Proteins or peptides relevant for user

II. COLUMN OPERATION

The most common causes of early column failure are inappropriate solvent selection, inadequate solvent preparation and insufficient cleanup of biological samples. These are all parameters which can be controlled with proper precautions.

A. Mobile-Phase Preparation

1. Dissolved Impurities

In HPLC, it is very important to use solvents of high purity (HPLC grade or spectroscopic grade) and filter them prior to use through a compatible microporous (0.2 - 0.45 µm) filter. Dissolved impurities in solvents can be retained on the stationary phase and either alter the phase characteristics or elute as spurious or "ghost" peaks. Glassware can also introduce impurities into the analysis, especially if it is not thoroughly cleaned or rinsed with HPLC-grade water (3).

Detection in the low UV (205 - 220nm) is often used for peptides and protein fragments because of the frequent absence of chromophores absorbing at 254 or 280nm. Many of the alcohol and acid mobile phase modifiers employed in peptide chromatography have UV cutoffs in this spectral range, producing some baseline offset in gradient elution (*see* Table 4.2). Solvent impurities can be a severe problem at these wavelengths, giving rise to noisy baselines or spurious peaks that may be incorrectly diagnosed as column or pump disorders.

Reversed-phase chromatography is particularly sensitive to mobile phase impurities, which are often hydrophobic and bind to the columns. Water is the most notorious source of impurities because it often contains trace organics that become concentrated on reversed-phase columns and appear as peaks during gradient elution. Even commercial HPLC-grade water may include interfering substances. This grade of water generally has a set limit for optical purity at 254nm with the actual value indicated for the absorbance at a lower wavelength like 204nm. Water purification systems, commercial HPLC-grade water, or sometimes bottled distilled water may or may not be satisfactory for use in HPLC. A blank gradient run is the best way to evaluate the purity of water or the mobile phase. If peaks appear, they can be traced to the water by pumping water through the column for varying time periods, followed by blank gradients. An increase in ghost peak height as a function of the volume of water used is indicative of impure water. Mobile phase additives, such as ion-pair agents, buffers and salts, can be checked individually in a similar manner. For example, a gradient from water to acetonitrile (0 - 100%) can be compared to one which is identical except for the addition of TFA or another agent. If trace impurities are found, substitution with a better quality of solvent or chemical is best. If that is not possible, the impurities may be removed by passing the water or aqueous mobile phase through an activated charcoal or reversed-phase column prior to use. Redistilled or recycled solvents should be carefully monitored because they may contain impurities.

Solvent impurities may also be a problem in ion-exchange chromatography, particularly when employing gradient elution with phosphate salts. Phosphates contain UV-absorbing impurities that cause severe baseline offset at high detector sensitivity. This problem is compounded by the tendency of these impurities to collect on the column and elute at high ionic strength; thus, the baseline offset may increase with column use. The best remedy is the use of high grades of salts and solvents and periodic stripping of adsorbed impurities from the column, as discussed later in this chapter. Solvent purification techniques for phosphate buffers using recrystallization, ion-exchange cleanup, or passage through a chelating resin are somewhat effective but do not completely eliminate the problem (1).

If impurities in the weak solvent of a gradient elution system can be removed by passage through an appropriate adsorbent bed, a stripper column packed with such an adsorbent can be placed in-line on the outlet side of the "A" pump in a multi-pump gradient HPLC system (1). However, if the breakthrough volume of the stripper column is exceeded, elution of impurities will occur during an analysis.

2. Dissolved Gases

When using pumps with ball and seat check valves, solvents must be thoroughly degassed to prevent loss of pump prime by cavitation. Oxidation of certain stationary phases like amino phases is also possible if oxygen is present in the mobile phase. If detection in the low UV (< 200 nm) is used, degassing will eliminate absorbance by dissolved oxygen which can yield noisy or drifting baselines. Another benefit of degassing mobile-phase solvents is minimization of bubble formation due to outgassing in the detector cell; this can also be prevented by creating 20 - 100psi resistance on the outlet side of the detector with a restrictor or a length of about 3 meters of 0.3mm ID tubing. Degassing also inhibits bacterial growth.

3. Particulates

Column failure may arise from the gradual accumulation of particulate material originating from either the wear of system seals or microbial growth in the mobile-phase solvents. The column can be protected by installation of an inline filter either before the injector or between the injector and the column. The former will protect from everything before the injector. The latter arrangement will additionally protect against particulates originating from the sample or the injector seal. In-line filters have easily replaceable frits in the 0.5 - 2μm range with only a minimal amount of dead volume. Pre-columns and guard columns also serve as effective solvent filters.

Premixed solvents should be filtered through a micropore filter before use in an HPLC. Insoluble particulates in the mobile phase will plug frits or filters, as well as inhibit check valve operation, producing either excessive or fluctuating operating pressure. Sometimes filtration can be a source of impurities if components of the filter are leached by the solvent. The filter must be matched to the solvent to avoid dissolving or leaching (4). Generally, nylon filters are appropriate for aqueous or buffer solutions and teflon for organic solvents. Filter vendors like Gelman provide selection guides for specific solvent compositions to ensure compatibility. It is sometimes advisable to discard the initial 25 - 50ml of filtrate to eliminate leachable impurities if a problem is suspected. Generally, HPLC-grade solvents have already been filtered, but it is imperative to filter all buffers and other aqueous mobile phases through 0.2μm filters before use. Aqueous mobile phases kept for more than one day must be filtered before each use, even after refrigeration, to eliminate particulates caused by microbial growth or precipitation (5). Aqueous mobile phases that contain a chemical which inhibits bacterial growth, such as trifluoroacetic acid, cyanide, or azide, usually do not have to be filtered after the initial preparation. Care must be taken not to cross-contaminate mobile phases with the filtering apparatus. It should be thoroughly washed and the filter membrane changed whenever contamination is possible.

When mixtures of an aqueous buffer and an organic solvent are used, as in reversed-phase chromatography, caution must be exercised to avoid precipitation of buffer salts in the system. Aqueous-organic mixtures proportioned from separate reservoirs by the HPLC pumping system may produce precipitation in hydraulic components, leading to reduced seal life or component failure. If precipitation occurs in the column bed, column failure is likely. The best means of avoiding buffer-organic solvent incompatibility is to use a premixed combination of buffer and organic modifier as the strong eluent. Buffers should never remain in a column or HPLC system during storage or downtime. When a system is converted from one chromatographic mode to another, e.g., from ion-exchange to reversed-phase chromatography, the entire system should be flushed with water before the new mobile phases and column are used.

4. Modification of Stationary Phase

A number of mobile-phase additives may cause column failure by irreversibly changing the stationary-phase characteristics or accelerating column degradation. Ion-pairing agents or detergents with bulky hydrophobic groups (e.g., sodium dodecylsulfate) may permanently bind to the stationary phase of size exclusion and reversed-phase columns, changing their effective phase chemistry, as well as reducing their pore volume. If applications require the use of such additives, it is recommended that the column be dedicated for this use. Similarly, anionic detergents are unsatisfactory for use with anion-exchange supports to which they will bind tenaciously.

5. Damaging Agents

As discussed in Chapter 5, most silica columns are degraded by use with basic pH. Cationic alkylamines (e.g., triethylamine, tetramethylammonium salts), which are used as ion-pairing agents or competing bases for silanol complexation, often accelerate the dissolution of silica at neutral pH. It is recommended that these agents be used only at pH less than 6 and that they be flushed from the column immediately after use. Biological buffers vary in their degradative properties towards silica-based packings, with phosphate being especially harmful (6, 7).

The operation of stainless steel HPLC pumping systems and columns with halide salts has long been a point of concern in protein purification, particularly since many ion-exchange procedures developed on carbohydrate-based gels employ chloride eluents with concentrations as high as 1 - 2 M. The 316 stainless steel used in most HPLC components is reasonably resistant to chloride at neutral pH, and component lifetimes may not be significantly reduced as long as such solutions are flushed from the system after use. Chloride at acidic pH should not be used in HPLC systems (1). If the solvent compatibility of a particular instrument or component is in question, the manufacturer should be consulted.

B. Sample Preparation

It is generally recommended that samples injected onto HPLC columns be free of contaminating material, both to minimize interference in detection and to prevent adsorption of undesirable sample components onto the column. This is frequently impossible for polypeptide samples from biological fluids or extracts; however, removal of certain contaminants is possible and advisable. Complex biological samples should always be filtered to remove particulates, and guard columns implemented to prevent strongly retained components from reaching the analytical column. Samples containing significant amounts of lipids should be extracted with a nonpolar solvent like ether or hexane prior to injection. High molecular weight components can be removed from samples by size exclusion or dialysis before injection onto ion-exchange or reversed-phase columns. Sample preparation is discussed at length in Chapter 13.

Chromatographic resolution is frequently affected by excessive sample volumes or concentrations. In size exclusion chromatography, the injection volume should be less than 2% of the column's permeation volume. For a standard 9.4mm x 25cm ID column, a sample volume of about 50µl and sample loads of 1 mg can typically be applied (8). In ion-exchange and reversed-phase chromatography using gradient elution, the sample must be applied in a weak solvent (typically the initial or A solvent) so that it will bind maximally. Sample volume is not as critical in these cases; therefore, several milliliters can often be applied to an analytical column, if necessary. Because of the high capacities of ion-exchange and reversed-phase HPLC supports, relatively large sample loads of at least 5 -15mg total protein can be injected onto columns with dimensions of 250 x 4.6mm ID. As a result, analytical columns can often be used for what might be considered semi-preparative applications, if gradient elution is employed. When columns are operated isocratically, resolution is more sensitive to sample volume, concentration, and matrix effects, particularly for compounds with low capacity factors (k).

C. Guard Columns

Guard columns are the most effective means of preserving and extending the lifetime of an analytical column (9). They are used to protect the analytical column from particulates and strongly retained contaminants in the sample matrix, as well as degrading factors in the mobile phase. Guard columns are placed between the injector and analytical column, being discarded or repacked often enough to ensure that bound contaminants are not accidentally eluted. The frequency of replacement depends on many factors including the column capacity, solvent strength, sample type, and sample load. Typically, the lifetime of a guard column is 10 - 50 injections. Because a separation will actually begin on the guard column, the selection of the appropriate packing and column configuration is based on two considerations. First, the stationary phase of the packing should be chemically identical to the analytical stationary phase so that the selectivity of the system is consistent. Unfortunately, this is not always possible because guard column analogs of certain analytical packings are unavailable and, when they are available, phase characteristics of the guard material may not be identical to the analytical packing. Sometimes it is feasible for the packing from an old column to be washed and packed into a guard column. In size exclusion chromatography, mismatches in guard column pore diameter or pore volume can distort protein elution profiles, due to differences in pore diameter, pore volume, and the bonded phase which may cause incompatibility with certain mobile phases. If a low capacity nonporous guard material is used, selectivity mismatch may not be as critical as with a higher capacity packing.

The second consideration in selecting a guard column is preservation of column efficiency; ideally, loss of plates should not exceed 10% to preserve overall resolution. This is extremely important when short, high efficiency analytical columns are used, because extra-column contributions to band broadening are more significant. Precautions must be taken with high capacity microparticulate guard columns because they effectively behave like an additional analytical column. The plate height (H) of a high-capacity guard column should be similar to that of the analytical column. For example, coupling a 3cm guard column packed with 10μm material to a 15cm analytical column with 5μm packing could result in a 30 - 40% loss in efficiency because the separation occurring on the guard column has low resolution. New developments in guard column technology, such as the guard disks supplied by Higgins Analytical, have low dead volume and only slightly diminish the column efficiency.

Two types of guard column packings are currently available: fully porous microparticulate packings and pellicular materials. Microparticulate guard columns are packed with the same material used in the corresponding analytical column and are designed for situations in which high efficiency and maximum loading capacity are required. Because they must be slurry-packed, they are usually obtained from the manufacturer as prepacked columns or cartridges. Some manufacturers offer disposable guard cartridges which fit into a reusable holder. They are available in a variety of 5μm and 10μm reversed-phase, ion-exchange, and size exclusion packings. Microparticulate guard columns must have a close selectivity match with the analytical column and minimal dead volume in connectors and transfer lines to reduce extra-column band broadening. When an exact selectivity match is not possible, it is best to use a guard column support that is less retentive than the analytical packing. In some cases, microparticulate guard columns can be regenerated off-line by washing with a series of strong solvents.

Pellicular packings consist of solid-core beads (usually 30 - 40μm) with the stationary phase bonded to or polymerized on the surface. Due to their large particle size, they are easily packed in dry form without any special equipment; therefore, replacement material for a 50 x 4mm ID guard column is inexpensive. Because the surface area of pellicular packings is low compared to porous microparticles, the loading capacity is also diminished. Consequently, the effect of such columns on selectivity is not significant, but they become saturated easily and require frequent repacking. Some manufacturers, such as Keystone Scientific, Scientific Systems, Inc. and Upchurch Scientific, offer low volume guard columns that couple directly to the analytical column. When these are packed with pellicular materials, efficiency losses of 5% or less are observed with analytical columns containing 5μm and sub-5μm particles (1). The need for frequent replacement is offset by the ease of repacking. The low capacity of pellicular sorbents usually render them unsatisfactory for complex biological samples which readily overload the guard column, thus contaminating the analytical column.

The relative utility of pellicular and porous guard column materials for biological samples was demonstrated in a study using crude ovalbumin from egg white (10). The integrity of the column was assessed using the resolution of a peptide standard. Resolution of the peptide components was observed before injection of 10mg of the protein mixture and after every five injections. The resolution of the column protected by the porous support was maintained for at least 20 injections, whereas the other dropped rapidly.

Guard columns can also be used for online concentration or cleanup of samples. In these applications, a guard column replaces the sample loop in a multiple port injection valve. After the sample is introduced through the injection port and concentrated on the guard column, the weakly retained contaminants can be eluted with mild conditions. Valve rotation can introduce the remainder of the sample onto the analytical column with a stronger solvent.

D. Saturator Columns

One method of extending the lifetime of silica-based supports in harmful mobile phases has been to saturate the mobile phase with dissolved silica to suppress degradation of the analytical support. This is usually accomplished by installing an in-line saturator or solvent-conditioning precolumn packed with porous silica before the injector. Large-particle (30 - 50μm) silica is generally used, due to packing ease and to low system pressure. Although it has been reported that use of a 300 x 4mm ID precolumn saturator may extend column life more than tenfold, lifetime extensions of 2- to 5-fold are more typical (11). The precolumn must be maintained at the same temperature as the analytical column and topped off periodically to replace dissolved packing.

Saturator pre-columns are rarely used due to several disadvantages. First, a silica-saturated mobile phase may precipitate if allowed to stand in the system. Second, dissolution of the silica packing can generate fines that may pass through the exit frit and then plug the injector, small bore tubing or the analytical column. Third, voids created by dissolution of the precolumn packing will increase the dead volume of the system and may lead to poor gradient reproducibility. Fourth, the precolumn can act as a trap, concentrating solvent impurities that may later elute when a stronger solvent is introduced. Finally, the presence of silicates in isolated sample fractions may alter biological activity or interfere with subsequent chemical characterization.

E. Equilibration

To achieve reproducible retention data, a column must be totally equilibrated in the mobile phase before sample injection. For isocratic runs, reproducibility of subsequent runs confirms equilibration but this is not relevant to gradient elution, which is employed in the majority of HPLC applications of protein and peptide isolation and characterization. Most commercial gradient HPLC systems have sufficiently high solvent proportioning precision to provide reproducibility of gradient retention time within a few percent. To achieve this level of performance, however, the column must be totally reequilibrated at initial conditions before the beginning of each analysis. The gradient elution protocol, including column regeneration and equilibration, must be the same for each run. This is an essential component of a rugged method.

When solvents of widely different strengths are used, the column is most efficiently regenerated with a reverse gradient to initial conditions, followed by equilibration with at least ten column volumes of the initial solvent. In other cases, immediate switching to the initial solvent is satisfactory. Regeneration and equilibration can proceed at elevated flow rates to reduce turnaround time. Usually 5 - 10 column volumes are adequate for reequilibration but occasionally column equilibration can require as many as fifty column volumes, especially if initial and final solvents differ in pH. It is not uncommon that ion-exchange columns, even when stored in the initial solvent, will exhibit a different retention time in the first analysis of the day than after stabilizing with successive runs. Operation of reversed-phase columns with ion-pairing agents or buffers may also require extended equilibration periods. For maximum reproducibility, the concentration of the ion-pairing agent should be kept constant across the solvent gradient. The minimum reequilibration volume can be determined by comparing retention during gradients after increasing column volumes (5, 10, 20, etc.) of initial mobile phase have passed through the column. If there is no change in retention, equilibration has been attained.

F. Operational Limits

Column supports and stationary phases differ in their limitations to pressure, flow rate, temperature, and pH. This information is supplied in installation instructions for the column and should always be noted to ensure that the column warranty is not voided. Column packings may be incompatible with certain mobile phases. For example, amine phases react with carbonyl groups; therefore, aldehydes or ketones cannot be used as mobile phase modifiers with such columns. If compatibility with an unusual solvent is not known, the column manufacturer should be contacted before use.

III. COLUMN STORAGE

For overnight or short periods, it is best to store silica-based columns in an organic solvent or, if salts have been used, mixtures of an organic solvent and water. Columns should never be stored in any potentially degrading solvents, such as trifluoroacetic acid, other acids or ion-pair agents. Buffers or salt solutions should always be purged from the column before storage. If it is necessary to leave buffer in the column for overnight equilibration, a low flowrate (0.1 ml/min for a 4.6mm ID column) should be maintained to prevent precipitation in either the column or the hydraulic components of the system. For long term storage, silica-based columns should be flushed with water if they have been used with buffers or salts, and then filled with an organic solvent (methanol or acetonitrile)/water mixture. The storage solvent should be indicated on a label attached to the column. Columns containing polystyrene resins or hydrophilic organic polymers should be stored with solvents recommended by the manufacturers. Because stationary phases can act as substrates for microbial growth, aqueous solvents are not recommended for storage unless they contain sodium azide or another bacterial inhibitor. Again, the manufacturer's recommendations should be followed. Column terminators should be tightly capped to prevent drying of the column bed which may cause changes in its geometry. Columns should be kept in a place where they will not be exposed to potential shock or extremes of temperature. Preparative columns ($\geq$ 10mm ID) should be stored vertically after use, because they are subject to void formation which can disrupt the whole bed if stored horizontally.

IV. COLUMN TESTING

A. Column Log

It is good practice to maintain a log in which column performance and use are detailed. This should include the type of column; its serial number; initial test data; the history of its analytical use, including sample type, mobile-phase conditions, and operator; test data from periodic performance checks; and details on column repair, as listed in Table 11.2. A log will enable rapid assessment of performance and the need for, or success of, maintenance procedures. It also allows accurate evaluation of column lifetime.

Table 11.2 Column Log

- Column Brand Name
- Serial Number
- Initial test data
- History of use
    samples
    mobile phases
    operators
- periodic performance checks
- maintenance/repairs

B. Troubleshooting

The most common problems encountered in the operation of HPLC columns, along with suggested diagnoses and treatments, are outlined in Chapter 15. Additionally, mode-specific troubleshooting is found at the end of Chapters 6 - 9. Several problems occur frequently in chromatographic separation of proteins and peptides, particularly in the reversed-phase mode. Spurious or "ghost" peaks can arise from mobile-phase impurities, from sample carryover in the injector or transfer lines, or from elution of adsorbed sample components in subsequent runs. Mobile-phase impurities are a common source of ghost peaks in gradient elution with detection in the low UV, and they typically originate from the water, as discussed previously. Occasionally, proteins that fail to elute quantitatively by the end of a gradient will appear as ghost peaks at the same elution position in subsequent blank gradients, with peak height decreasing progressively. In such instances, the column can be washed between runs by pumping or injecting up to several milliliters of a strong solvent between analyses.

Tailing peaks and split peaks occur frequently with HPLC of biological samples. Peak tailing can arise from sample overload, a poorly packed or degraded column, buildup of adsorbed contaminants on the stationary phase, or extra-column effects. If the sample concentration or viscosity is too high, peaks may be deformed by splitting or tailing. In reversed-phase chromatography of polypeptides, tailing is often a sign of interaction between residual silanol groups on the stationary phase and basic amino acid side chains. This effect can be minimized by operating at acidic pH to suppress silanol ionization, by adding a competing base such as triethylamine or a tetramethylammonium salt to the mobile phase, or by using an endcapped column with high surface coverage. As discussed in Chapter 7, high surface coverage does not necessarily produce decreased tailing or increased performance.

V. COLUMN REPAIR

Careful column operation can prolong the lifetimes of some columns to a year or longer. Column performance will gradually degrade, as strongly adsorbed sample components accumulate on the stationary phase and as a void is formed by the dissolution or compression of the support. Split peaks or doublets may appear, arising from the formation of multiple flow paths through the column - usually voids or channels formed by settling or degradation of the column bed. Peak doublets can also occur from partial plugging of the inlet frit or column head by sample contaminants, creating partial flow resistance. This effect may be accompanied by a gradual rise in operating pressure and can be relieved by frit replacement or, as a last resort, removing and repacking the top millimeters of the column bed. Column lifetimes can often be extended for old or abused columns by simple operations such as frit replacement, washing, backflushing, and bed repair (10). Whenever a column has a void or blockage which needs to be removed, it is advisable to order a new column, because the old column is near the end of its lifetime.

A. Frit Replacement

High pressure which has been isolated to the column is most frequently caused by plugging of the inlet frit. To remove a column frit, tightly secure the column tube vertically and cautiously remove the top terminator fitting, being careful not to disturb the column bed. If a replaceable frit remains in the end fitting, it can usually be dislodged with the tip of a spatula blade or by passing a 20 - 24 gauge wire through the terminator inlet. The end fitting should be thoroughly cleaned with water and/or solvent to remove silica or other contaminants before use. It is advisable to install a new frit; however, replaceable stainless steel frits may sometimes be cleaned by immersing them in a sonic bath containing 3 - 6N nitric acid. When refitting the end fitting, the new frit should be aligned and the surfaces cleaned of packing material to permit proper seating. If the ferrule on the column has moved or become distorted, it may be difficult to achieve a leak-free connection with a new frit; in such cases, installation of PEEK-encased or double frits will sometimes allow a tight seal to be formed.

Some end fittings have pressed-in frits. These can be cleaned by pumping a nitric acid solution through the terminator after removal from the column. If this procedure is ineffective, as indicated by continued high back pressure, the end fitting can be installed on an empty guard column and pumped in the reverse direction to backflush the frit. Whenever an end fitting is removed from a column, such as during frit cleaning, the analytical column should be capped with a spare end fitting to prevent contamination, drying, or disruption of the bed.

B. Column Washing

With extended use, the buildup of strongly retained material, such as lipids or proteins, will cause increased operating pressure and altered chromatographic behavior on an HPLC column. Column performance can often be recovered by stripping adsorbed material with one or more strong solvents. The key in rejuvenating a contaminated column lies in knowing the nature of the contaminants and finding an effective strong solvent to remove them. Lipids can be removed by washing with organic solvents such as methanol, acetonitrile, or tetrahydrofuran. There is no solvent or series of solvents that will universally strip all adsorbed molecules from HPLC stationary phases, but Table 11.3 lists several strong eluents or solubilizing agents that have been used to strip proteins. Some columns, particularly those based on organic polymeric supports rather than silica, are not compatible with these solvents, and, if solvent compatibility is unknown, the column manufacturer should be consulted.

Table 11.3 Wash Solvents for HPLC Columns

Solvent	Composition
Reversed-Phase and Size Exclusion Columns	
acetic acid	1% in water
trifluoroacetic acid (TFA)	1% in water
0.1% TFA/propanol[b]	40/60
TEAP[c]/propanol[b]	40/60
aqueous urea or guanidine	5 - 8M
aqueous sodium chloride, sodium	
phosphate or sodium sulfate	0.5 - 1M
DMSO/water[a]	50/50
Ion-Exchange Columns	
acetic acid	1% in water
phosphoric acid	1% in water
aqueous sodium chloride, sodium	
phosphate or sodium sulfate	1 - 2M

[a] consult manufacturer for column compatibility.
[b] high viscosity solvent; pump at reduced flowrate to avoid high pressure.
[c] triethylamine-phosphoric acid; adjust 0.25N phosphoric acid to pH 2.5 with triethylamine.
Reprinted from *Meth. Enzymol.* 104, p. 133 (1) with permission of Academic Press.

In most cases, neat organic solvents such as acetonitrile or methanol do not dissolve peptides and proteins and, therefore, are not effective stripping solvents for HPLC columns. Mixtures of organic solvents with a buffer, organic acid, or ion-pairing agent often serve as good cleaning agents for reversed-phase and size-exclusion columns. High salt and/or aqueous trifluoroacetic acid (TFA) usually clean ion-exchange columns effectively. Repeated up and down gradients between aqueous trifluoroacetic acid (TFA) and TFA-propanol can frequently regenerate contaminated HPLC columns (10). Detergent cleanup of size exclusion columns should be used as a last resort and must be followed by extensive washing with water and methanol. Detergents such as sodium dodecylsulfate and Triton are good protein solvents, but they tend to be strongly retained on many HPLC stationary phases, and may irreversibly change column characteristics

Several precautions should be observed during column cleanup. If the initial wash solvent is not compatible or miscible with the mobile phase, the column should first be flushed with water or a totally miscible solvent. Similarly, sets of wash solvents used in series must be sequentially miscible or, if not, interspersed with a mutually compatible flushing solvent. For example, after washing with high salt, urea, or guanidine, the column must be purged with at least ten column volumes of water before the introduction of any organic solvent. Similarly, if nonpolar solvents such as hexane or methylene chloride are used to strip lipids from a reversed-phase column, a ten column volume propanol purge is necessary before the introduction of aqueous solvents. To avoid shocking the column bed, it is advisable to introduce wash solvents with gradients over 5-10 column volumes. Viscous solvents, such as dimethylsulfoxide (DMSO)/water, methanol/water, and propanol/water mixtures, should be pumped at low enough flowrates to maintain moderate pressures.

Successful regeneration of a contaminated column can be a time-consuming process. With microprocessor-controlled HPLC systems, multi-step washing sequences can be programmed and run automatically overnight. Alternatively, column regeneration can be carried out off-line with an inexpensive low-pressure pump at reduced flowrates.

C. Column Backflushing

Most columns have arrows or other indications for the normal direction of column flow and they should be operated in that direction. When a column exhibits excessive operating pressure due to partial plugging of the inlet frit or column head, it is sometimes possible to dislodge the material by reversing the direction of flow and backflushing with the mobile phase or a wash solvent. This simple remedy avoids the hazards of opening a terminator; however, it may have other negative consequences. Several precautions about backflushing should be observed. First, some manufacturers do not recommend backflushing of columns; the column installation instructions or manufacturer should be consulted prior to performing the operation. Second, if the pressure rise is accompanied by a loss in efficiency, it may be indicative of voids in the bed. In this case, backflushing will disturb the bed and prevent any further repair by topping off procedures. Third, the column should be disconnected from the detector during backflushing to prevent contaminants or particulates from entering the flow cell. Occasionally, increasing pressure can arise from blockage of the outlet frit, particularly when fines are generated by degradation of silica supports during operation at high pH, high temperature, or with cationic ion-pair agents. Under such conditions, backflushing will produce only a partial or transient reduction in pressure. The outlet frit can sometimes be changed; however, it is likely that the column will need to be replaced in the near future.

D. Repair of Voids and Bed Irregularities

Symptoms such as peak broadening, peak tailing, or split peaks may indicate a void or channel in the column bed. If these are observed with a new column or during the first few injections of normal operation, the column may be damaged or poorly packed. The column manufacturer should be consulted for technical assistance and/or exchange procedures. Voids that develop with extended use can sometimes be repaired by topping off with the identical column packing (12). Original performance is rarely regained, but the improvement in resolution is often acceptable, particularly during gradient elution. Column hardware which allows axial compression for void repair is an excellent solution, especially for preparative ($\geq$ 1cm ID) columns. Large bore columns ($\geq$ 1cm ID) packed with microparticulate column packings are subject to void formation due to column expansion caused by high pressures.

To repair a void, the column should be secured vertically and the inlet terminator removed. Packing material should be level and flush with the face of the column tube. Even small depressions in the bed can result in significant band broadening and should be filled; voids of a centimeter or more probably cannot be repaired successfully. A light gray or brown discoloration of the bed surface is normal after extended use, but dark discoloration, as seen in Fig. 11.2A, indicates contaminated packing material which should be removed before addition of new support.

Figure 11.2 Repairing a Column Bed

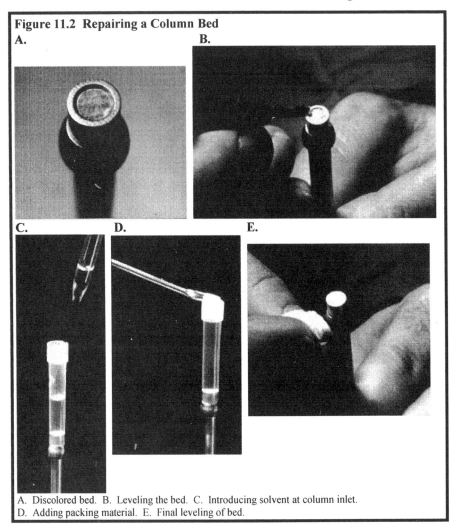

A.

B.

C.   D.   E.

A. Discolored bed. B. Leveling the bed. C. Introducing solvent at column inlet.
D. Adding packing material. E. Final leveling of bed.

Before filling the void, the bed should be leveled with a square blade, as shown in Fig. 11.2B; a small flat-tip laboratory spatula trimmed to the column internal diameter works well. It is best if the void is filled with the same microparticulate packing or, alternatively, a pellicular guard column packing with the same stationary phase. Underivatized glass beads are not recommended because they will adsorb proteins and many other compounds. If a microparticulate packing material is used, it should be prepared as a slurry in a suitable organic solvent and added drop-wise to the column, allowing excess solvent to permeate the bed (Fig. 11.2C). A plastic pipettor tip can be trimmed and fit to the column tube for use as a solvent reservoir. It will probably be necessary to reconnect the end fitting, pump the column, and refill one or more times. Because microparticulate packings have high capacity, the stationary phase should be the same as that of the column to prevent selectivity changes. In many cases, a pellicular packing will serve as the better repair material because it can be added dry, and the loss in efficiency due to a poorly repacked void is not as significant. Once the void is filled, the bed surface should be leveled flush with the face of the tubing (Fig. 11.2D), excess packing material removed from the tubing surface, and the end fitting replaced. If the repaired column does not yield satisfactory efficiency when tested, the end fitting should be removed and the bed checked for settling. If settling has occurred, the top-off procedure should be repeated. Void repair should be considered a temporary measure of prolonging column lifetime and the necessity of column replacement to be imminent.

VI. REFERENCES

1. C.T. Wehr, *Meth. Enzymol.* 104 (1984) 133.
2. G. Wieland, K. Cabrera, and W. Eymann, *LC-GC* 15 (1997) 98.
3. C.T. Mant and R.S. Hodges, "Mobile phase Preparation and Column Maintenance" in *High-Performance Liquid Chromatography of Peptides and Proteins* (ed. C.T. Mant and R.S. Hodges), CRC Press, Boca Raton (1991) 37.
4. J.W. Dolan and L.R. Snyder, *Troubleshooting LC Systems*, Humana Press, New Jersey (1989) 154.
5. M.L. Mayer, *Amer. Lab.,* Jan. (1997) 34.
6. H.A. Claessens, M.A. van Straten, and J.J. Kirkland, *J. Chromatogr. A* 728 (1996) 259.
7. J.J. Kirkland, *Current Issues in HPLC Technology* (May 1997) 46.
8. *Zorbax HPLC Columns for Analytical Biochemistry*, #96051TB, MAC-MOD Analytical.
9. R. Henry, "HPLC Guard Columns" in *High-Performance Liquid Chromatography of Peptides and Proteins* (ed. C.T. Mant and R.S. Hodges), CRC Press, Boca- Raton (1991) 47.
10. H.H. Freiser, M.P. Nowlan, M.N. Schmuck, D.L. Gooding and K.M. Gooding, *The use of a stabilized support with a short alkyl chain for reversed phase preparative chromatography of proteins,* MICRA Scientific, (1988) PB 10.
11. R. Majors, *LC Magazine* 2 (1984) 16.
12. C.T. Wehr, "Sample Presentation and Column Hygiene" in *HPLC of Proteins, Peptides and Polynucleotides* (ed. M.T.W. Hearn), VCH Publishers, New York (1991) 37.

CHAPTER 12
CAPILLARY ELECTROPHORESIS

I. INTRODUCTION

Capillary electrophoresis (CE) is a relatively new separation technology which combines aspects of both gel electrophoresis and HPLC. Like gel electrophoresis, the separation depends upon differential migration in an electrical field. Unlike conventional electrophoresis, however, the separations are usually performed in free solution without the requirement for a casting a gel. Since its first description in the late 1960's, capillary electrophoretic techniques analogous to most conventional electrophoretic techniques have been demonstrated: zone electrophoresis, displacement electrophoresis (isotachophoresis), isoelectric focusing, and molecular sieving separations. As in HPLC, detection is accomplished as the separation progresses, with resolved zones producing an electronic signal as they migrate past the monitor point of a concentration-sensitive detector, such as absorbance or fluorescence, eliminating the need for staining and destaining. Data presentation and interpretation are also similar to HPLC; the output (peaks on a baseline) can be displayed as an electropherogram and integrated to produce quantitative information in the form of peak area or height. In CE as in HPLC, a single sample is injected at the inlet of the capillary and multiple samples are analyzed in serial fashion. This contrasts with conventional electrophoresis in which multiple samples are frequently run in parallel as lanes on the same gel. This limitation of CE in sample throughput is compensated by the ability to process samples automatically using an autosampler. Compared to its elder cousins, CE is characterized by high resolving power, sometimes higher than electrophoresis or HPLC. The implementation of narrow-bore capillaries with excellent heat dissipation properties enables the use of very high field strengths (sometimes in excess of 1000 V/cm), which decreases analysis time and minimizes band diffusion. Electroosmotic flow (EOF), a phenomenon caused by a high-density charged surface, such as that of an uncoated capillary tube, is an additional force with which analytes can be transported down the capillary in the presence of an electrical field. Separations performed in the presence of EOF often exhibit plug-flow characteristics, resulting in high efficiency. This contrasts with the laminar flow properties of liquid chromatography, where resistance to mass transfer can reduce separation efficiency.

Because of its many advantages, CE shows great promise as an analytical tool in the characterization of biological materials. In some cases, it may replace HPLC and electrophoresis, but more often will be used in conjunction with existing techniques, providing a different separation selectivity, improved quantitation or automated analysis. One anticipated benefit of performing protein separations in open-tubular capillaries was thought to be the reduced potential for surface interactions. In fact, this proved not to be the case; the high surface-to-volume ratio of the capillaries and the high surface activity of the fused silica capillary wall has proven to be a major problem in applying CE to protein separations. Much of the research in separation chemistries and capillary wall modifications has been directed towards improving CE performance in protein separations.

This chapter provides an introduction to the separation mechanisms in capillary electrophoresis, a description of CE instrumentation, and some guidelines in selecting conditions for a CE separation. More detailed presentations of CE theory and practice can be found in references 1-7.

II. BASIC CONCEPTS

As the name implies, capillary electrophoresis separates species within the lumen of a small-bore capillary filled with an electrolyte. A schematic of a CE system is presented in Fig. 12.1.

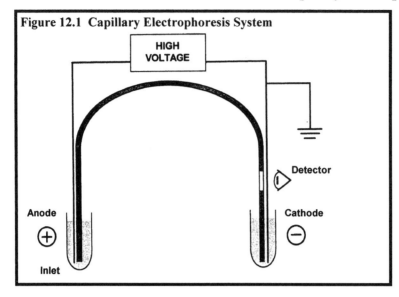

Figure 12.1 Capillary Electrophoresis System

The capillary is immersed in electrolyte-filled reservoirs containing electrodes connected to a high voltage power supply. A sample is introduced at one end of the capillary (the inlet) and analytes are separated as they migrate through the capillary towards the outlet end. As separated components migrate through a section at the far end of the capillary, they are sensed by a detector and an electronic signal is sent to a recording device.

As in conventional gel electrophoresis, the basis of separations in capillary electrophoresis is differential migration of molecules in an applied electric field. The electrophoretic migration velocity (u_{ep}) will depend upon the magnitude of the electric field (E) and electrophoretic mobility (μ_{ep}) of the analyte:

$$u_{ep} = \mu_{ep}E \qquad\qquad\qquad\qquad \text{Eq. 12.1}$$

In a medium of a given pH, the mobility of an analyte is given by the expression:

$$\mu_{ep} = \frac{z}{6\pi\,\eta\,r} \qquad\qquad\qquad\qquad \text{Eq. 12.2}$$

where z is the net charge of the analyte, η is the viscosity of the medium, and r is the Stoke's radius of the protein. Mobility will increase directly with increasing charge and inversely with molecular weight because the Stoke's radius is related to molecular mass. Mobility can be determined from the migration time and field strength by:

$$\mu_{ep} = \left(\frac{L}{t_m}\right)\left(\frac{L_{tot}}{V}\right) \qquad\qquad\qquad\qquad \text{Eq. 12.3}$$

where L is the distance from the inlet to the detection point (termed the effective length of the capillary), t_m is the time required for the analyte to reach the detection point (migration time), V is the applied voltage, and L_{tot} is the total length of the capillary.

In contrast to most forms of gel electrophoresis, the velocity of an analyte in capillary electrophoresis will also depend upon the rate of electroosmotic flow (EOF), also known as electroendosmotic flow. This phenomenon is observed when an electric field is applied to a solution contained in a capillary with fixed charges on the capillary wall. Charged sites are typically created by ionization of silanol groups on the inner surface of the fused silica. Silanols are weakly acidic and ionize at pH values above about 3. Hydrated cations in solution associate with ionized SiO⁻ groups to form an electrical double layer - a static inner Stern or Helmholtz layer close to the surface and a mobile outer layer, termed the Gouy-Chapman layer (8). Upon application of the field, hydrated cations in the outer layer move toward the cathode, creating a net flow of the bulk liquid in the capillary, as seen in Fig. 12.2.

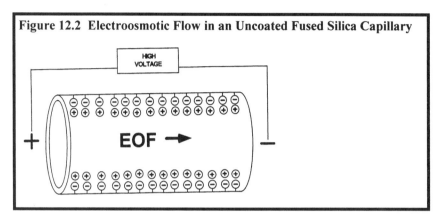

Figure 12.2 Electroosmotic Flow in an Uncoated Fused Silica Capillary

The rate of movement is dependent upon the field strength and the charge density of the capillary wall. The population of charged silanols is a function of the pH of the medium; therefore, the magnitude of EOF increases directly with pH until all available silanols are fully ionized, as shown in Fig. 12.3.

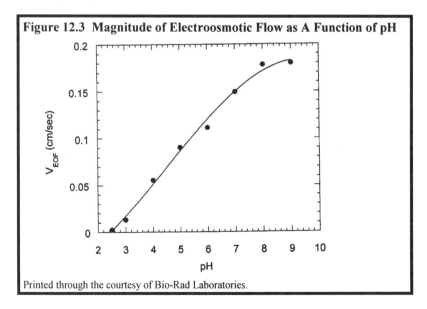

Figure 12.3 Magnitude of Electroosmotic Flow as A Function of pH

Printed through the courtesy of Bio-Rad Laboratories.

The velocity of electroendosmotic flow (u_{eo}) can be expressed as:

$$u_{eo} = \mu_{eo}E \qquad\qquad\qquad \text{Eq. 12.4}$$

where μ_{eo} is the electroendosmotic mobility, defined as:

$$\mu_{eo} = \frac{\varepsilon_r \zeta}{\eta}$$ Eq. 12.5

Where, ζ is the zeta potential of the capillary wall, ε_r is the dielectric constant and η is the viscosity of the medium. Electroosmotic mobility can be determined experimentally by injecting a neutral species and measuring the time t_n when it appears at the detection point:

$$\mu_{eo} = \left(\frac{L}{t_n}\right)\left(\frac{L_{tot}}{V}\right)$$ Eq. 12.6

The apparent velocity (u) of an analyte in an electric field will therefore be the combination of its electrophoretic velocity and its movement in response to EOF:

$$u = u_{ep} + u_{eo} = (\mu_{ep} + \mu_{eo})E$$ Eq. 12.7

In the presence of electroosmotic flow, analytes which possess net positive charge will migrate faster than the rate of EOF; analytes which are isoelectric will be carried toward the cathode at the rate of EOF; and anionic analytes will migrate towards the cathode at a rate which is the difference between their electrophoretic velocity and u_{eo}. If the magnitude of EOF is sufficiently great, all analytes, regardless of their charge state, will migrate past the detection point. In this regard, performing separations in the presence of EOF is highly desirable. However, achieving reproducible separations requires that EOF be constant, and this in turn requires that the surface characteristics of the capillary wall remain consistent from run to run. Proteins are notorious for interacting with silica surfaces and changing the level of EOF. A great variety of capillary surface treatments and buffer additives have been developed for reducing protein adsorption and controlling EOF; these are described in detail in Section V.

Separation efficiency (N) in capillary electrophoresis is given by the following expression:

$$N = \frac{\mu V}{2D_M}$$ Eq. 12.8

where D_M is the diffusion coefficient of the analyte and μ is the apparent mobility in the separation medium. This equation predicts that efficiency is only diffusion-limited and increases directly with field strength. Capillary electrophoresis separations are usually characterized by very high efficiency, often as high as several hundred thousand plates. Efficiency in CE is much higher than in HPLC because there is no mass transfer between phases and the flow profile in EOF-driven systems is flat (approximating plug flow), in contrast to the laminar-flow profiles characteristic of pressure-driven flow in chromatography columns, as shown in Fig. 12.4.

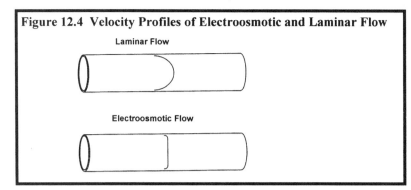

Figure 12.4 Velocity Profiles of Electroosmotic and Laminar Flow

Laminar Flow

Electroosmotic Flow

According to equation 12.8, proteins should exhibit excellent separation efficiencies due to their low diffusion coefficients relative to small molecules. Resolution (R_s) in CE is defined as:

$$R_s = \frac{1}{4}\left(\frac{\Delta\mu_{ep}N^{1/2}}{\mu_{ep}+\mu_{eo}}\right)$$
Eq. 12.9

This implies that resolution will be greatest when μ_{ep} and μ_{eo} are of similar magnitude, but of opposite sign. From Eq. 12.9, it can be seen that high resolution will be at the expense of lower velocity and increased analysis time. It is also evident from Eq. 12.9 that, although Eq. 12.8 suggests that the use of higher field strengths is the most direct route to high efficiency, a doubling of voltage yields only a 1.4-fold increase in resolution, at the expense of Joule heat.

III. BAND BROADENING

Band broadening and the resultant reduction in resolution can arise from several contributing factors. If the total band broadening is expressed as plate height (H), the contributions to band broadening due to the initial zone width, diffusion and electrodispersion, Joule heating, and adsorption can be expressed qualitatively (9) as:

$$H = H_{inj} + H_{diff+cond} + H_{joule} + H_{ads}$$
Eq. 12.10

A. Initial Zone Width (H_{inj})

Optimum resolution will only be obtained if the initial sample zone is kept as small as possible. The starting zone length should not exceed 5% of the total capillary length. In electrokinetic injection, the injection zone length (L_{inj}) can be estimated from the injection time (t_{inj}), the injection and separation field strengths (E_{inj}) and (E_{sep}), and the migration time (t_m) of the peak of interest:

$$L_{inj} = \frac{(L_{eff})(E_{inj})(t_{inj})}{(E_{sep})(t_m)}$$
Eq. 12.11

Sharpening of the sample zone can be achieved by preparing the sample in an electrolyte of lower conductivity than the analysis buffer (10-11). Under these conditions, there is a discontinuity in field strength at the sample/buffer boundary such that ions migrating rapidly from a region of higher field strength become focused at the boundary. This focusing or stacking not only produces narrow zones for increased resolution, but also increases zone concentration for enhanced sensitivity. Sensitivity enhancement is directly proportional to the ratio of buffer to sample conductivity.

In displacement injection, the injection zone volume (V_{inj}) in milliliters can be estimated from the Poiseuille equation:

$$V_{inj} = \frac{\pi P t_{inj} r_c^4}{8 L_{tot} \eta}$$ Eq. 12.12

where P is the pressure in dynes/cm^2, t_{inj} is the injection time in seconds, r_c is the capillary radius in cm, L_{tot} is the total length of the capillary in cm, and η is the viscosity of the electrolyte in poise. Pressure in psi can be converted to dynes/cm^2 using the value of 68947.6 dynes/cm^2/psi. Stacking effects will be observed in displacement injection if the sample electrolyte and analysis buffer differ in conductivity, but the effect is less pronounced than seen in electrokinetic injection. An alternative method of stacking employs differences in sample and buffer pH (12). The sample is prepared in an alkaline solution in which analytes are negatively charged. The sample is pressure-injected into a capillary containing buffer at a lower pH and, when voltage is applied, sample ions stack at the anodic end of the sample zone. Following the stacking phase, the sample electrolyte diffuses away from the zone, and the analytes become positively charged and migrate towards the cathode.

B. Diffusion and Electrodispersion ($H_{diff + cond}$)

Axial diffusion (H_{diff}) can broaden bands in CE, but shortening the analysis time by operating at high field strengths with short capillaries will reduce the effect. Axial diffusion is minimal for polypeptides, which possess small diffusion coefficients.

Electrodispersive band broadening (H_{cond}) arises from conductivity differences between the zone and the background electrolyte. If this difference is large, diffusion at one boundary of the zone is negligible due to the zone sharpening effect caused by the discontinuity in field strength. In this case, the peak will be asymmetric - the peak will exhibit fronting or tailing depending on whether the conductivity of the zone is greater or less than that of the background electrolyte, as illustrated in Fig. 12.5.

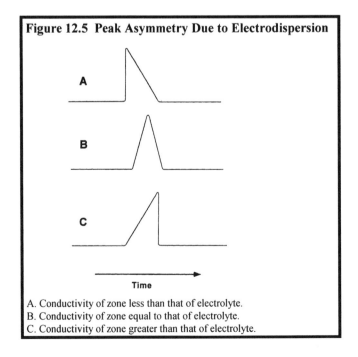

Figure 12.5 Peak Asymmetry Due to Electrodispersion

A
B
C

Time

A. Conductivity of zone less than that of electrolyte.
B. Conductivity of zone equal to that of electrolyte.
C. Conductivity of zone greater than that of electrolyte.

This phenomenon is readily observable when analyzing highly-charged small molecules such as inorganic ions. Satisfactory peak shapes can only be obtained by carefully matching sample and electrolyte conductivities. The effect is much less noticeable in most other applications.

C. Joule Heating (H_{joule})

When an electric field is applied to a capillary containing an electrolyte, Joule heat is generated uniformly across the circumference of the tube. Because heat can only be removed at the margin of the tube, a temperature gradient exists across the radius. As noted in Eq. 12.2, mobility is inversely related to viscosity, which decreases with temperature, resulting in an increase in mobility of approximately 2.5% for each degree rise in temperature. Consequently, the temperature gradient creates a mobility gradient across the tube radius, which contributes to band broadening, as shown in Fig. 12.6.

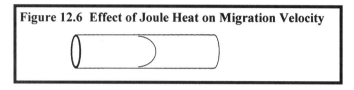

Figure 12.6 Effect of Joule Heat on Migration Velocity

The small internal diameters of fused silica capillaries have been the key to high resolution CE separations. The smaller the capillary bore, the greater the surface-to-volume ratio and the more efficiently heat is removed from the tube. The optimal capillary inside diameter (ID) for most applications is 50 - 75 µm. Capillaries with smaller diameters are subject to plugging and their high surface-to-volume ratios increase the risk of analyte adsorption. Capillaries with larger internal diameters may exhibit significant loss in resolution due to thermal effects, and thus may require operation at low field strengths or with low conductivity buffers.

D. Adsorption (H_{ads})

Analyte-wall interactions have been the greatest obstacle to achieving satisfactory resolution and reproducibility in protein analysis by capillary electrophoresis, especially zone electrophoresis and isoelectric focusing. Approaches to minimizing or eliminating protein adsorption include the use of buffer additives and dynamic or covalently-coupled capillary coatings. These various approaches are discussed extensively in Section V.

IV. INSTRUMENTATION

Commercial CE instruments are available in a wide range, from simple modular systems consisting of a power supply, detector and injection device to fully integrated automated systems under computer control. This section will focus on the features of automated CE instrumentation.

A. Power Supply

Power supplies capable of delivering constant voltage at high precision up to 30kV are standard throughout the industry. Most systems additionally offer constant current operation up to 300µA; constant current operation may be desirable in systems without adequate temperature control. High voltage is applied at the capillary inlet, with the outlet or detector end at ground potential. When separations are performed in uncoated fused silica capillaries in the presence of EOF, the inlet (high voltage) electrode is the anode and EOF carries analytes towards the cathode; this positive-to-negative is termed "normal" polarity. Separations performed in coated capillaries without EOF may require "reversed" polarity. In some separations, the direction of EOF is reversed by using capillaries coated with positively-charged polymers or by using osmotic flow modifying additives in the background electrolyte. In these cases, reversed polarity is also used. All commercial instruments have reversible polarity; however, for frequent switching among different applications, polarity reversal through software is more convenient than manual reversal.

B. Injection

Sample injection in CE requires the introduction of very small amounts of analyte at the capillary inlet with high precision. All commercial instruments offer electromigration injection and at least one type of displacement injection.

1. Electrokinetic

Electromigration is the simplest injection method in CE; the capillary inlet is immersed in the sample solution and high voltage is applied for a brief period of typically a few seconds. If no EOF is present, sample ions enter the capillary by electrophoretic mobility alone. If EOF is present, sample ions will be introduced by a combination of electrophoretic mobility and EOF; this mode is generally termed electrokinetic injection.

Electrophoretic injection offers two advantages. First, only species of like charge will enter the capillary, discriminating against compounds of opposite charge and simplifying the separation problem. Second, zone sharpening can be achieved using the stacking principle described in Section IIIA. Unfortunately, these advantages are countered by two major limitations: sample ions will enter the capillary based on mobility; therefore, low-mobility ions will be present in lower concentrations and have decreased detector response. More importantly, the presence of non-analyte ions in the sample will reduce injection efficiency, making electrophoretic injection very sensitive to the presence of salts or buffers in the sample matrix. The disadvantages of electrophoretic injection argue against its use in routine analysis except in cases where displacement injection is not possible, such as in capillary gel electrophoresis (CGE). Electrokinetic injection suffers from the further disadvantage that many sample matrices contain components like proteins which adsorb to the capillary wall and change the magnitude of EOF.

2. Displacement

Displacement is usually the preferred method of injection because analyte ions are present in the sample zone in proportion to their concentration in the bulk sample, making injection efficiency less sensitive to variations in sample ionic strength. However, it should be noted that the presence of high salt with displacement injection can affect detector response. Additionally, changes in the sample viscosity due to temperature variations or the presence of viscosity-modifying components can affect displacement injection efficiency.

Two modes of displacement injection have been employed in commercial CE instruments: application of positive pressure at the capillary inlet and application of vacuum at the capillary outlet. The former method can employ pressurization of the sample headspace by gas (pressure injection) or application of hydrostatic pressure by elevating the capillary inlet relative to the capillary outlet (gravity injection). Gravity injection has been reported to provide reproducible injection of very small sample zones. On the other hand, pressure injection will allow the introduction of larger sample zones; this can be an advantage when using injection pneumatics to introduce a chiral selector or to perform on-column concentration. Moderate injection pressures are useful when capillaries contain analysis buffers with viscous agents like sieving polymers. In some instruments, high precision in displacement injection is achieved by integrating the pressure signal during the injection process. The operator programs a time-pressure product as a constant, and the instrument applies pressure to the sample until the measured time-pressure product matches the programmed constant. This injection method compensates for variations in seal compliance and sample-to-sample headspace.

C. Temperature Control

Temperature control of the capillary environment is essential for attaining satisfactory reproducibility because inadequate temperature control results in variable migration times. In CE, peak area depends upon the residence time of the component in the detector light path, and therefore, it is dependent upon migration velocity. If migration times vary because of inadequate temperature control, peak area precision will be poor. Control of capillary temperature above or below ambient temperature may be desirable in special applications, such as kinetic studies, on-column enzyme assays, or protein folding experiments. Operation at higher temperatures in capillary zone electrophoresis decreases analysis time and may improve peak shape; however, at high temperatures the risk of protein denaturation and precipitation is increased. Low temperature operation generally has no advantage in most modes of capillary electrophoresis.

Capillary temperature control can be achieved by forced air or a circulating liquid coolant. Forced air control is less efficient, but permits the use of free-hanging capillaries. Liquid cooling requires that the capillary be enclosed in a sealed cartridge; this cartridge format provides for automatic alignment of the capillary in the detector light path and reduces time required for capillary installation when changing methods.

The effectiveness of capillary thermostatting can be determined by the variation in current as a function of voltage. According to Ohm's Law, this should be a linear relationship; deviation from linearity in an Ohm's Law plot is indicative of poor efficiency in heat dissipation by the capillary temperature control system.

D. Detectors

1. Absorbance

As in HPLC, absorbance detection is used in the vast majority of CE applications; all commercial CE systems employ UV or UV-VIS absorbance as the primary mode of detection. The simplest approach is the use of line-source lamps or continuum-source lamps with wavelength selection by filters. Better flexibility is obtained using a continuum source (e.g. deuterium lamp) with wavelength selection by a grating monochromator. Because the low output of a deuterium lamp above 360nm limits sensitivity, a secondary tungsten source is desirable for detection at visible wavelengths.

All commercial CE absorbance detectors employ on-tube detection: a section of the capillary itself is used as the detection cell. This permits detection of separated zones with no loss in resolution. Usually capillaries used for CE are coated with a polymer (usually polyimide) which protects the fused silica and provides it with mechanical stability. Because the polymer is not optically transparent, it must be removed from the detection segment of the capillary to form a "window" which must be accurately positioned in the optical path to achieve good sensitivity. This segment of bare capillary is very fragile and is subject to breakage during manipulation and installation. Capillaries with a UV-transparent coating which eliminates this problem are available from Polymicro Technologies; however, the coating is not resistant to some coolants (e.g. fluorinated hydrocarbons) used in liquid-cooled CE systems.

In on-tube detection, the internal diameter of the capillary forms the detection light path. In accordance with Beer's law, the sensitivity of a concentration-sensitive detector is a direct function of the length of the light path. Consequently, in comparison to an HPLC detector with a 1cm path length, detector signal strength is reduced 200-fold in a CE system equipped with a 50μm ID capillary. Concentration sensitivity can be improved by employing focusing lenses to collect light at the capillary lumen, detecting at low wavelengths (where most analytes have greater absorbance) and using sample-focusing techniques during the injection process. Even under ideal conditions, however, the concentration limit of detection (CLOD) is about 10^{-6}M. Detection sensitivity can be improved by extending the light pathlength at the detection point. In one approach, the capillary ID is increased at the window; these "bubble cell" capillaries can increase sensitivity about threefold. Another approach couples the separation capillary to a detection capillary in which source light is directed axially down a segment of the tube. These "Z-cells" can increase sensitivity 20 - 50 times, depending on the length of the detection segment. Both bubble cells and Z-cells can compromise resolution.

Several commercial CE systems incorporate scanning absorbance detectors, as discussed in Chapter 4. Scanning detection enables on-the-fly acquisition of spectra as analytes migrate through the detection point. This information can assist in the identification of peaks based on spectral patterns, detection of peak impurities by variation in spectral profiles across a peak or determination of the absorbance maximum of an unknown compound. Two different designs are used for scanning detection in CE instruments. In photodiode array (PDA) detectors, after the capillary is illuminated with full-spectrum source light, the light passing through the capillary is dispersed by a grating onto an array of photodiodes, each of which samples a narrow spectral range. In fast-scanning detectors, monochromatic light is collected from the source using a moveable grating and slit assembly and directed to the capillary; light transmitted by the capillary is detected by a single photodiode. Scanning is accomplished by rapidly rotating the grating through an angle to "slew" across the desired spectral range.

2. Fluorescence

Fluorescence detection (see Chapter 4) offers the possibility of high sensitivity and improved selectivity for certain complex samples. This mode of detection requires that the analyte exhibit native fluorescence or contain a group to which a fluorophore can be attached by chemical derivatization. The number of compounds that fall into the former category is small. While many analytes contain derivatizable groups (e.g. amino, carboxyl, hydroxyl), most derivatization chemistries are limited by one or more disadvantages, such as slow reaction kinetics, complicated reaction or cleanup conditions, poor yields, interference by matrix components, derivative instability, interference by reaction side products, or unreacted derivatizing agent. Fluorescence detection of proteins usually requires derivatization because amino acids exhibiting native fluorescence (Tyr, Trp) may be absent or sparsely represented in the protein. Unfortunately, most proteins possess multiple reactive sites which, if incompletely derivatized, can yield a family of products varying in the number of fluorophores. The reactive sites are usually side-chain amino groups, and derivatization products which vary in mass and charge can elute as multiple peaks or as a single broad peak.

When compared to fluorescence detectors for HPLC, the design of a fluorescence detector for CE presents some technical problems. In order to obtain acceptable sensitivity, it is necessary to focus sufficient excitation light on the capillary lumen. This is difficult to achieve with a conventional light source, but is easily accomplished using a laser. The most popular source for laser-induced fluorescence (LIF) detection is the argon ion laser which is stable and relatively inexpensive. The 488nm argon ion laser line is close to the desired excitation wavelength for several common fluorophores. The CLOD for a laser-based fluorescence detector can be as low as 10^{-12}M.

3. On-line Coupling With Mass Spectrometry

With the increasing need to obtain absolute identification of separated components and the gradual price reduction of mass spectrometers (MS), there is a growing demand for direct coupling of CE with MS. The most common configuration is introduction of the capillary outlet into an electrospray interface (ESI) coupled to the mass spectrometer (13). In this configuration, the outlet electrode of the CE is eliminated and the MS becomes the ground. Because the volumetric flow out of the capillary is negligible, separated components are usually transported from the capillary to the electrospray using a liquid sheath flow. The major limitation in CE-ESI/MS is the requirement for volatile buffers which narrows the choice of CE separation modes and resolving power.

E. Preparative Capillary Electrophoresis

Because of its high resolving power, CE is often considered for micropreparative isolation. Many of the commercially available CE systems have the capability of automatic fraction collection. Unfortunately, the desirability of using CE as a preparative tool has to be carefully weighed against the problems encountered in fraction collection. When using narrow-bore (e.g. 50μm ID) capillaries, the volume injected into the capillary is quite small, typically a few nanoliters. Unless the analyte is in very high concentration, recovery of sufficient material will require numerous repetitive injections of the same sample and highly reproducible run-to-run migration times to ensure accurate collection. Additionally, the recovered analyte must be stable for the time required to collect the desired amount of material (often several hours) under the collection conditions. An alternative strategy is the use of larger-diameter capillaries (≥ 75μm ID); however, thermal effects may compromise resolution, and low voltages or low-conductivity buffers may be necessary to prevent excessive heating.

V. CONTROLLING ELECTROOSMOTIC FLOW

Controlling electroosmotic flow (EOF) has been the major challenge in achieving reproducible separations in capillary electrophoresis. At pH values above about 2, the weakly acidic silanol groups on the capillary surface become ionized, and the charge density on the wall increases with pH to a maximum at about pH 10 where the silanol groups are fully dissociated, as was illustrated in Fig. 12.3. This characteristic of fused silica has two consequences for the CE separation. First, sample species which are cationic or have cationic groups can have electrostatic interactions with ionized silanols. Proteins are particularly susceptible to wall interactions due to the presence of basic amino acid residues on their surface. Protein adsorption at the capillary wall can result in band broadening, tailing, and, in the case of strong interaction, reduced detector response or complete absence of peaks. In addition, changes in the state of the wall either during an analysis or run-to-run can alter the magnitude of EOF, resulting in variance in analyte migration times and peak areas. Protein adsorption can alter the zeta potential of the capillary wall, changing the EOF and degrading reproducibility. Three strategies have been employed to minimize protein-wall interactions: operation at extremes of pH, use of buffer additives, and use of wall-coated capillaries.

A. Operation at pH Extremes

The simplest approach to minimizing protein-wall interaction is to use a buffered pH at which interactions do not occur. Under alkaline conditions, neutral and acidic proteins (pI ≤ 7) will carry net negative charges and the capillary surface will also be anionic due to silanol ionization; therefore, protein-wall adsorption will be prevented by Coulombic repulsions. The same approach can be used for basic proteins (pI 7 - 10) using very alkaline buffers (pH 11 - 12) (14), but inadequate selectivity and the risk of protein degradation at high pH has prevented it from being a general strategy for protein separations. An alternative strategy is operation at very low pH (pH < 3). Under these conditions, the degree of silanol ionization and thus, the magnitude of EOF, is very low. At such low pH, most proteins will bear a net positive charge, and in the absence of significant EOF, migrate electrophoretically towards the cathode. Phosphate buffers have proven to be very effective for operation in this pH range due to their high buffering capacity and low UV absorbance. In addition, complexation of phosphate groups with surface silanols may contribute to reduced EOF, as well as reduced polypeptide adsorption (15).

Although use of low pH conditions has been remarkably successful for separations of complex mixtures of peptides, the approach has been less successful for proteins due to limited selectivity. In the case of protein analysis, successful application of CE to most separation problems requires the ability to operate under conditions where protein charge differences are greatest and, preferably, where the proteins are in their native states. Consequently, there have been tremendous efforts to develop conditions for protein analysis at physiological pH in which protein adsorption is minimized, and EOF is either minimized or adequately controlled.

B. Use of Buffer Additives

The use of bare fused silica capillaries has advantages in simplicity, low cost, and good capillary lifetime. For this reason, many investigators have searched for buffer components which permit protein separations to be achieved with uncoated capillaries.

Monovalent alkali metal salts have been used to reduce protein adsorption (16), but this approach has limitations. The effective concentration of salt is usually high enough to generate both excessive Joule heat and high background UV absorbance. These problems can be circumvented by use of capillaries with small internal diameter or operation at lower field strengths, but these approaches compromise detection sensitivity and analysis time.

A variety of amine-containing organic bases have been used as additives to reduce protein adsorption. Alkylamines added to the electrophoresis buffer under conditions where they are protonated interact with silanols on the capillary wall, stabilizing the state of the double layer and the level of EOF (17, 18). Protein-wall interactions are thus reduced, resulting in improved peak shape and separation efficiencies. At very low pH, alkylamine additives reverse the direction of EOF towards the anode (19, 20) and increase the operating current. Sometimes protein-additive interactions may advantageously affect the separation, as in the analysis of protein glycoforms (21, 22).

Zwitterions, such as sarcosine and betaine, have been used successfully as buffer additives for CE (23) because they reduce protein-wall interactions without increasing the conductivity of the electrophoresis buffer. Ideally, such additives should be zwitterionic over a wide pH range, have good solubility to enable use at high concentrations, produce a stable EOF, and exhibit low UV absorbance.

C. Use of Wall-Coated Capillaries

The most widely employed strategy for control of EOF is the use of physically adsorbed or covalently attached coatings on the capillary wall (24, 25). Neutral hydrophilic polymers, such as polyethylene glycol, polyvinyl alcohol, and alkylated celluloses (methylcellulose, hydroxyethylcellulose, hydroxypropylmethylcellulose) are typically used for physically adsorbed coatings. These materials are applied by purging the capillary with an aqueous solution of the polymer prior to use. To prevent exposure of sites caused by coating bleed, a low concentration of the polymer is typically added to the electrolyte used for separations so that the coating is continuously replaced. Capillaries coated with the same materials used for gas chromatography columns (e.g. DB-1, DB-17 and DB-WAX) have also been used for CE applications (26).

In some cases, charged polymers are used as capillary coatings for CE. These are typically polyamine compounds which bind tenaciously to the fused silica surface (27). Under acidic to neutral conditions, a charge reversal is produced along the capillary wall, switching the direction of EOF towards the anode. This technique has been used for separation of basic proteins which do not adsorb to the cationic coating.

Covalently-attached coatings have the advantage of improved stability relative to adsorbed coatings, eliminating the problem of bleed and the necessity for polymer additives in the separation electrolyte. Some covalent coatings are composed of short-chain ligands with interactive groups, such as alkylhydrocarbons or alkylamines. More typically, they are neutral hydrophilic polymers which reduce both EOF and analyte adsorption. Formation of a polymeric coating is usually a two-step process in which a bifunctional silane is reacted with silanol groups on the wall surface using the same chemistry employed for synthesis of HPLC stationary phases (*see* Chapter 5). The second functional group is then used to graft the desired polymer, which can be crosslinked to add stability. An alternative approach employs a silane with a vinyl group which can be reacted with a monomer, such as acrylamide, to form a linear polyacrylamide (LPA) coating (28); this can be crosslinked by adding N,N'-methylene-bis-acrylamide to the reaction. Polymeric coatings have two advantages. First, they form a viscous layer at the wall surface, which is more effective than thinner layers in reducing EOF. Second, the thick polymer layer tends to protect the siloxane bond anchoring the coating to the wall and blocks access of analytes to residual adsorptive sites on the wall.

A hybrid coating procedure, in which an adsorbed coating was applied to an octyldecyl ligand attached to the wall to form a hydrophobic covalent coating, was described by Towns and Regnier (29). In this procedure, a neutral surfactant like Brij® was adsorbed to the covalent coating by hydrophobic interactions, producing a dynamic hydrophilic coating by the polar surfactant head groups on the interior surface of the capillary. For this coating, a low concentration of surfactant was required in the separation electrolyte to maintain the level of the adsorbed coating.

A disadvantage of covalent coatings is their instability under harsh conditions. The siloxane anchoring bond is unstable under alkaline conditions, and even the most stable coatings will eventually degrade after prolonged operation at high pH. The amide bond of polyacrylamide will also hydrolyze at high pH to generate acrylic acid groups. This has led to the use of other monomers such as N-acryloylaminoethoxyethanol (AAEE) and N-acryloylaminopropanol (AAP) to form alkaline-resistant polymeric coatings (30, 31).

VI. SEPARATION MODES

One of the major advantages of CE as a separation technique is the wide variety of separation modes available. Analytes can be separated on the basis of charge, molecular size or shape, isoelectric point, or hydrophobicity. The same CE instrument can be used for zone electrophoresis, isoelectric focusing, sieving separations, isotachophoresis, and chromatographic techniques such as micellar electrokinetic chromatography and capillary electrokinetic chromatography. This section provides a description of each separation mode with an emphasis on those used for separation of proteins and other biological macromolecules.

A. Capillary Zone Electrophoresis (CZE)

Capillary zone electrophoresis (CZE) is a simple and straightforward method which offers several advantages over other CE separation modes. In most cases, a single buffer is used throughout the capillary and electrode vessels, and the sample is introduced as a zone or plug at one end. Capillary preparation often involves merely filling the capillary with the separation buffer, although uncoated capillaries generally require prior washing or conditioning steps. The technique is inexpensive in comparison to capillary gel electrophoresis or isoelectric focusing because only low-cost buffers and salts are used. Separation selectivity can be easily varied by manipulation of buffer pH or use of additives. For these reasons, CZE is the most widely-used CE mode.

1. Method Development

The development of a CZE separation is typically a three-step process: determination of the optimum buffer composition, selection of the appropriate capillary type, and determination of high-voltage power supply parameters.

a. Buffer selection

i. pH

Selection of the analysis buffer is dictated by the pH required to achieve satisfactory analyte mobility and resolution. In the separation of proteins, the pH of the buffer should be at least one pH unit above or below the isoelectric point (pI) of the analyte protein; at pH values closer to the protein pI, low mobility will result in long analysis times, peak broadening, and increased risk of protein-wall interactions. To achieve satisfactory resolution, the buffer pH should provide satisfactory differences in the mass-to-charge ratios of the sample proteins so that they exhibit significant differences in mobility. In many applications, a sample may contain several proteins with such a wide range of isoelectric points that some components may take a very long time to reach the detection point, or may (in the absence of EOF) migrate in the opposite direction. In such cases, two strategies can be used: 1. the separation can be performed at extremes of pH where all proteins possess net positive charge (pH 2 - 2.5) or net negative charge (pH 10 - 11); or 2. the separation can be run twice, at different polarities. When operating under alkaline conditions with uncoated capillaries, the level of EOF will sweep acidic proteins past the detection point, allowing separations of mixtures of acidic and basic proteins using positive-to-negative polarities. In this case, however, adsorption of basic proteins to the silica surface may alter the level of EOF and lead to poor reproducibility.

ii. Buffer

The buffer for CZE should have enough buffering capacity at the selected pH to provide good reproducibility, at low ionic strength to minimize Joule heat. Low conductivity buffering ions are also preferable to minimize thermal effects.

Buffers with low absorbance in the UV region must be used to achieve low background noise and satisfactory detection limits. For conventional gel electrophoresis where proteins are stained following the separation, absorbance is not a limitation in buffer selection. Because CE employs on-tube absorbance detection, many of the buffers commonly used in gel electrophoresis are unsuitable for CE due to their unacceptably high background at 200nm. In some cases, satisfactory detection limits can be achieved by the use of slightly longer wavelengths (210 - 220nm) to reduce background noise significantly without a serious decrease in analyte signal.

A list of buffer systems with characteristics relevant to their use for CZE separations is provided in Table 12.1.

Table 12.1 Buffer Systems for CE						Absorbance	
Buffer Ion	Counter Ion	Molarity (mM)	pH	pK of Buffer Ion	Current	200nm (mAU)	220nm (mAU)
Phosphate	Sodium	100	2.0	2.12	66	3	3
Phosphate	Tris	50	2.0	2.12	37	3	0
Citrate	Citrate	100	3.0	3.06	25	807	47
Formate	Tris	50	4.0	3.75	24	342	70
Acetate	Tris	50	5.0	4.75	21	612	46
MES	Sodium	100	6.0	6.15	17	2458	273
MES	Tris	100	6.0	6.15	16	2356	134
MOPS		100	7.0	7.20	27	2564	344
Taurine	Sodium	100	8.0	8.95	9	504	28
Taurine	Sodium	200	8.0	8.95	17	1008	28
Phosphate	Tris	100	8.0	8.30	35	1000	0
Bicine		100	8.0	8.35	20	3260	2368
HEPES	Tris	100	8.0	8.00	27	2752	1105
Boric Acid	Sodium	100	8.5	9.24	18	-0.15	-0.34
CHES		100	9.0	9.50	17	2025	270
Phosphate	Tris	200	9.0	8.30	19	2363	41
Beta Alanine	Sodium	100	10.0	10.19	20	2272	160
CAPS	Sodium	100	10.0	10.40	13	1908	281
Glycine	TEA	40	10.2	9.60	20	1612	192
CAPS		50	11.0	10.40	25	2429	351

Current measured using a 24 cm x 50 μm capillary at 10,000 V
Absorbance measured against water using a 24 cm x 50 μm capillary
Printed through the courtesy of Bio-Rad Laboratories.

The current and absorbance values in this table were determined under typical CZE analysis conditions (24cm x 50μm ID capillary operated at 10 kV). The properties of these buffers illustrate the difficulties in selecting a satisfactory CZE buffer for a specific pH range. The zwitterionic buffers MES, CAPS, CHES, β-alanine, and bicine are useful for their low conductivity, but they exhibit high absorbance in the low UV region. In the pH 2 - 3 range, phosphate buffers are excellent due to their low UV absorbance and low conductivity, but they exhibit high conductivity at pH 6 - 7. Taurine buffers offer satisfactory buffering capacity, relatively low UV absorbance, and low conductivity in the pH 8 - 8.5 range. Borate buffers are excellent choices for the pH 8 - 9 range because of their low UV absorbance, good buffering capacity, and low conductivity.

iii. Additives

It is often necessary to include additional components in the analysis buffer to achieve a satisfactory separation. Additives such as neutral salts, zwitterions, alkylamines, and neutral polymers can be used to reduce or control EOF and protein adsorption, as described above. Other additives may be used to solubilize proteins or to reduce protein aggregation, especially in the analysis of transmembrane, structural, or large proteins with significant hydrophobic surface areas. Solubilizing agents may also be necessary when separating core fragments generated by cyanogen bromide cleavage. Chaotropes like guanidinium chloride or urea are sometimes used in chromatographic mobile phases to improve protein solubility and recovery, but guanidinium salts are impractical in CE because they cause excessive Joule heating.

Urea can be used in concentrations up to 8M; however, high concentrations of urea introduce high background absorbance levels. In such cases, it may be necessary to increase the detection wavelength to 215 - 220nm to reduce the baseline absorbance. An alternative approach is to use 7 - 8M urea in the sample to solubilize the proteins and to include a lower concentration (4 - 6M) in the analysis buffer to maintain protein solubility. Use of urea in CE systems requires great care to achieve good instrument performance. When used at such high concentrations, urea readily contaminates CE components, producing current leakage and arcing. Attention to system cleanliness is extremely important to avoid these problems. The high voltage electrodes and capillary housing should be inspected daily, and if urea crystallization is observed, the contaminated surfaces should be cleaned with deionized water and dried. Exposure of buffer-containing capillaries to the air will result in urea crystallization and capillary plugging. After use, capillaries should be immediately purged with water or stored with the tips immersed in buffer. Because urea is very water-soluble, plugged capillaries can often be recovered by immersing the capillary tips in warm (70°C) deionized water for 30min or in water at ambient temperature overnight. If this treatment is unsuccessful, the capillary tips can be immersed in a sonic bath containing deionized water for several minutes.

Surfactants, which can be added to the buffer to improve the solubility of proteins, should be selected for their solubilizing power and low UV background. Because ionic surfactants may contribute to Joule heating, zwitterionic or neutral surfactants may be preferable. Brij 35 and Triton X-100 have been used with good success at concentrations up to 1% (w/v). Only the reduced (hydrogenated) form of Triton X-100 should be used so that there is low baseline noise; this form exhibits similar surfactant properties to unreduced Triton X-100 (32). A number of the common surfactants used for solubilizing proteins, with characteristics relevant to their use in CZE, are listed in Table 12.2.

Table 12.2 Surfactant Performance Characteristics for CZE

Surfactant	Molecular Weight	Aggregation Number	CMC (mM)	Current	Absorbance 200 nm (mAU)	220 nm (mAU)
Brij 35	1200	40	0.05-0.1	5.19	5.2	1.5
CHAPS	615	4-14	6-10	0.40	102.5	3.7
Cholate, Sodium	431	2	9-15	11.63	17.6	0.4
Deoxycholate, Sodium	415	3-12	2-6	12.81	17.3	0.2
Dodecylsulfate, Sodium	289	62	7-10	11.20	-0.7	-0.6
Octylglucoside	292	84	20-25	0.61	-0.4	-0.6
Taurodeoxycholate, Sodium	522	6	1-4	9.23	109.4	4.2
Triton X-100, Reduced	631	-	0.25	2.20	13.8	2.5
Tween 20	1228	16.7	0.059	0.38	5.8	2

Current and absorbance of 1% aqueous solution measured using a 24 cm x 50 μm capillary at 10,000 V
From manufacturers' literature and through the courtesy of Bio-Rad Laboratories.

Organic solvents may be used as additives for hydrophobic species which have poor solubility in aqueous electrolytes. Glycerol can be added in concentrations up to 20%. Because this additive will increase the buffer viscosity, it may require adjustments to capillary purge volumes and injection parameters when using displacement injection (pressure, vacuum or gravity). Alcohols such as methanol, ethanol, propanol, or isopropanol may also be used as buffer modifiers. Like glycerol, they will change the buffer viscosity. Additionally, short-chain alcohols exhibit significant vapor pressure and will evaporate if buffer reservoirs are not tightly capped. Solvent evaporation can result in variability, both in injection volumes due to changing viscosity and in migration times due to changing buffer conductivity.

Resolution of stereoisomers can be achieved by CZE by using additives which modulate the migration rate of one enantiomer relative to the other; the use of such chiral selectors is described in more detail in Section VI.A.2.c.

Buffers should be filtered and degassed prior to use by vacuum filtration or by centrifugation for 2 - 5min in a microcentrifuge. It is advisable to store buffers at 4°C to prevent microbial growth because even slight contamination by microorganisms can cause capillary plugging and detector interference.

b. Capillary Selection

Fused silica capillaries are almost universally used in capillary electrophoresis. The inside diameter (ID) of fused silica capillaries varies from 20 - 200µm and outside diameter (OD) varies from 150 - 360µm. The capillary is coated externally with polyimide to provide mechanical stability but this opaque polymeric coating must be removed at the detection point. This is most easily accomplished by heating the detection segment in a flame, by dripping hot concentrated sulfuric acid on the coating, or by scraping with a sharp blade such as a scalpel. This detection "window" is very fragile; therefore, after the window is created, the capillary must be handled with great care to prevent breakage. Capillaries coated with a UV-transparent cladding are commercially available and can be used directly without removal of the coating. However, the coating is not stable in high concentrations of organic solvents and is attacked by the fluorocarbon liquid used as a capillary coolant in some CE instruments.

Selection of the capillary ID is a tradeoff between resolution, sensitivity and capacity. The best resolution is achieved by reducing the capillary diameter to maximize heat dissipation; however, maximum sensitivity and capacity are achieved with large internal diameters. A capillary ID of 50µm is optimal for most applications. Diameters of 75 - 100µm may be needed for high sensitivity or micropreparative applications; however, those greater than 75µm exhibit poor heat dissipation and may require use of low-conductivity buffers and low field strengths to avoid excessive Joule heating. Small bore capillaries (ID ≤ 25 µm) are prone to plugging and may increase the risk of protein-wall interactions due to their high surface-to-volume ratios.

Capillaries with outside diameters of 360µm are the most widely used, although smaller OD capillaries are commercially available. Capillaries with smaller OD and large ID (e.g. 50µm ID x 150µm OD) exhibit great fragility due to reduced wall thickness and may fracture when subjected to high voltage. Short capillaries with large ID are susceptible to siphoning which may compromise reproducibility and introduce laminar flow band broadening.

Selection of capillary length is dictated by the type of capillary and the required resolution. Analysis on coated capillaries with insignificant EOF may only require relatively short effective lengths (20 - 30cm, inlet to detection point). Uncoated capillaries, using conditions where there is appreciable EOF, may necessitate longer lengths (≥ 50 cm) to achieve a separation, particularly for basic proteins which are migrating towards the detector under the combined forces of EOF and electrophoresis. Capillary coatings, described in detail in Section V.C., overcome these problems by being neutral to suppress protein adsorption, or positively-charged to reverse the direction of EOF and reduce adsorption of basic proteins.

c. High Voltage Parameters

High voltage parameters include the mode of operation, field strength, and polarity. Commercial CE systems can be operated in constant voltage, constant current or, in some instruments, constant power mode. The vast majority of protein separations reported in the literature have been performed in constant voltage mode. With well-designed liquid or forced-air temperature control systems, constant voltage operation provides good reproducibility; however, in systems without adequate temperature control, constant current operation may be preferable. The constant power mode allows separation time to be minimized without excess heat generation, but there are very few reports of separations achieved in this manner.

Selection of field strength is a tradeoff between resolution and analysis time. Operation at high field strength reduces analysis time but increases band broadening due to thermal effects; operation at low field strength reduces heating, but increases analysis time and band broadening due to diffusion and adsorption. Generally, operation at 400 - 600 V/cm provides optimal separations in terms of speed and resolution. Selection of the polarity depends on the sample composition and capillary type. For uncoated capillaries under neutral to alkaline conditions, "normal" polarity (positive-to-negative) is used. When uncoated capillaries or capillaries coated with a neutral material are used with an acidic analysis buffer (pH 2 - 3), positive-to-negative polarity should also be used because all the proteins will carry a net positive charge. If coated capillaries are used at pH 3 - 9, polarity selection will depend on the isoelectric points of the proteins of interest. When using capillaries coated with cationic functionalities, the direction of EOF will be reversed, requiring operation with negative-to-positive ("reversed") polarity.

2. Applications

a. Peptides

Application of CZE to peptide separations has been remarkably successful. Under acidic conditions, peptides are usually very stable and exhibit a net positive charge so that they can be analyzed with normal polarity in the absence of EOF. Because the wall is essentially uncharged under acidic conditions, adsorption is seldom a problem. Separation of peptides with subtle structural differences is often possible, and since the separation selectivity (mass/charge) is different from that of reversed phase HPLC (hydrophobicity), CZE is often used as a complementary technique for determination of peptide purity or characterization of peptide digests. Fig. 12.7 illustrates the analysis of a tryptic digest by CZE.

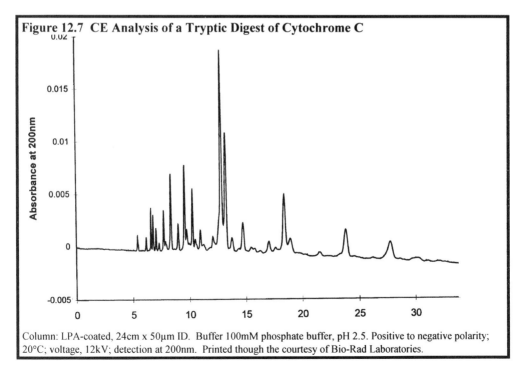

Figure 12.7 CE Analysis of a Tryptic Digest of Cytochrome C

Column: LPA-coated, 24cm x 50μm ID. Buffer 100mM phosphate buffer, pH 2.5. Positive to negative polarity; 20°C; voltage, 12kV; detection at 200nm. Printed though the courtesy of Bio-Rad Laboratories.

b. Proteins

Successful application of CZE to the separation of proteins has been limited by the problems of wall interactions, as discussed previously. In some cases, operation at extremes of pH to minimize the interactions has been successful, but poor protein stability under these conditions has limited this as a general strategy for protein separations. In most cases, inclusion of electrolytes with ionic or zwitterionic additives or coated capillaries are necessary for success. In spite of these limitations, there is a large and growing literature on the use of CZE for protein analysis (25). Some applications include characterization of milk and dairy proteins (33), protein profiling of cereal grain extracts for cultivar identification (34), and separation of metalloproteins (35). Fig. 12.8 shows the comparison of milk components from two cows fed the same controlled diet.

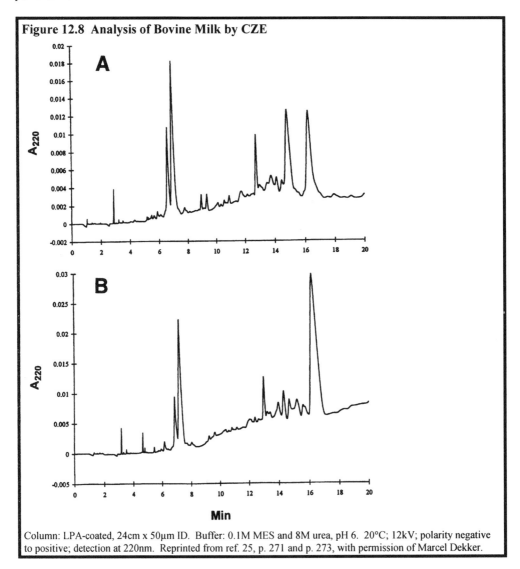

Figure 12.8 Analysis of Bovine Milk by CZE

Column: LPA-coated, 24cm x 50µm ID. Buffer: 0.1M MES and 8M urea, pH 6. 20°C; 12kV; polarity negative to positive; detection at 220nm. Reprinted from ref. 25, p. 271 and p. 273, with permission of Marcel Dekker.

Fig. 12.9 compares protein components of sister lines of the wheat cultivar, Rawhide. The sister lines shown in traces 1-5 carry the 1BI.1RS rye protein chromosomal translocations; those in traces 6-10 do not carry it. Protein peaks characteristic of the translocation migrate as a doublet around 13min.

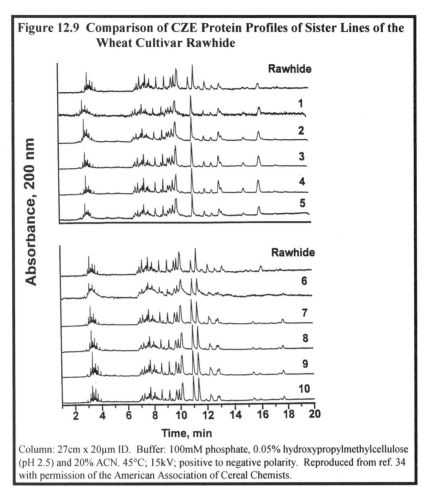

Figure 12.9 Comparison of CZE Protein Profiles of Sister Lines of the Wheat Cultivar Rawhide

Column: 27cm x 20μm ID. Buffer: 100mM phosphate, 0.05% hydroxypropylmethylcellulose (pH 2.5) and 20% ACN. 45°C; 15kV; positive to negative polarity. Reproduced from ref. 34 with permission of the American Association of Cereal Chemists.

A very important application in the development and quality control of biopharmaceutical products is the separation of protein glycoforms, such as that seen in Fig. 12.10 (22, 36). The labeled peaks indicate individual glycoforms of the dimeric protein, which differ by one mannose residue at the single glycosylation site on each monomer. Glycoforms which possess the same number of sugar residues were not resolved under these conditions, resulting in the separation of the 15 glycoforms into nine peaks.

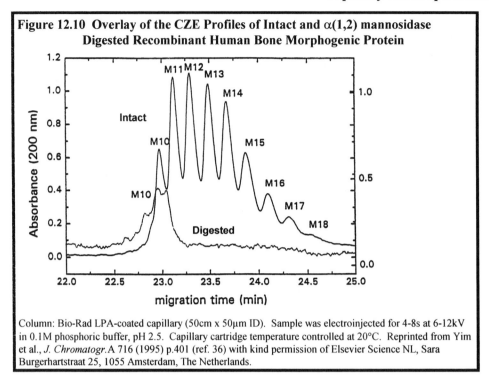

Figure 12.10 Overlay of the CZE Profiles of Intact and α(1,2) mannosidase Digested Recombinant Human Bone Morphogenic Protein

Column: Bio-Rad LPA-coated capillary (50cm x 50μm ID). Sample was electroinjected for 4-8s at 6-12kV in 0.1M phosphoric buffer, pH 2.5. Capillary cartridge temperature controlled at 20°C. Reprinted from Yim et al., *J. Chromatogr.* A 716 (1995) p.401 (ref. 36) with kind permission of Elsevier Science NL, Sara Burgerhartstraat 25, 1055 Amsterdam, The Netherlands.

Affinity capillary electrophoresis has been used to separate receptor-ligand complexes from free receptor and ligand by changes in mass-charge ratios in the complex. This technique, which has been applied to protein receptors and drug ligands (37, 38), as well as antibody-antigen interactions (39), is useful in the determination of binding kinetics and for quantitative assay of ligand concentration. On-column assay of enzyme activity has been demonstrated by electrophoretically migrating a substrate through a zone of enzyme to form a detectable product at the monitor point of the capillary (40-42). CZE has also been used to monitor protein folding, taking advantage of the change in migration behavior induced by conformational changes (43).

c. Chiral Separations

CZE is often used to separate low molecular weight ionic drugs from impurities or metabolites. It is being investigated in clinical research laboratories as a technique for therapeutic drug monitoring and determination of drugs of abuse in biological fluids. In the pharmaceutical industry, CZE is being evaluated for chiral separation of drugs. Resolution of enantiomers in CE is accomplished by incorporating into the electrolyte an additive which complexes with the analytes and selectively moderates the migration of one enantiomer relative to the other (44). The most frequently used chiral selectors are cyclodextrins, cyclic oligosaccharides which have a central hydrophobic cavity. Certain enantiomers form inclusion complexes within the cavity, retarding the isomer with higher affinity for the cyclodextrin. The separation is optimized by selecting a cyclodextrin with properties such as a cavity diameter tailored to the dimensions of the analyte, or functional groups on the cavity rim which enhance stereoselectivity. Charged cyclodextrins with carboxyl or sulfate groups can complex with neutral molecules, enabling their electrophoretic separation as inclusion complexes. Other chiral selectors used for CE are macrocyclic antibiotics such as vancomycin, amino acids, crown ethers, and proteins like cellulase and albumin. Fig. 12.11 shows the separation of catecholamine enantiomers using β-cyclodextrin as the chiral selector.

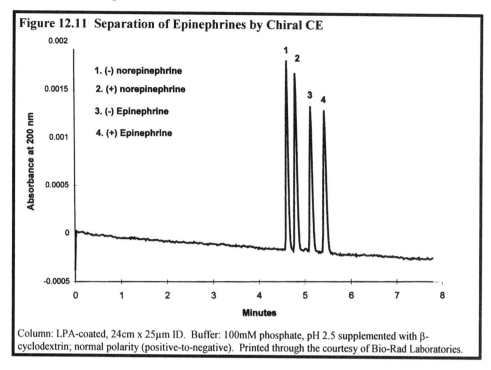

Figure 12.11 Separation of Epinephrines by Chiral CE

1. (-) norepinephrine
2. (+) norepinephrine
3. (-) Epinephrine
4. (+) Epinephrine

Column: LPA-coated, 24cm x 25μm ID. Buffer: 100mM phosphate, pH 2.5 supplemented with β-cyclodextrin; normal polarity (positive-to-negative). Printed through the courtesy of Bio-Rad Laboratories.

d. Ions

Analysis of inorganic ions by CE is a challenge because of their high mobility and low detectability. When injected into an uncoated capillary under high EOF conditions, inorganic anions exhibit high mobility towards the anode and negligible net mobility towards the detector. To circumvent this problem, a cationic surfactant which binds to the silica surface of the capillary and reverses the charge on the wall is added to the electrolyte (45). Under these conditions, the direction of EOF is towards the anode and anions migrate towards the detection point under the summed influence of EOF and their native electrophoretic mobility. To circumvent the lack of UV absorbance, the technique of indirect detection is used (45): a UV-absorbing co-ion of the same charge as the analyte is used as the electrolyte. Displacement of the co-ion forms a negative peak with a magnitude proportional to the analyte ion concentration. Negative peaks can be inverted electronically prior to integration. Fig. 12.12 illustrates the resolution of eleven common ions using pyromellitic acid (1,2,4,5 tetracarboxylic benzene) as the co-ion.

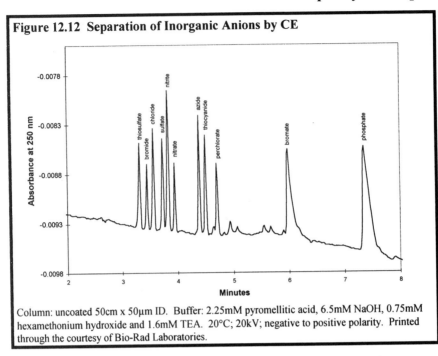

Figure 12.12 Separation of Inorganic Anions by CE

Column: uncoated 50cm x 50µm ID. Buffer: 2.25mM pyromellitic acid, 6.5mM NaOH, 0.75mM hexamethonium hydroxide and 1.6mM TEA. 20°C; 20kV; negative to positive polarity. Printed through the courtesy of Bio-Rad Laboratories.

Organic acids can similarly be analyzed with indirect detection and reversed EOF. The common co-ions used are chromate and aromatic carboxylic acids such as pyromellitic acid or phthalic acid. Inorganic cations can be analyzed by indirect detection using aromatic amines as electrolyte co-ions (46); EOF reversal is unnecessary in this case. For adequate separation of inorganic cations, as shown in Fig. 12.13, complexing agents like hydroxyisobutyric acid, lactic acid or crown ethers may be required.

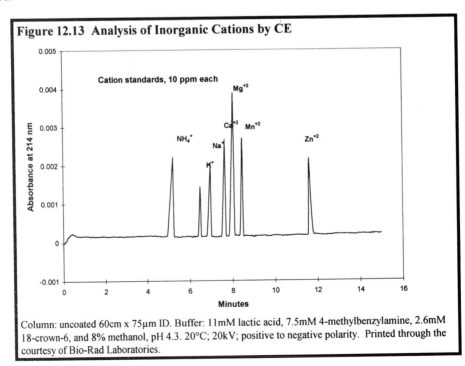

Figure 12.13 Analysis of Inorganic Cations by CE

Column: uncoated 60cm x 75µm ID. Buffer: 11mM lactic acid, 7.5mM 4-methylbenzylamine, 2.6mM 18-crown-6, and 8% methanol, pH 4.3. 20°C; 20kV; positive to negative polarity. Printed through the courtesy of Bio-Rad Laboratories.

B. Capillary Isoelectric Focusing (CIEF)

Capillary isoelectric focusing (CIEF) is similar in concept to conventional gel IEF - using carrier ampholytes, a stable pH gradient is formed in the capillary and proteins become focused at their isoelectric points. The major difference in performing IEF in the capillary format is the requirement for moving the focused protein zones past the detection point (47, 48). Thus, CIEF procedures typically include three steps: sample preparation, focusing, and mobilization.

1. Sample Preparation

Sample preparation consists of mixing the sample with a solution of ampholytes of the appropriate composition and concentration. For screening experiments or for separation of complex mixtures of proteins over a wide range of isoelectric points, a broad range ampholyte solution (pI 3 - 10) should be used. For high resolution separation of proteins with similar pI values, a narrow range of ampholytes covering 1 - 2 pI units is more appropriate. The final ampholyte concentration should be 1 - 2% (w/v). Unlike other modes of CE in which the sample is introduced as a zone or plug at the inlet end of the capillary, in CIEF, most or all of the capillary is filled with the protein-ampholyte solution. At the conclusion of focusing, individual proteins are confined in bands of only a few millimeters in length. Because the proteins can be concentrated up to several hundred-fold during the focusing process, the risk of precipitation is great if protein concentration in the original sample is too high. Protein concentration should not exceed 200 - 500µg/ml per component. The presence of salt in the sample may interfere with the focusing process; therefore, a desalting step employing ultrafiltration or dialysis against ampholytes is recommended prior to analysis.

2. Focusing

After injection of the sample-ampholyte mixture into the capillary, the capillary inlet is immersed in an acidic anolyte solution (typically dilute phosphoric acid) and the capillary outlet is immersed in a basic catholyte solution (typically dilute sodium hydroxide). Upon application of high voltage (positive-to-negative polarity), ampholytes and proteins migrate within the field, establishing a stable pH gradient at equilibrium. Proteins reach the point in the gradient where their net charge is zero (isoelectric point) and form a highly focused zone; movement out of the zone by diffusion results in acquisition of charge which forces the protein molecule back into the zone. It is this focusing effect that yields extremely sharp peaks and high resolving power. Similarly, the anolyte and catholyte solutions form a pH "cage," confining the gradient within the capillary so that at steady state, the only ions moving in or out of the capillary are protons or hydroxyls. Because the ampholyte gradient spans the entire capillary under these conditions, any proteins focusing between the monitor point and the capillary outlet will be undetected during mobilization. If these are proteins of interest, they can be caused to focus "upstream" from the monitor point by including in the mixture a basic spacer such as N,N,N',N'-tetramethylethylenediamine (TEMED), which at steady state occupies this "blind" segment of the capillary (49). The TEMED concentration must be carefully controlled so that it does not appreciably over- or under-fill the blind segment.

Focusing is typically performed at field strengths of 300 - 600 V/cm and is complete in 4 - 6min. The current drops exponentially during focusing, as ampholytes and proteins cease to become current carriers. The low currents at steady state allow rather high field strengths to be used for rapid focusing without excessive Joule heating; however, extended focusing times increase the risk of protein precipitation.

The presence of salt can severely compromise resolution in CIEF by compressing the zones of the gradient in the focusing step. During focusing, salt ions exit the capillary to be replaced by protons at the anodic end and hydroxyls at the cathodic end. Consequently, at the conclusion of focusing, there are segments of strong acid and strong base at the capillary margins whose segment lengths will be proportional to the original salt concentration in the sample. Compression of the ampholyte zones between these segments reduces resolution and increases the risk of protein precipitation.

3. Mobilization

The mobilization step following completion of focusing is designed to move the focused zones past the detection window without changing their order or band width. This is achieved by applying an electrophoretic, hydraulic or electrokinetic mobilizing force, while maintaining the electric field so that zones do not defocus during transport to the monitor point.

a. Electrophoretic

In this technique, variously termed "electrophoretic," "chemical," and "ion addition" mobilization, proteins are induced to acquire charge and thus move electrophoretically past the detection window. To accomplish this, a non-hydroxyl anion is added to the catholyte after focusing, resulting in mobilization towards the cathode. Alternatively, a non-proton cation can be added to the anolyte to effect anodic mobilization. In practice, the most common method is cathodic mobilization, by addition of a neutral salt such as sodium chloride to the catholyte. As indicated above, at steady state there is an electrical balance with proton flux from the anolyte equal to hydroxyl flux from the catholyte. With the addition of the mobilizing salt, chloride migration from the catholyte reduces hydroxyl movement into the capillary outlet, thereby reducing pH at the cathodic end of the capillary. This pH reduction causes focused proteins to acquire charge and migrate towards the cathode (50). As mobilization proceeds, chloride continuously migrates down the capillary, shifting the pH gradient with time. The net result is migration of all protein zones past the detection zone in order of decreasing pI values, as illustrated in Fig 12.14.

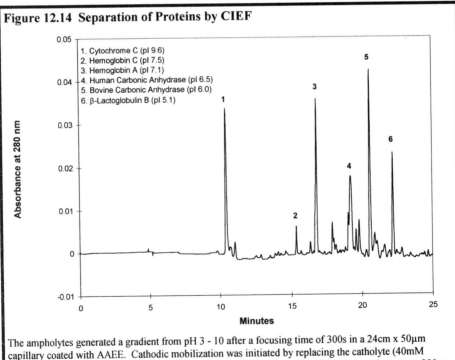

Figure 12.14 Separation of Proteins by CIEF

1. Cytochrome C (pI 9.6)
2. Hemoglobin C (pI 7.5)
3. Hemoglobin A (pI 7.1)
4. Human Carbonic Anhydrase (pI 6.5)
5. Bovine Carbonic Anhydrase (pI 6.0)
6. β-Lactoglobulin B (pI 5.1)

The ampholytes generated a gradient from pH 3 - 10 after a focusing time of 300s in a 24cm x 50μm capillary coated with AAEE. Cathodic mobilization was initiated by replacing the catholyte (40mM NaOH) with an alkaline zwitterion solution. Polarity was positive-to-negative and detection was at 280nm. Printed through the courtesy of Bio-Rad Laboratories.

Under optimized conditions, mobilization times can be correlated with isoelectric points. A calibration curve constructed with standard proteins, as seen in Fig. 12.15, can be used to estimate the pI of an unknown protein.

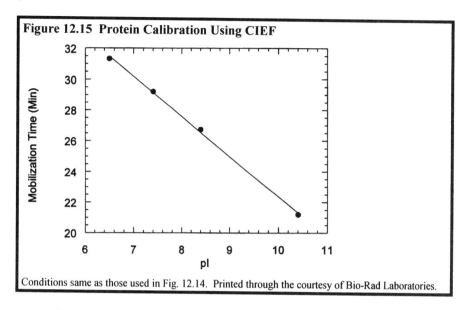

Figure 12.15 Protein Calibration Using CIEF

Conditions same as those used in Fig. 12.14. Printed through the courtesy of Bio-Rad Laboratories.

Refinements on electrophoretic mobilization include use of salts other than sodium chloride and use of zwitterions to enhance the mobilization efficiency of acidic proteins (49).

b. Hydraulic

Hydraulic mobilization can be effected by applying either pressure at the capillary inlet (51) or vacuum at the capillary outlet (52). In either case, the key is to keep the mobilization force low to reduce resolution loss caused by laminar flow mixing. Pressure mobilization requires application of pressures up to 0.5psi. A simple and convenient method of hydraulic mobilization is adjustment of the height of the liquid levels of the catholyte and anolyte so that a small siphoning pressure moves zones towards the cathode or anode. This method, termed gravity mobilization, can be optimized by selecting the capillary dimensions and adding viscosity-modifying polymers to the ampholytes (53).

c. Single-step CIEF

As the name implies, in single-step CIEF, the focusing and mobilization processes occur simultaneously (54, 55). The capillary is prefilled with catholyte, and the sample/ampholyte mixture is introduced as a plug at the capillary inlet. This plug moves towards the detector while the electric field is applied; the motive force can be hydraulic or electrokinetic. In either case, the rate of movement must be carefully controlled to insure that protein zones become fully focused by the time they reach the detection point. For hydraulic mobilization, gravity provides a very modest and easily-controlled force to achieve single-step mobilization. In electrokinetic mobilization, EOF provides the motive force. Attempts have been made to perform electrokinetically-mobilized CIEF in uncoated capillaries, but variations in the magnitude of EOF, induced by both pH shifts during the separation and protein adsorption to the wall, have limited its success. Better results have been obtained using coated capillaries in which EOF is modest and relatively constant across the pH range of the gradient.

Single-step CIEF may also be performed by monitoring nascent zones formed at the initiation of focusing as they move past the detection point during formation of the pH gradient. Since zones are detected before achievement of steady state conditions, this technique is sometimes termed nonequilibrium CIEF.

4. Limitations

The number of variables that impact resolution and reproducibility have limited the success of CIEF as a robust and reliable separation method. Nonetheless, improved understanding of the effects of the sample matrix, focusing, and mobilization conditions has improved the performance of the technique to the point that it has found some utility in quality control of protein therapeutics. Key requirements for successful CIEF are control of sample salt content, use of stable capillary coatings, use of high-quality ampholyte preparations, and normalization of protein mobilization times to internal standards. The ultimate limitation to successful CIEF is the nature of the particular proteins in the sample. Isoelectric focusing conditions inherently increase the risk of precipitation for several reasons:

1. proteins are less soluble at their isoelectric points;
2. the focusing process strips salt from the proteins;
3. proteins can be brought to extraordinarily high concentrations during focusing.

All of these conditions favor protein-protein interactions, aggregation, and precipitation. In conventional gel IEF, precipitation can result in poor separation patterns, but in CIEF it can terminate the analysis by plugging the column. In many cases, protein precipitation can be reduced by the incorporation of surfactants like Triton or Brij or chaotropes like urea into the sample/ampholyte mixture.

C. Capillary Sieving Techniques

Sieving techniques are used to separate species like SDS-protein complexes and nucleic acids which have no difference in mass-to-charge ratio and therefore would co-migrate in a CZE system. Sieving separations can be achieved by using crosslinked or linear polymeric gels cast in the capillary or by employing replaceable polymer solutions.

1. Gel-filled Capillaries

Capillaries containing crosslinked gels, such as polyacrylamide, provide the highest resolution of all capillary sieving systems (56, 57) but the problems with their preparation, operation and lifetimes result in their use only when other sieving systems do not provide satisfactory performance.

In order to immobilize the gel within the capillary during use, it is necessary to anchor the gel to the capillary wall. This is accomplished using the same chemistry employed to attach polymer coatings: reaction of a bifunctional silane with silanols to form an anchoring siloxane bond. The other functionality, such as a vinyl group, is reacted with an acrylamide monomer and bis-acrylamide in the presence of TEMED and ammonium persulfate to form the crosslinked gel. The total amount of acrylamide (%T) and the proportion of crosslinker (%C) determine the sieving strength of the gel. For oligonucleotides in the range of 10 - 60 bases, a 5%T, 5% C gel provides satisfactory sieving strength, while for double-stranded DNA in the range of 100 - 1000 base pairs, a 3%T, 3%C gel is recommended. During the polymerization process, the volume of the gel changes, causing voids which prevent current flow through the gel during operation. To compensate for this volume change, special casting techniques, such as carrying out the polymerization under pressure or initiating the polymerization sequentially along the capillary axis, have been developed (58).

Once the gel is cast in the capillary, several precautions are recommended to prevent damage to the matrix during installation and use. If the gel is to be stored at low temperature prior to use, cooling and warming must be done slowly to prevent damage due to temperature effects. Because drying of the gel at the capillary ends damages the matrix, the ends must be kept wet at all times. The best approach is to trim the capillary to length immediately before installing it in the instrument and to keep the ends immersed in buffer. Exposure to drying conditions even momentarily can destroy the gel. Once installed in the instrument, the gel is usually brought slowly to the operational voltage and then preconditioned for a period to achieve a stable baseline.

Because of their fragile nature, gel-filled capillaries typically exhibit short lifetimes. With clean samples like synthetic oligonucleotides, more than 100 analyses are possible. With biological samples, contamination of the gel with sample matrix components such as proteins will shorten capillary lifetimes. It is possible to remove contaminants by trimming a segment of the capillary inlet, but this will also change the performance.

Use of gel-filled capillaries is limited primarily to DNA sequencing applications and separation of antisense oligonucleotides. Application of polyacrylamide gel-filled capillaries to protein separations is rarely possible because the absorbance of the gel matrix in the low UV limits their use to wavelengths above 250 nm, where proteins have poor detectability.

2. Replaceable Polymer Solutions

The vast majority of sieving separations implement replaceable polymer solutions because of their ease of use with good reproducibility. In this technique, the electrolyte contains a hydrophilic polymer which, under the appropriate conditions, mimics the effect of a gel and causes retardation of sample molecules based on their molecular size (59). At a particular concentration (termed the entanglement threshold), the polymer chains interact to form an entangled network with a dynamic pore-like structure. This technique has been variously termed "nongel sieving," "entangled polymer sieving," "dynamic sieving," or "physical gel sieving." Typical sieving polymers are alkylated celluloses, such as methylcellulose (MC), hydroxymethylcellulose (HMC), and hydroxyethylcellulose (HEC), polyethylene glycols (PEG), and dextrans.

The procedure for carrying out an entangled polymer sieving separation is very similar to performing a CZE separation: the capillary is filled with the polymer solution, the sample is injected, and voltage is applied. The polymer solution is replenished at the beginning of each separation. Because of this, the separations are very reproducible and, typically, hundreds of runs can be obtained from a single capillary. The polymer solutions are quite transparent in the low UV, allowing their use for protein separations at 200 nm, as well as nucleic acid analysis at 260 nm.

Optimization of a polymer solution system necessitates selection of a polymer with the appropriate composition and chain length for the analytes; typically the chain length is similar to the molecular size of the analyte. The polymer concentration is adjusted to a level to achieve the desired sieving strength, usually around the entanglement threshold. Stronger sieving can be achieved by increasing the polymer concentration, but at some point the viscosity of the solution will prevent easy introduction into the capillary. Viscous polymer solutions may adhere to the capillary and electrode surfaces, causing carryover problems. In this case, it is advisable to rinse the capillary and electrode with water or buffer after the polymer replacement step.

Polymer sieving systems have been widely applied to purity determination of synthetic oligonucleotides. Fig. 12.16 illustrates the resolution of a mixture of oligonucleotides by this technique.

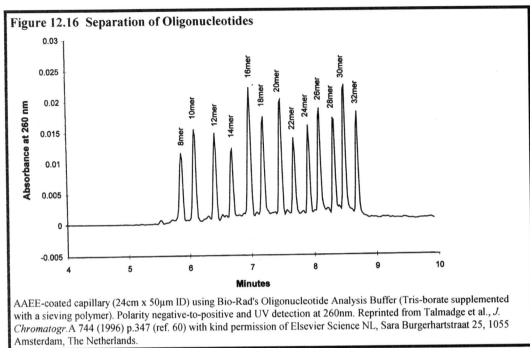

Figure 12.16 Separation of Oligonucleotides

AAEE-coated capillary (24cm x 50μm ID) using Bio-Rad's Oligonucleotide Analysis Buffer (Tris-borate supplemented with a sieving polymer). Polarity negative-to-positive and UV detection at 260nm. Reprinted from Talmadge et al., *J. Chromatogr.* A 744 (1996) p.347 (ref. 60) with kind permission of Elsevier Science NL, Sara Burgerhartstraat 25, 1055 Amsterdam, The Netherlands.

Fig. 12.17 shows the analysis of the components of an unpurified synthetic oligonucleotide using entangled polymer gel sieving.

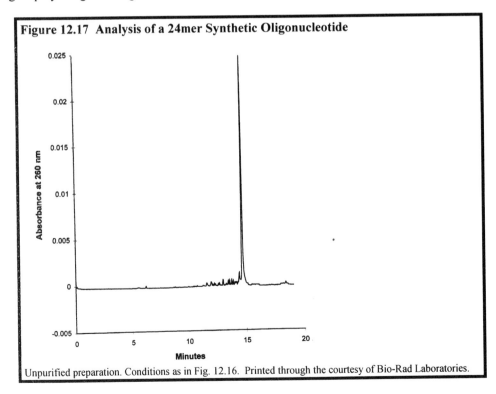

Figure 12.17 Analysis of a 24mer Synthetic Oligonucleotide

Unpurified preparation. Conditions as in Fig. 12.16. Printed through the courtesy of Bio-Rad Laboratories.

Fluorescence detection can be used to quantitate double-stranded DNA sequences, such as PCR products, which have been separated by capillary sieving by adding an intercalating dye like ethidium bromide or thiazole orange to the polymer solution, as illustrated in Fig. 12.18 and 12.19.

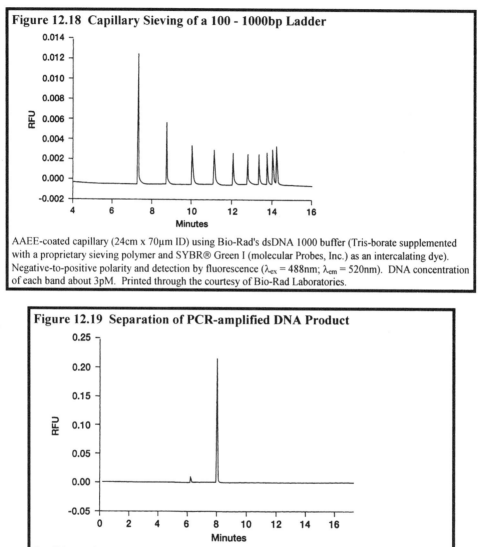

Figure 12.18 Capillary Sieving of a 100 - 1000bp Ladder

AAEE-coated capillary (24cm x 70μm ID) using Bio-Rad's dsDNA 1000 buffer (Tris-borate supplemented with a proprietary sieving polymer and SYBR® Green I (molecular Probes, Inc.) as an intercalating dye). Negative-to-positive polarity and detection by fluorescence (λ_{ex} = 488nm; λ_{em} = 520nm). DNA concentration of each band about 3pM. Printed through the courtesy of Bio-Rad Laboratories.

Figure 12.19 Separation of PCR-amplified DNA Product

Conditions as in Fig. 12.18. PCR-amplified 150bp length of M13mp18 DNA product. After 15 cycles of amplification, the sample was diluted 1000-fold for analysis. Printed through the courtesy of Bio-Rad Laboratories.

These CE systems have excellent potential for automated DNA forensic analyses and mutation screening.

Resolution and sensitivity for protein mixtures by CE are not as good as that achieved with silver-stained SDS-PAGE gels; however, the advantages of capillary sieving in terms of automation and quantitative analysis make this technique a promising one for industrial protein chemistry laboratories. Fig. 12.20 shows the separation of SDS-protein complexes by such a system.

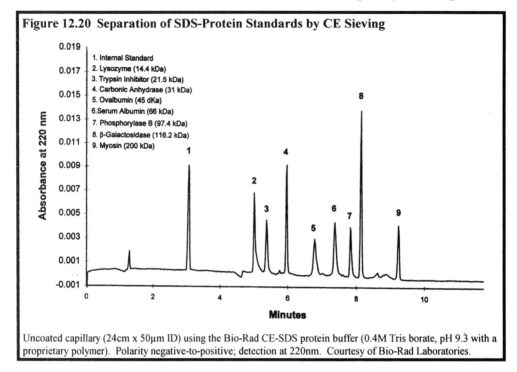

Figure 12.20 Separation of SDS-Protein Standards by CE Sieving

1. Internal Standard
2. Lysozyme (14.4 kDa)
3. Trypsin Inhibitor (21.5 kDa)
4. Carbonic Anhydrase (31 kDa)
5. Ovalbumin (45 dKa)
6. Serum Albumin (66 kDa)
7. Phosphorylase B (97.4 kDa)
8. β-Galactosidase (116.2 kDa)
9. Myosin (200 kDa)

Uncoated capillary (24cm x 50µm ID) using the Bio-Rad CE-SDS protein buffer (0.4M Tris borate, pH 9.3 with a proprietary polymer). Polarity negative-to-positive; detection at 220nm. Courtesy of Bio-Rad Laboratories.

D. Isotachophoresis (ITP)

Isotachophoresis is based on work by Kohlrausch which showed that concentration of ions at a migrating boundary between two salt solutions were related to their effective mobilities (61). The sample is injected into the capillary between a leading electrolyte which contains only one ionic species, with ion mobility greater than that of all sample components, and a terminating buffer containing one ionic species, with ion mobility less than that of all sample components. Zones migrate at equal velocity towards the detection point where they are detected as contiguous steps with zone lengths proportional to concentration. If UV-transparent spacers are added to the sample, ITP zones will appear as isolated peaks and the detector trace will resemble a CZE electropherogram. ITP resolves proteins as contiguous zones which migrate in order of mobility. It is rarely used as a separation method for proteins, but is occasionally used as a preconcentration technique. When used for protein preconcentration, protein zones are stacked at the column inlet between a terminating and a leading electrolyte in an initial transient isotachophoretic step. After preconcentration, the terminating buffer is replaced by the leading buffer, and proteins proceed to be separated by a conventional CZE mechanism (62). Fig. 12.21 illustrates the increased sensitivity for β-lactoglobulin B and A obtained by this method. For the separation by CZE alone (A), the sample was injected with pressure for 2 psi*sec, providing an injection zone of 1cm. For the ITP preconcentration procedure (B), the capillary was prefilled with sample and preconcentration was carried out for 20min at 1kV with terminating buffer at the cathode and leading buffer at the anode. A height difference of about 2cm between the terminating and leading buffer levels was used to generate gravity counterflow. Following preconcentration, the terminating buffer was replaced by leading buffer at the same level.

Figure 12.21 Analysis Using CZE Alone vs. CZE With ITP Preconcentration

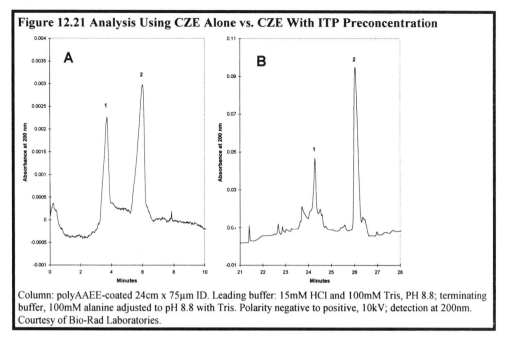

Column: polyAAEE-coated 24cm x 75μm ID. Leading buffer: 15mM HCl and 100mM Tris, PH 8.8; terminating buffer, 100mM alanine adjusted to pH 8.8 with Tris. Polarity negative to positive, 10kV; detection at 200nm. Courtesy of Bio-Rad Laboratories.

E. Micellar Electrokinetic Chromatography (MEKC)

As the name implies, micellar electrokinetic chromatography (MEKC) or capillary micellar electrochromatography (CMEC) is a chromatographic technique in which samples are separated by differential partitioning between two phases. The technique was pioneered by Terabe and coworkers (63-65), expanding the utility of CE instrumentation to the separation of neutral, as well as ionic, species. MEKC is usually performed in uncoated capillaries under alkaline conditions to generate a high EOF. The background electrolyte contains a surfactant at a concentration above its CMC (critical micelle concentration) where surfactant monomers are in equilibrium with micelles. The most widely used MEKC system employs sodium dodecylsulfate (SDS) as the surfactant. The sulfate groups of SDS are anionic; therefore, both surfactant monomers and micelles have electrophoretic mobility counter to the direction of EOF. Sample molecules are distributed between the bulk aqueous phase and the micellar phase, depending upon their hydrophobic character, as diagrammed in Fig. 12.22.

Figure 12.22 Schematic of an MEKC Separation

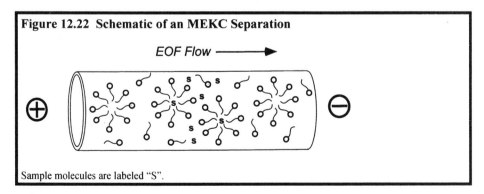

Sample molecules are labeled "S".

Hydrophilic neutral species with no affinity for the micelle will remain in the aqueous phase and reach the detector in the time required for EOF to travel the effective length of the column (t_0 in Fig. 12.23).

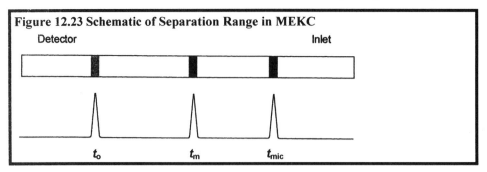

Figure 12.23 Schematic of Separation Range in MEKC

Hydrophobic neutral species will spend varying amounts of time in the micellar phase depending on their hydrophobicity; therefore, their migration will be somewhat retarded by the anodically-moving micelles (t_m in Fig. 12.23). Very hydrophobic species will be totally retained in the micellar phase and migrate at its rate (t_{mic}). Charged species will display more complex behavior because they have the potential for electrophoretic migration and electrostatic interaction with the micelles in addition to hydrophobic partitioning. MEKC is used almost exclusively for small molecules such as drugs and metabolites; Fig. 12.24 shows the analysis of caffeine in coffee using the technique.

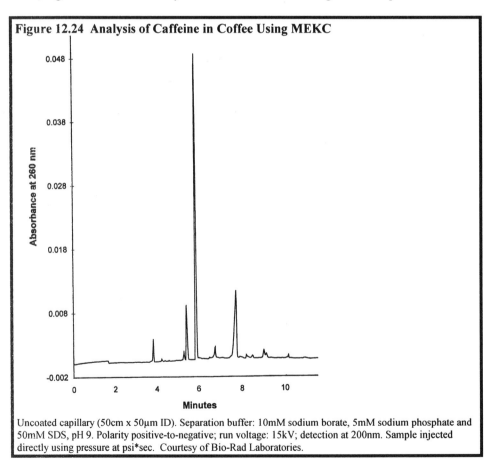

Figure 12.24 Analysis of Caffeine in Coffee Using MEKC

Uncoated capillary (50cm x 50µm ID). Separation buffer: 10mM sodium borate, 5mM sodium phosphate and 50mM SDS, pH 9. Polarity positive-to-negative; run voltage: 15kV; detection at 200nm. Sample injected directly using pressure at psi*sec. Courtesy of Bio-Rad Laboratories.

The selectivity of MEKC can be expanded with the introduction of chiral selectors or chiral surfactants to the system. MEKC has been occasionally used for peptides, but the mechanism is not suitable for proteins because they are too large to partition into a surfactant micelle; rather, they bind surfactant monomers tenaciously to form SDS-protein complexes.

1. Principles

If an anionic surfactant such as SDS is used as the micellar phase, micelles move towards the anode - counter to the direction of EOF. By adjusting the buffer pH and ionic strength to generate an EOF higher than the electrophoretic velocity of the micelles, the micelles and their contents eventually flow past the detection point (t_{mic} in Fig. 12.23). By convention, mobility towards the cathode is assigned a positive value, while mobility towards the anode is assigned a negative value. A neutral analyte is retarded in its movement towards the detection point by its partitioning into the micellar phase; this can be described by its capacity factor (k) where

$$k = \frac{n_{mic}}{n_{aq}} = \phi K \qquad \text{Eq. 12.13}$$

where n_{mic} and n_{aq} are the number of moles of analyte in the micellar phase and the aqueous phase, respectively, ϕ is the phase ratio (ratio of the volume of the micellar phase to the volume of the aqueous phase), and K is the partition coefficient of the analyte between the micellar and aqueous phase. The capacity factor of an analyte can be determined empirically from its retention time (t_m), the time required for an unretained species to reach the detector (t_{eo}), and the time required for the micelle to reach the detector (t_{mic}):

$$k = \frac{t_m - t_{eo}}{t_{eo}\left[1 - \left(\dfrac{t_m}{t_{mic}}\right)\right]} \qquad \text{Eq. 12.14}$$

Note that this is similar to the expression for k in conventional packed-bed chromatography:

$$k = \frac{t_R - t_0}{t_0} \qquad \text{Eq. 12.15}$$

The additional term in the denominator in Eq. 12.14 accounts for the fact that the hydrophobic phase in MEKC is not stationary, but moving within the capillary. For this reason, it is referred to as a "pseudophase". Two features of MEKC are apparent from the above discussion. First, retention of an analyte can be increased by increasing the phase ratio. In conventional column chromatography, this requires replacing the column with one having a stationary phase with higher capacity. In MEKC, this is simply accomplished by increasing the concentration of surfactant in the buffer, thereby augmenting the population of micelles. Second, all components which partition strongly into the micellar phase (in essence, having $k = \infty$) co-migrate at t_{mic}. In conventional chromatography, this separation problem can be overcome with gradient elution. In MEKC, it requires a means of extending the elution range to resolve strongly retained species.

2. Extending the Elution Range

Several strategies for extending the elution range in MEKC have been employed, including the use of organic modifiers, chaotropic agents, cyclodextrins, and mixed-micelle systems. Organic modifiers, such as short-chain alcohols or acetonitrile, can improve the resolution of analytes that coelute with the micellar phase. These additives primarily act by decreasing t_m and k, but they may also alter the level of EOF. Chaotropic agents like urea can have a similar effect by increasing the solubility of a lipophilic analyte in the aqueous phase.

Selectivity in MEKC can be modulated by using mixed micelle systems, for example, combining SDS with a neutral surfactant such as Brij-35®. To resolve chiral analytes, chiral surfactants have been used, either alone or in combination with nonchiral surfactants.

Cyclodextrins can also modify retention in MEKC. As described earlier, cyclodextrins can form inclusion complexes by hydrophobic partitioning of the analyte into the cavity of the cyclic polysaccharide. In an MEKC system, this complexation can compete with hydrophobic partitioning into the micellar phase. Because the analyte-cyclodextrin complex is uncharged, it will be carried towards the detector by EOF at a higher velocity than that of the anodically-moving micellar complexes.

F. Capillary Electrochromatography (CEC)

Like MEKC, capillary electrochromatography (CEC) is a chromatographic technique performed with CE instrumentation, primarily to separate small molecules (66, 67). It employs fused silica capillaries packed with 1.5 - 7μm microparticulate porous silica beads, usually derivatized with a hydrophobic ligand such as C18. Mobile phases are similar to those used for conventional reversed phase HPLC - mixtures of aqueous buffers and an organic modifier like acetonitrile. The silica surface of the derivatized beads has sufficient density of ionized silanol groups to generate a high EOF when a voltage is applied to the system, causing mobile phase to move through the column. Because EOF is plug-like rather than laminar in nature, efficiencies in CEC are often much higher than in HPLC, as illustrated in Fig. 12.25.

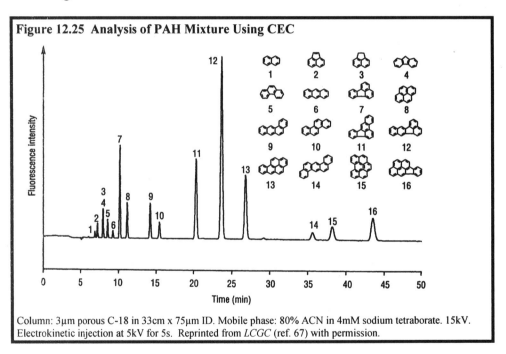

Figure 12.25 Analysis of PAH Mixture Using CEC

Column: 3μm porous C-18 in 33cm x 75μm ID. Mobile phase: 80% ACN in 4mM sodium tetraborate. 15kV. Electrokinetic injection at 5kV for 5s. Reprinted from *LCGC* (ref. 67) with permission.

VII. STRATEGIES FOR DEVELOPING A CE METHOD

Method development in CE is often simpler and less time-consuming that in HPLC because it usually involves rapid screening of a series of buffers, which can be accomplished with a single automation sequence using a standard buffer library. A general discussion of CE method development has recently been published (68); this section focuses on method development for biomolecules.

A. Selecting CE As An Analytical Technique

As an analytical tool, CE shares with HPLC the advantages of high resolution, rapid analysis, quantitative results, and ease of automation. Moreover, many of the separations performed by HPLC can also be done by CE. CE may be the preferred technique for a variety of reasons. First, the sample type or analytical goal may be uniquely suited for CE. For example, the separation of proteins by pI can only be accomplished by capillary isoelectric focusing. Similarly, separation of SDS-protein complexes is easily accomplished using entangled-polymer sieving CE. Second, HPLC techniques may not provide satisfactory resolution of species with similar characteristics, such as glycoforms, which are often readily resolved by CZE and IEF. Third, the different separation selectivity of CE may be desirable for confirming purity of a sample previously analyzed by HPLC. In this case, the two techniques are used in concert. Fourth, limitations in the amount of available sample may rule out other separation techniques. Finally, the reduced requirements for toxic and expensive solvents may make CE the desired choice.

B. Sample Preparation

The sample preparation considerations for capillary electrophoresis are similar to those for HPLC. Preliminary sample cleanup may be required to remove particulates or soluble components which would plug the capillary or adsorb to the capillary wall, to remove interfering compounds, or to simplify the separation problem. One requirement which is unique to capillary electrophoresis is removal of salt. Salts and buffer ions reduce injection efficiency (particularly for electrokinetic injection) and can compromise resolution, quantitative accuracy, and reproducibility. Typical desalting techniques include dialysis, ultrafiltration, and size exclusion. Since sample volumes are often small, the desalting apparatus must be suitable for handling microliter volumes. Microdialysis devices and low-volume size exclusion columns are available from several commercial sources.

C. Selecting the CE Separation Mode

Once the decision is made to employ CE for an analysis, selection of the appropriate separation mode is the next step. As in HPLC, more than one separation mode can often be used to achieve satisfactory results. Usually, additional considerations such as reagent cost, analysis time, method robustness and reproducibility, and instrument capability enter into the decision. For some applications, selection of the separation mode is obvious. For the separation of proteins by isoelectric point or for the determination of protein pI, capillary isoelectric focusing is the mode of choice (pI determination using CZE by plotting mobility *vs.* pH is time consuming and requires extensive buffer preparation). For separation of species with identical charge densities (SDS-protein complexes, synthetic oligonucleotides and double-stranded DNA, such as restriction fragments and PCR products) capillary gel electrophoresis or entangled-polymer sieving CE is employed. For analysis of neutral species, such as nonionic drugs, MEKC is required. For many of these applications, commercial reagents and analysis kits are available. A flowchart for selecting the appropriate CE mode is presented in Fig. 12.26.

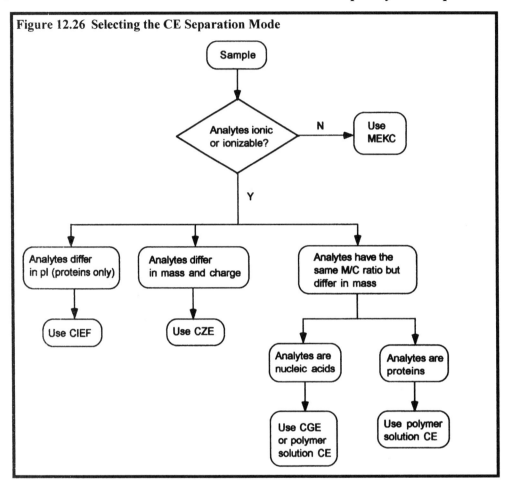

Figure 12.26 Selecting the CE Separation Mode

D. Developing a CE Method

Capillary electrophoresis method development was discussed in detail for the various CE separation modes in Section VI. Guidelines for each of the modes are summarized in Fig. 12.27 - 12.30.

1. Capillary Zone Electrophoresis

Controlling protein-wall adsorption and EOF are key considerations in developing a robust CE method for peptides and proteins, and the decision to use coated or uncoated capillaries dictates the strategies for method development. Peptides and some protein mixtures can be resolved on uncoated capillaries under acidic conditions (e.g. phosphate buffer at pH 2-3) with minimal EOF and adsorption. Capillaries coated with cationic materials (e.g. amines or polyamines) can be used under similar conditions with reversed EOF. Proteins may be analyzed on uncoated capillaries under basic conditions in the presence of EOF, although adsorption may be a problem for basic proteins or proteins with accessible basic residues. At intermediate pH values (pH 6 - 9), uncoated capillaries can be used with normal polarity, but additives such as salts, alkyl amines, zwitterions, or polymers will probably be necessary to suppress adsorption and control EOF. Many applications in this pH range may require the use of neutral coated capillaries. If such capillaries have negligible EOF, polarity selection will depend on the ionic characteristics of the analytes. Additives may still be required to reduce any residual adsorption, maintain protein solubility, or tailor selectivity (e.g. for separation of glycoforms). A flowchart for CZE method development is presented in Fig. 12.27.

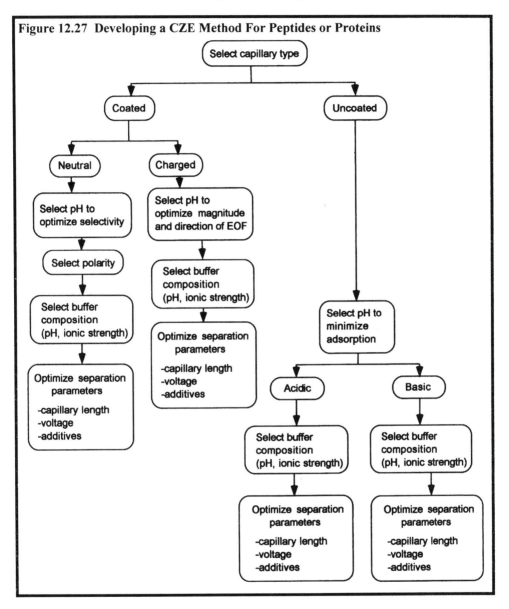

Figure 12.27 Developing a CZE Method For Peptides or Proteins

2. Capillary Isoelectric Focusing

The first step in developing a CIEF method is selection of the ampholyte range necessary for satisfactory resolution. Broad-range ampholyte mixtures (pH 3-10) are satisfactory for screening or separating complex protein samples, but highest resolution may require mixing broad-range with narrow-range ampholytes. Spacers may also be needed in the ampholyte mixtures for two-step IEF. The second decision in CIEF method development is selecting single- or two-step IEF. In single-step CIEF, the entire capillary is filled with the ampholyte-protein mixture and the resolved zones are detected as they migrate past the detection window from the outlet segment of the capillary. In this case, the analysis is performed under nonequilibrium conditions. Alternatively, the ampholyte-protein mixture can be injected as a plug at the inlet end of the capillary and displaced towards the detection window by hydraulic or electrokinetic forces during the focusing process. This approach has the potential for higher resolution because steady state is achieved by the time the zones are detected. Two-step mobilization, in which complete focusing is achieved before a separate mobilization step, often provides higher resolution than single-step CIEF. The mobilization force can be displacement or electrophoretic; electrophoretic mobilization with ion addition often yields highest resolution in CIEF. Capillary IEF methods which employ EOF for zone transport require capillaries with significant and controlled EOF, while CIEF methods which use displacement or electrophoretic mobilization require coated capillaries with negligible EOF. An additional CIEF method development requirement is the selection of additives (surfactants, chaotropes) to reduce protein precipitation. A flowchart for CIEF method development is presented in Fig. 12.28.

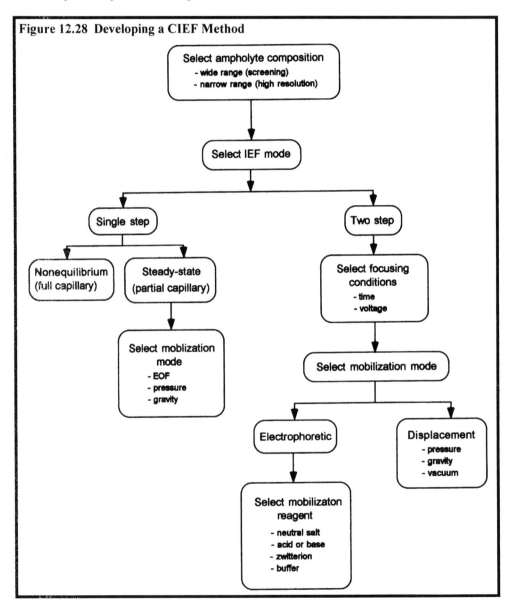

Figure 12.28 Developing a CIEF Method

3. Sieving Separations

Separations of nucleic acids, protein aggregates, and SDS-protein complexes require the use of a sieving medium. Nucleic acids can be analyzed using capillary gel electrophoresis or replaceable polymer solutions, although polymer solutions are currently preferred for the majority of applications because of ease of use and method robustness. The key to method development is achieving satisfactory sieving strength for the desired separation, which requires selecting the correct gel concentration and crosslinking in CGE or the correct polymer type and concentration for replaceable polymer solutions. Nucleic acid separations using polymer solutions usually require coated capillaries, while analysis of SDS-protein complexes has been achieved with coated and uncoated capillaries. Use of uncoated capillaries generally necessitates preconditioning of the capillary or capillary wash steps between analyses. A flowchart for development of nucleic acid sieving separations is presented in Fig. 12.29. The development of a sieving method for SDS-protein complexes is diagrammed in Fig. 12.30.

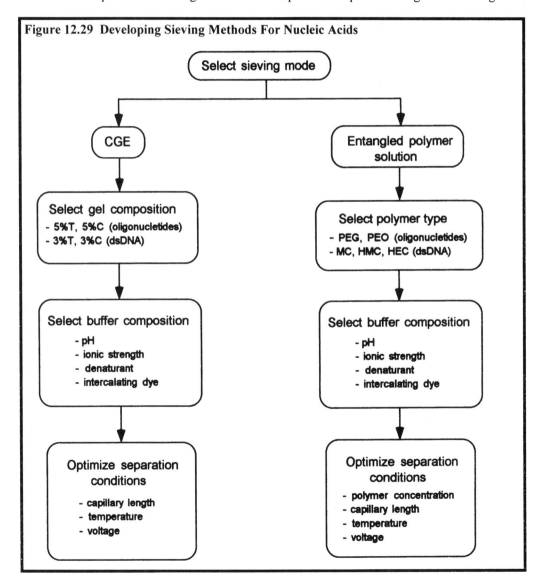

Figure 12.29 Developing Sieving Methods For Nucleic Acids

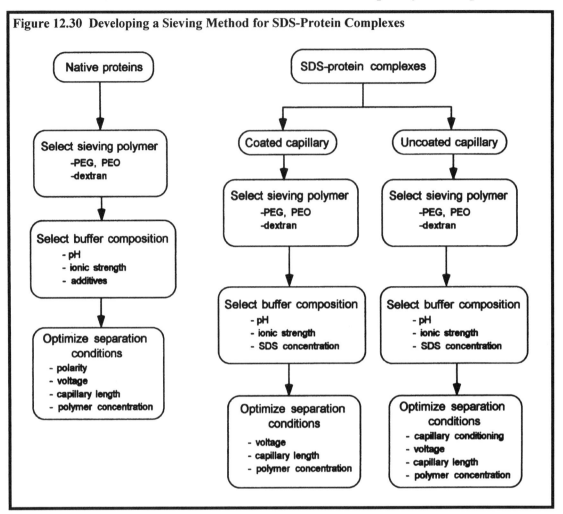

Figure 12.30 Developing a Sieving Method for SDS-Protein Complexes

VIII. REFERENCES

1. *Capillary Electrophoresis, Theory and Practice* (ed. P.D. Grossman and J. C. Colburn), Academic Press, San Diego, 1992.
2. S.F.Y. Li, "Capillary Electrophoresis, Principles, Practice, and Applications", *J. Chromatogr. Library*, Vol. 52, Elsevier Science Publisher, Amsterdam, 1992.
3. R. Weinberger, *Practical Capillary Electrophoresis*, Academic Press, Boston, 1992.
4. *Handbook of Capillary Electrophoresis, Second Edition* (ed. James P. Landers), CRC Press, Boca Raton, 1996.
5. *Capillary Electrophoresis, Theory and Practice* (ed. P. Camilleri), CRC Press, Boca Raton, 1993.
6. N. A. Guzman, *Capillary Electrophoresis Technology*, Marcel Dekker, New York, 1993.
7. J. Vindevogel, *Introduction to Micellar Electrokinetic Chromatography*, Hüthig, 1992.
8. J.H. Knox, *J. Chromatogr. A* 689 (1994) 3.
9. S. Hjertén, *Nucleosides and Nucleotides* 9 (1990) 319.
10. S. Hjertén, S. Jerstedt, and A. Tiselius, *Anal. Biochem.* 11 (1965) 219.
11. D.S. Burgi and R.-L. Chien, *Anal. Chem.* 63 (1991) 2042.
12. R. Aebersold and H.D. Morrison, *J. Chromatogr.* 516 (1990) 79.
13. R.D. Smith, H.R. Udseth, J.H. Wahl, D.R. Goodlett, and S.A. Hofstadler, "Capillary Electrophoresis-Mass Spectrometry" in *Meth. Enzymol.* 271 (1996) 448.

14. H. Lauer and D. McManigill, *Anal. Chem.* 58 (1986) 166.
15. R.M. McCormick, *Anal. Chem.* 60 (1988) 2322.
16. J.S. Green and J.W. Jorgenson, *J. Chromatogr.* 478 (1989) 63.
17. L. Song, Q. Ou, and W. Yu, *J. Liq. Chromatogr.* 17 (1994) 1953.
18. N. Cohen and E. Grushka, *J. Chromatogr. A* 678 (1994) 167.
19. D. Corradini, A. Rhomberg, and C. Corradini, *J. Chromatogr. A* 661 (1994) 305.
20. D. Corradini, G. Cannarsa, E. Fabbri, and C. Corradini, *J. Chromatogr. A* 709 (1995) 127.
21. J. P. Landers, R.P. Oda, B.J. Madden, and T.C. Spelsberg, *Anal. Biochem.* 205 (1992) 115.
22. R.P. Oda, B.J. Madden, T.C. Spelsberg, and J.P. Landers, *J. Chromatogr. A* 680 (1994) 85.
23. M.M. Bushey and J.W. Jorgenson, *J. Chromatogr.* 480 (1989) 301.
24. T. Wehr, *LC-GC* 11 (1993) 14.
25. T. Wehr, R. Rodriguez-Diaz, and C.-M. Liu, "Capillary Electrophoresis of Proteins," *Advances in Chromatogr.* 37 (ed. P.Brown and E. Grushka), Marcel Dekker, New York, 1997, pp. 237-361.
26. K. Ulfelder, K.W. Anderson, H.E. Schwartz, F.J. Sunzeri, M.P. Busch, and R.G. Brownlee, Poster PT-5 presented at the *3rd International Symposium on Capillary Electrophoresis*, San Diego, CA, 1991.
27. J.E. Wiktorowicz and J. C. Colburn, *Electrophoresis* 11 (1990) 769.
28. S. Hjertén, *J. Chromatogr.* 347 (1985) 191.
29. J.K. Towns and F.E. Regnier, *Anal. Chem.* 63 (1991) 1126.
30. M. Chiari, C. Micheletti, M. Nesi, M. Fazio, and P.G. Righetti, *Electrophoresis* 15 (1994) 177.
31. P.G. Righetti and C. Gelfi, *Anal. Biochem.* 244 (1997) 195.
32. J. Neugebauer, *A Guide to the Properties and Uses of Detergents in Biology and Biochemistry*, Calbiochem Corp., 1988.
33. R. Rodriguez-Diaz, M. Zhu, V. Levi, R. Jimenez, and T. Wehr, presented at the *7th Symposium on Capillary Electrophoresis*, Würzburg, Germany, 1995.
34. G.L. Lookhart, S.R. Bean, R. Graybosch, O.K. Chung, B. Morena-Sevilla, and S. Baenziger, *Cereal Chem.* 73 (1996) 547.
35. M.P. Richards and J.H. Beattie, *J. Cap. Elect.* 1 (1994) 196.
36. K. Yim, J. Abrams, and A. Hsu, *J. Chromatogr. A* 716 (1995) 401.
37. Y.H. Chu, L.Z. Avila, H.A. Biebuyck, and G.M. Whitesides, *J. Med. Chem.* 35 (1992) 2915.
38. L.Z. Avila, Y.H. Chu, E.C. Blossey, and G.M. Whitesides, *J. Med. Chem.* 36 (1993) 126.
39. T.J. Pritchett, R.A. Evangelista, and F.-T.A. Chen, *J. Cap. Elec.* 2 (1995) 145.
40. J. Bao and F.E. Regnier, *J. Chromatogr.* 608 (1992) 217.
41. D. Wu and F.E. Regnier, *Anal. Chem.* 65 (1993) 2029.
42. B.J. Harmon, D.H. Patterson, and F.E. Regnier, *Anal. Chem.* 65 (1993) 2655.
43. V.J. Hilser, G.D. Worosila, and E. Freire, *Anal. Biochem.* 208 (1993) 125.
44. S. Fanali, An Introduction to Chiral Analysis by Capillary Electrophoresis, Bio-Rad Laboratories, Hercules, CA, 1995.
45. P. Jandik, W.R. Jones, A. Weston, and P.R. Brown, *LC-GC* 9 (1991) 634.
46. Y. Shi and J.S. Fritz, *J. Chromatogr. A* 671 (1994) 429.
47. T. Wehr, M. Zhu, and R. Rodriguez-Diaz, "Capillary Isoelectric Focusing," in *Meth. Enzymol.* 270 (ed. B.L. Karger and W.S. Hancock), Academic Press, San Diego, CA, 1996, 358-374.
48. R. Rodriguez-Diaz, T. Wehr, M. Zhu, and V. Levi, "Capillary Isoelectric Focusing," in *Handbook of Capillary Electrophoresis, Second Ed.* (ed. J. Landers), CRC Press, Boca Raton, 1997, 101-188.
49. M. Zhu, R. Rodriguez, and T. Wehr, *J. Chromatogr.* 559 (1991) 479.
50. S. Hjertén, J.L. Liao, and K. Yao, *J. Chromatogr.* 387 (1987) 127.
51. S. Hjertén and M. Zhu, *J. Chromatogr.* 346 (1985) 265.
52. S.M. Chen and J.E. Wictorowicz, *Anal. Biochem.* 206 (1992) 84.
53. R. Rodriguez and C. Siebert, presented at the *6th International Symposium on Capillary Electrophoresis*, San Diego, CA, 1994.

54. R. Mazzeo and I.R. Krull, *Anal. Chem.* 63 (1991) 2852.
55. W. Thormann, J. Caslavska, S. Molteni, and J. Chmelik, *J. Chromatogr.* 589 (1992) 321.
56. S. Hjertén, *Electrophoresis '83*, H. Hirai (ed.), Walter de Gruyter & Co., New York (1994) 71-79.
57. A.S. Cohen and B.L. Karger, *J. Chromatogr.* 397 (1987) 409.
58. D.J. Rose and R.R. Holloway, poster 207 presented at the *Second International Symposium on High Performance Capillary Electrophoresis*, San Francisco, 1990.
59. M. Zhu, D.L. Hansen, S. Burd, and F. Gannon, *J. Chromatogr.* 480 (1989) 311.
60. K.W. Talmadge, M. Zhu, L. Olech, and C. Siebert, *J. Chromatogr.* A 744 (1996) 347.
61. F. Kohlrausch, *Ann. Phys Chem.* 62 (1987) 209.
62. F. Foret, E. Szoko, and B.L. Karger, *J. Chromatogr.* 608 (1992) 3.
63. S. Terabe, K. Otsuka, K. Ichikawa, A. Tsuchiya, and T. Ando, *Anal. Chem.* 56 (1984) 111.
64. S. Terabe, K.Otsuka, and T. Ando, *Anal. Chem.* 57 (1985) 834.
65. S. Terabe, "Micellar Electrokinetic Chromatography," in *Capillary Electrophoresis Technology* (ed. N. Guzman), Marcel Dekker, New York, 1993, p. 65.
66. M.M. Dittmann, K. Wienand, F. Bek and G.P. Rozing, *LC-GC* 13 (1995) 800.
67. R. Dadoo, C. Yan, R.N. Zare, D.S. Anex, D.J. Rakestraw, and G.A. Hux, *LC-GC* 15 (1997) 630.
68. D. Heiger, R.E. Majors, and R.A. Lombardi, *LC-GC* 15 (1997)14.

CHAPTER 13
SAMPLE PREPARATION

I. INTRODUCTION

An important consideration in the design of cost-effective separation methods is that the best HPLC column performance will only be obtained if the sample is introduced in a suitable form. In the worst cases, inadequate sample preparation can lead to equipment failure and reduced column lifetime. For "clean" samples, such as relatively pure polypeptides, preparation may only require simple procedures such as filtration, concentration, or dilution. When HPLC methods are used early in purification schemes, however, the complexity of the samples requires more sophisticated manipulations. In these cases, sample preparation can be the most time-consuming and labor-intensive element of the purification process; hence, sample handling techniques that lend themselves to automation are highly desirable, especially in manufacturing environments. An additional consideration in purification schemes for peptides, proteins, and nucleic acids is prevention of inactivation, degradation, or modification of the biomolecules during the process (1). Oxidation can be prevented by the addition of compounds like β-mercaptoethanol or dithiothreitol to the samples. Proteolysis can be controlled with reduced temperatures and addition of proteolytic inhibitors.

Although most HPLC applications utilize one or more sample preparation procedures, it is best to keep sample preparation as simple as possible (2). Some basic procedures that have found success for protein and peptide samples are precipitation and chromatographic methods. This chapter reviews the commonly used methods and devices for sample preparation in chromatography of biomolecules, particularly proteins. The proper choice and execution of the methods can greatly affect the success of the analysis, especially in terms of recoveries (1, 3-4). Planning the sequence of procedures can minimize time and maximize efficiency; a flow chart for sample preparation strategies was recently presented in *LC-GC* (5). Table 11.1 summarizes some of the typical steps in the preparation of a sample for injection onto an HPLC column.

TABLE 13.1 Objectives in Sample Preparation

- Prefractionation
- Extraction
- Removal of contaminants
- Solubilization
- Detergent removal
- Concentration
- Desalting
- Dilution
- Removal of particulates
- Preservation of structure
- Preservation of biological activity

Some traditional methodologies for these operations are laborious, frequently providing poor recoveries. Fortunately, many improved procedures which are rapid and simple, with increased yields, are available. In some cases, commercial products allow parallel processing of multiple samples, sometimes on-line as part of an automated chromatographic method.

II. PREFRACTIONATION

The first step in sample preparation for HPLC is to ensure that the sample is in a suitable physical form, that is, in a liquid state. After that is achieved, the objectives of sample prefractionation are twofold. The first goal is to simplify the mixture so that the separation is within the peak capacities of available HPLC columns. This can be attained by separating the target class of biomolecules from extraneous molecules or fractionating a complex protein mixture into subclasses for separate processing. The other goal of sample prefractionation is to eliminate sample components that would bind irreversibly to the stationary phase, changing column selectivity or otherwise degrading chromatographic performance. The sample prefractionation techniques which accomplish these aims are listed in Table 11.2.

TABLE 13.2 Prefractionation Techniques

- Liquid Extraction
    - Liquid-Solid
    - Liquid-Liquid
- Solid Phase Extraction (SPE)
    - Batch
    - Cartridge
- Precipitation

A. Liquid Extraction

1. Liquid-Solid

The first step in preparing a solid sample for HPLC is usually homogenization, in preparation for extraction (5). Extraction of the relevant components of the solid can be accomplished by a variety of techniques including Soxhlet extraction, accelerated extraction methods, and supercritical flow extraction (6). Soxhlet extraction, which leaches a specific component from a sample by refluxing in an appropriate solvent, is the benchmark classical technique for extraction of small molecules from a matrix (5, 6).

Supercritical fluid extraction (SFE) is gaining popularity as a prefractionation technique because it can reduce extraction time with concomitant good recoveries and ease of automation (7). SFE frequently employs carbon dioxide, which is converted into a supercritical state by pressure and temperature. Adjustment of these parameters and addition of modifiers and cosolvents yield selective extractions of specific classes of molecules (7-9).

2. Liquid-Liquid

When solutions contain classes of compounds with radically different solubilities, they can often be isolated with classical extraction techniques. In liquid-liquid extraction methods, two solvents which are not miscible are vigorously agitated, usually by shaking or stirring. These procedures can be manual or automated to include simple separatory funnel extractions, counter current chromatography, and on-line procedures (10). Liquid extraction is especially useful for very hydrophobic proteins, protein aggregates, or lipids which are primarily soluble in organic solvents or aqueous/organic mixtures. In some cases, a single solvent extraction can be used to remove the desired protein or protein class; alternatively, a series of extractions with different organic solvents or aqueous/organic mixtures can be employed to effect class separations. This latter approach is often used for fractionation of mixtures of hydrophobic membrane proteins and structural proteins. For example, the four major classes of cereal proteins (albumins, globulins, prolamins, and glutelins) can be sequentially extracted with water, dilute salt, and ethanol/water (4). Similarly, the four small subunits of yeast cytochrome c oxidase (a seven-subunit oligomeric membrane protein) have been selectively extracted from the holoenzyme with water/acetonitrile (60/40), followed by water/acetonitrile/isopropanol (50/25/25) (11).

B. Solid-Phase Extraction (SPE)

1. Batch

Batch extraction with solid-phase sorbents has been widely used for prefractionation of biological mixtures containing small proteins and peptides (4). The technique is inexpensive, can be applied to large volumes, and is amenable to handling viscous samples that would plug solid-phase cartridges or analytical HPLC columns. The steps in batch extraction are listed in Table 2-3.

TABLE 13.3 Steps in Solid-Phase Extraction

- Sorbent selection
- Sorbent conditioning
- Sample binding
- Extraneous substance extraction
- Sample elution

As with most solid-phase extraction techniques, the selected sorbent must first be conditioned with a suitable solvent to wet the phase. Methanol is commonly used for bonded-phase sorbents and hexane for unbonded silica. The conditioned sorbent is then mixed with the sample so that specific molecules can bind to it, typically requiring gentle stirring in the cold for several hours. Characteristically, low molecular weight species (< 40,000 Da) bind more tightly than large species. Binding can be followed by one or more extraction steps with suitable solvents, such as aqueous salt solutions or aqueous/organic solvent mixtures, to remove the undesirable components. Final elution of the biomolecules of interest is achieved batchwise in a small volume of a strong eluent. Alternatively, the sorbent can be packed into a column and the desired molecules eluted in a very small volume with step or gradient elution. Solid-phase batch extraction has been applied to purification of human immune interferon (12), gibbon interleukin-2 (13), and murine B-cell stimulatory factor (14).

2. Cartridge

Solid-phase extraction employing fully porous microparticulate silica or polymers, with or without covalently bonded stationary phases, has been used in the analysis of low molecular weight compounds in clinical, environmental, food and agricultural samples for over a decade. Typically, up to a few hundred milligrams of 20 - 50μm porous support is packed into a minicolumn, a syringe barrel, or a plastic cartridge that can be fitted to a tuberculin or Luerlock-type syringe. Sample application, extraction, and elution are achieved by passage of liquids through the cartridge bed with positive pressure or vacuum. The advantages of solid-phase extraction are its ease of use and the wide variety of commercially available sorbents and cartridge types, permitting selection of the appropriate phase for any particular sample preparation problem. Also, vacuum and pressure manifolds are available for simultaneous processing of multiple samples (4). This rapid fractionation is advantageous when bioactive peptides or intermediates in peptide metabolism might be lost or degraded by more extensive isolation procedures.

The steps in solid-phase cartridge extraction are the same as those listed in Table 11.3 for batch extraction. The choice of the sorbent is dependent on the nature of the fractionation desired. Compounds may be discriminated on the basis of charge, using cation- or anion-exchange sorbents, or on relative surface hydrophobicity, using one of a variety of alkyl, aromatic, or nitrile phases. The most commonly used phases are silica, octadecyl, octyl, and ion-exchange supports, although many others have also been implemented, including mixed phases (2, 15).

Approaches to developing a solid-phase sample cleanup method for low molecular weight compounds are generally applicable to biopolymers (4). It is best to have the analytical chromatographic method, including column, mobile phase, and gradient parameters established so that the effectiveness of the extraction method can be assessed. Using all available information about the sample and matrix, the general cleanup strategy, such as selective retention of sample or interfering compounds, should be determined. At this point, any constraints on the extraction method dictated by the sample chemistry or HPLC method should be identified. These might include the ultimate concentration or volume of the sample or the solvent composition, including pH, ionic strength, organic modifier concentration, and volatility. The final sample volume should be small (a minimum of two bed volumes is generally required to elute sample from the sorbent) and the sample solvent should be weak in strength relative to the HPLC mobile phase.

The selection of the appropriate sorbent depends on the amino acids on the interactive surface of the desired peptide or peptide class relative to the functional groups of interfering substances. Chromatographic data on the sample are useful in predicting its behavior in solid-phase extraction. For example, polypeptides with basic side chains can be selectively retained on a sulfopropyl cation-exchange phase and selectively excluded from an anion-exchanger. For peptides with significantly hydrophobic contact regions, octyl or phenyl phases would be appropriate. Very hydrophobic polypeptides that are soluble in organic solvents can sometimes be retained on polar sorbents in the presence of nonpolar organic solvents.

Composition of the sample matrix may also affect the choice of sorbent, particularly if the extraneous substances are in high enough concentration to interfere with sample binding. For example, the presence of concentrated salts would reduce retention on ion-exchange sorbents and enhance reten-tion on nonpolar sorbents; desalting or sample dilution of such solutions would be required prior to SPE by ion-exchange. Certain detergents can reduce polypeptide binding by competing for active sites on ion-exchange sorbents while others may reduce surface tension on nonpolar sorbents. The HPLC method may suggest that a sorbent different from that of the analytical column is preferable for reasons of solvent strength. For example, because the aqueous salt solutions used to elute samples from ion-exchange sorbents are weak solvents in reversed-phase systems, sample fractions can be introduced directly onto a reversed-phase HPLC column with no intermediate manipulations.

Once the sorbent type has been selected, extraction of the sample should be confirmed. The cartridge should be conditioned with 1 - 2 bed volumes of a solvent such as methanol, then prepared for sample introduction by passage of 10 - 20 bed volumes of the sample solvent. If an ion-exchange phase is used, it should first be converted to the proper counterion, then equilibrated at the appropriate pH. Because of the slow diffusion rates of macromolecules, samples should be applied at relatively low flowrates, preferably less than 10 ml/min; a cartridge syringe device can generate flowrates of 200 ml/min (3). After sample application, the bed should be washed twice with 10 - 20 bed volumes of sample solvent, followed by analysis of both washes by HPLC. Absence of sample components in the chromatogram indicates adequate extraction by the sorbent, whereas appearance of samples in the wash implies that a more retentive sorbent is needed.

The last steps in SPE are to wash out extraneous bound substances with suitable solvents and, finally, to elute the sample in a small volume with a strong solvent. Typical wash solvents and sample elution solvents are given in Table 11.4.

TABLE 13.4 Solvents for Sorbent Extraction

Sorbent Class	Weak Solvents (Wash)	Strong Solvents (Sample Elution)
Nonpolar	water low ionic strength buffers low % organic in water or buffer	high % polar organic (methanol, propanol or acetonitrile) in water or buffer
Polar	nonpolar organic (hexane or methylene chloride)	polar organic solvents aqueous buffers or acids
Ion-exchange	water low ionic strength buffers cation-exchange, pH < pK anion-exchange, pH > pK	cation-exchange, 2 pH units > pK anion-exchange, 2 pH units < pK high ionic strength buffers

Elution conditions for SPE columns follow the same principles as analytical columns of the same mode. Because proteins interact at multiple sites with chromatographic supports, the transition from complete sorption to full desorption occurs over an extraordinarily narrow range of solvent strengths; therefore, in many cases it is possible to find elution conditions where the desired protein or protein class is recovered, while unwanted material is separated by significantly stronger or weaker retention on the sorbent. Similarly, sequential elution steps can be used to fractionate a complex protein sample into subfractions. In both ion-exchange and nonpolar interactions, retention is strongly dependent on the ionic state of the sample; therefore, manipulation of solvent pH provides a powerful tool for modulating retention of both interfering substances and the molecules of interest. Release of strongly retained samples from ion-exchange sorbents may be achieved by increasing ionic strength, changing pH, or using a counterion with stronger selectivity. For example, a protein strongly retained in chloride buffer may be eluted with phosphate. Reversed-phase sorbents which use silica as the support material may excessively retain peptides or proteins containing basic side chains because of silanol interactions. In such cases, addition of a competing base such as triethylamine or tetramethylammonium salt to the eluent may improve recovery. Sorbents for SPE with pore diameters in the 60 - 100Å range will possess low loading capacities for proteins and peptides, which are excluded from most of the sorbent surface.

A number of automated sample processing systems for SPE are available. In some cases, sample wash and elution steps are performed automatically off-line and eluted samples are transferred to HPLC autosampler vials. In other systems, the processor functions as an on-line autosampler by employing cartridges compatible with high pressure operation. This permits samples to be eluted from the sorbent directly onto the HPLC column via an automated injection valve. Conditioning and wash steps may be manually performed off-line or automatically by the processor or an autosampler interfaced to it. In on-line methods, the choice of sorbent and extraction solvents will depend strongly on the HPLC method because the initial mobile phase composition of the HPLC method must quantitatively transfer the sample from the sorbent to the head of the analytical column. It must be both a strong eluent for the SPE sorbent and a weak eluent for the HPLC column; thus coordination of the SPE sorbent to the HPLC column chemistry is required. Ion-exchange prefractionation should be ideal for automated processing prior to a reversed-phase separation because the typical initial mobile-phase composition for RPC (dilute aqueous acid with low percentages of organic modifier) will effectively strip sample from an IEC sorbent. If a reversed phase SPE support is desired for use with an octadecyl analytical column, a less hydrophobic sorbent, such as, C_8 or C_3, would be required and quantitative transfer would need to be confirmed. The requirement for correct matching of SPE sorbents and analytical columns suggests that all manual off-line methods may not be easily adapted to on-line processing.

Applications of SPE for peptide and protein samples abound in the literature. In one interesting example, a series of bonded-phase sorbents was used to fractionate pituitary peptides from biological tissues (16). Initially, an acid extract was passed through reversed-phase (C_{18}) cartridges to remove hydrophobic proteins. The pH of the eluent was then adjusted and the sample was run through tandem cation- and anion-exchange cartridges, which bound basic peptides and acidic peptides, respectively; neutral peptides passed unretained through the cartridge set. Peptides retained on each cartridge could be eluted with appropriate solvents. Four fractions of a complex tissue extract, which could each be conveniently analyzed by RPC without interference or damage to the column by unwanted substances, were thus obtained. Optimum fractionation in this kind of system is highly dependent on both pH and phase selection. In this case, at pH 7, weakly basic peptides failed to bind to the cation-exchanger and eluted unretained with the neutral fraction.

On-line multi-column prefractionation is an efficient but complex method of isolating fractions for HPLC analysis (1). Switching devices can implement the transfer of fractions to other extraction columns or to an analytical column. Each solvent used for extraction must be compatible with the subsequent step. Such on-line procedures minimize sample loss from manual manipulations. On-line multi-column techniques are most effective for high volume repetitive procedures, where the time involved in method development is inconsequential due to throughput. If cross-contamination must be avoided, columns or cartridges must be discarded after each use; in other cases, regeneration is possible.

SPE has also provided a direct approach for removal of contaminants from proteins excised from two-dimensional electrophoresis gels (17). An empty cartridge containing the gel material was connected to a short reversed-phase cartridge column. With passage of 0.1% aqueous TFA through the coupled columns, the protein was transferred from the gel and bound to the reversed-phase column, whereas gel contaminants, buffer, and stain eluted in the void volume. Proteins were analyzed by gradient elution with TFA/acetonitrile. This procedure is well-suited for recovering small peptides that are often lost in electroblotting.

C. Precipitation

Precipitation is a very effective means of removing proteins from a complex solution (18). Acetonitrile containing either 0.1 - 1% trifluoroacetic acid or trichloroacetic acid precipitates many proteins. Cold methanol containing acids functions similarly. High concentrations of ammonium sulfate or guanidinium chloride may precipitate other classes of proteins.

III. SOLUBILIZATION

Any molecules which are separated by HPLC must first be soluble in the mobile phase. For some biomolecules, particularly membrane proteins, additives must be used to achieve this (19). Integral membrane proteins are a very hydrophobic species that lie embedded in the lipid bilayers of cell membranes. They are composed of relatively hydrophilic regions, which are located on the membrane surfaces and one or more hydrophobic sequences, which are associated with the interior hydrophobic portions of the membrane lipid. Disruption of this organization during purification usually leads to aggregation and precipitation unless detergents are added. Detergents compete with the lipids for the hydrophobic regions of the protein, thus maintaining its soluble state in aqueous environments, usually in the form of protein-detergent or mixed protein-lipid-detergent micelles. In a typical purification scheme, loosely associated hydrophilic peripheral proteins are first removed from the membrane with salt or chaotropic agents, and then integral membrane proteins are liberated by the addition of the detergent.

The choice of detergent used for solubilization of membrane proteins depends on both the protein and the subsequent method of purification. Some detergents with applications in membrane protein purification are listed in Table 11.5.

TABLE 13.5 Detergents Used for Solubilization of Membrane Proteins

Detergent	CMC (mM)
Ionic	
sodium dodecyl sulfate	8.13 (water)
	2.3 (0.05M NaCl)
	0.51 (0.5M NaCl)
sodium cholate	13 - 15
sodium deoxycholate	4 - 6
sodium taurodeoxycholate	2 - 6
Nonionic	
Triton X-100	0.24 - 0.3
Nonidet P40	0.29
Triton X-114	0.2
Tween 80	0.012
Emulphogen BC-720	0.087
Octylglucoside	25
Brij 35	0.091
dodecyl dimethylamineoxide	2.2
Amphoteric	
CHAPS	4 - 6
Zwittergent 3-12	3.6

Reprinted from Welling et al., *J. Chromatogr.* 418 (1987) p.223 (ref. 20) with kind permission of Elsevier Science NL, Sara Burgerhartstraat 25, 1055 Amsterdam, The Netherlands.

A systematic approach to the design of a solubilization procedure includes selecting the detergent, buffer and temperature, and testing for solubilization (20). Major considerations in detergent selection are the constraints imposed by the subsequent chromatographic method and any requirements for recovery of biological activity. The critical micelle concentration (CMC) is the concentration of a detergent monomer at which micelles, i.e., spherical aggregates of detergent molecules, begin to form. Detergents with high CMC are easier to remove by dialysis than those with low CMC. Some detergents, such as Triton and Nonidet, are UV-active, and thus interfere with absorbance detection at wavelengths less than 280nm. A reduced (hydrogenated) form of Triton X-100 is commercially available from Sigma. It has similar properties to the nonreduced form and no absorption in the mid- to low UV (21). If ion-exchange chromatography is to be used, ionic detergents should be avoided. Most detergents will bind to reversed-phase and size exclusion columns via hydrophobic interactions and, if not removed prior to chromatography, are likely to permanently alter the bonded phase and thus, the selectivity of the column. If an HPLC column is used for the analysis of detergent-containing materials, it is best if it is dedicated to this application.

Sodium dodecyl sulfate (SDS) binds strongly to proteins and is one of the most effective solubilizing agents; however, it tends to irreversibly denature proteins with loss of biological activity. Bile salts have been found to be more effective than nonionic detergents in dissociating protein complexes (4). Bile salts like sodium cholate and deoxycholate and their derivatives, such as CHAPS, are known for their gentleness, and are often used when solubilization without denaturation is desired; however, they form aggregates which precipitate at acidic pH (19). If divalent cations are required for maintenance of biological activity, sodium cholate should be avoided (4).

In general, solubilization is best achieved at high ionic strength (0.1 - 0.5M salt) in a 25mM buffer with a pH close to the pK of the buffer. The salt and buffer must be compatible with the detergent; for example, SDS will precipitate in potassium phosphate, but is soluble in sodium phosphate. If ion-exchange chromatography is to be used subsequently, the buffer and salt ions should not have greater strength than the mobile phase counterions. Usually, solubilization is carried out at reduced temperatures. After the detergent, salt, and buffer conditions have been selected, the effectiveness of solubilization can be determined by centrifuging after incubation to separate solubilized from insoluble material for subsequent assay. A range of detergent concentrations (0.1 - - 3.0% in the protein/salt/buffer solution) should be tested with gentle stirring in the cold for one hour. Successful solubilization may depend on finding the optimal detergent/protein ratio by this procedure. Protein solubilization occurs at or near the CMC for most detergents. If biological activity is lost during detergent solubilization, addition of a stabilizer such as glycerol, dithiothreitol, or EDTA may be necessary.

Effective solubilization of a membrane protein may require a detergent that is incompatible with subsequent steps in the purification process. For example, competition with membrane lipids to form protein-detergent micelles may require a high concentration of a detergent with low CMC and high aggregation number, such as Triton. Removal of such detergents by dilution or dialysis is often difficult; however, it may be necessary due to its UV absorption or other properties. It is sometimes possible to exchange the detergent for one more compatible, such as a bile salt (22). In such a strategy, protein-detergent micelles are first separated from detergent micelles, detergent-lipid micelles, and detergent monomers by SEC. Detergent exchange is then achieved by running the detergent-protein fraction through an SEC column with a mobile phase containing a detergent of high CMC and low aggregation number. The mobile phase detergent will disperse the solubilization detergent into mono- mers, as well as replace it in the protein-detergent micelle.

IV. DETERGENT REMOVAL

A. General

Removal of detergents from extracted or solubilized hydrophobic proteins may be necessary prior to subsequent purification steps (23). Detergent micelle formation may complicate concentration procedures or detergents may degrade column performance and interfere with UV detection in HPLC separations. Techniques for removal of detergents from proteins are listed in Table 11.6.

TABLE 13.6 Methods for Detergent Removal
• Dialysis
• Solvent extraction
• Ion-pair extraction
• Adsorption on polystyrene resin
• Adsorption on ion-retardation resin
• Binding to affinity matrix

Dialysis can be used to remove detergents with high CMC values from proteins after dilution of the detergent below the CMC, but this is generally a time consuming technique.

B. Extraction

Chloroform extraction is a method which has been effective in removing SDS from proteins electroeluted from SDS polyacrylamide gels (24). After elution, the aqueous fractions were mixed with methanol and extracted with chloroform. The SDS was removed by the chloroform and the extracted protein was washed extensively with methanol, and then dried to remove any residual chloroform.

Ion-pair extraction has also been used to remove SDS from detergent-protein complexes (25). Dry samples were extracted with a solution of ion-pairing agent in an organic solvent, such as acetone/triethylamine/acetic acid/water or heptane/tributylamine/acetic acid/butanol. Sufficient water to promote formation of the alkylammonium-SDS ion pair was required in the extractant or protein sample; therefore, in the second solvent system, water had to be added to about 1%. A single extraction removed up to 95% of the total SDS; protein was recovered as a precipitate that could be separated from the extractant by low-speed centrifugation. Washing with acetone or heptane removed any residual extractant or SDS. Salts in the sample had to be removed prior to extraction due to interference with SDS removal. This procedure can be adapted to extraction of small volumes of aqueous protein-SDS solutions by 1/20 dilution using extractant prepared without water. A two step extraction quantitatively removes the detergent with recovery of protein that generally exceeds 80%. This extraction procedure also removes the Coomassie blue stain from proteins recovered from SDS polyacrylamide gels.

C. Adsorption

Adsorption on neutral polystyrene resins by batch or by column is a popular technique for removal of surfactants from aqueous solutions of proteins. Because this technique only removes detergent micelles, the detergent concentration will only be reduced to the CMC. Amberlite XAD-2 resin and Biobeads SM have been used for removal of Triton X-100 by both batch and column methods (3).

Ion-retardation resins, which consist of acrylic acid polymerized inside a strong anion-exchange resin on a polystyrene divinylbenzene matrix (26), are also effective for removal of SDS from proteins. Passage of a protein-SDS complex through the resin results in complete retention of SDS and elution of protein with 80-90% recovery (27). The capacity of the resin for SDS is more than 2.2mg/g, which effectively reduces the SDS level to less than one molecule of SDS per protein molecule. Because SDS binds tenaciously to the resin, it cannot be removed and the resin must be discarded after use. In the presence of buffers, adsorption of SDS by an ion-retardation column is reduced, resulting in incomplete removal of detergent from the protein. This can be circumvented by prior removal of buffer by SEC or, more conveniently, by the addition of a few grams of size exclusion gel to the head of the ion-retardation resin bed to retard the buffer (4).

D. Affinity Binding

Removal of detergents using Extractigel (Pierce Chemical, Rockford, IL), an affinity matrix developed for detergent removal, is based on interaction with proprietary affinity ligands after selective permeation of the detergent into the gel (28). Proteins larger than 10,000 Da are excluded and pass through in the void volume; therefore, operation in the column mode rather than batch format is required. The technique cannot be used for polypeptides below the 10,000 Da exclusion limit because they permeate the matrix and bind to the ligand. Dilute protein solutions ($\leq 50\mu g/ml$) may yield poor recoveries due to nonspecific binding. In such cases, it is recommended that a protein like bovine serum albumin be added to the sample to saturate nonspecific binding sites. This affinity product has been used for removal of both ionic and nonionic detergents (4).

V. CONCENTRATION

A. General

Biological samples are frequently very dilute by the time HPLC procedures are employed; therefore, concentration is necessary for detection, as well as general volume reduction. Commonly used techniques for sample concentration are listed in Table 11.7.

```
┌─────────────────────────────────────────────────────────────────┐
│  TABLE 13.7  Sample Concentration Techniques                      │
│                                                                   │
│     •   Lyophilization                                            │
│     •   Vacuum evaporation                                        │
│     •   Precipitation                                             │
│     •   Dialysis against concentrator resins                      │
│     •   Ultrafiltration                                           │
│     •   Chromatography                                            │
└─────────────────────────────────────────────────────────────────┘
```

Conventional techniques, such as lyophilization and evaporative concentration, are lengthy and may cause chemical modification or loss of sample by nonspecific adsorption if the sample is taken to near dryness. Ultrafiltration and chromatographic concentration are more rapid and gentle, minimizing the problems encountered in conventional approaches.

B. Precipitation

Fractional precipitation is a useful tool for biomolecule purification because it can be very selective for specific classes of molecules. Precipitation of proteins may be initiated by pH manipulation; high salt concentration, most notably with ammonium sulfate; high concentration of organic; or addition of certain chemicals like perchloric acid (1, 18). Similarly, RNA can be specifically precipitated with lithium chloride (1). The success of such techniques depends on the goals. If the objective is to remove the proteins or nucleic acids and analyze the components of the supernatant, these can be very effective procedures. If the aim is to analyze the precipitated molecules, success will depend on their quantitative reconstitution and preservation of biological activity. Sample loss can occur during precipitation by formation of insoluble aggregates.

C. Ultrafiltration

Ultrafiltration is based on the selective passage of low molecular weight sample components through a porous membrane filter of appropriate porosity, while high molecular weight species, such as proteins which cannot penetrate the pores, become concentrated. Centrifugation is the driving force. Filters are available with molecular weight cutoffs in the range of 10,000 - 30,000 Da (29, 30). Ultrafiltration devices that are designed for use in fixed angle rotors permit unrestricted solvent flow through the membrane, and prevent the sample from reaching complete dryness. These devices permit milliliter samples to be rapidly concentrated up to 80-fold with low nonspecific adsorption and generally good recoveries.

D. Chromatography

In chromatographic concentration, the strategies for choosing columns and packings are similar to those employed for solid phase extraction. The sample, diluted in a large volume of a weak solvent, is concentrated onto the column and then eluted in a small volume of strong solvent. For example, low pressure ion-exchange gels packed into mini-columns were used to concentrate samples up to 30-fold prior to SEC (31). The necessity of using a solvent of low ionic strength to allow quantitative binding of protein to an ion exchanger may require appropriate dilution before application to the column.

Short (3 - 10cm) HPLC columns can also be used for rapid sample concentration. Microbore reversed-phase columns have been used for preconcentration of proteins prior to microsequence analysis, after collection and dilution of milliliter fractions from conventional HPLC columns (33). Gradient elution at low flowrates (0.1 - 0.2 ml/min for 2mm ID or 0.02 - 0.04 ml/min for 1mm ID columns) permits recovery of proteins in volumes as small as 25μl, with concentration factors up to 80-fold. The high capacities of porous microparticulate packings enable 50 - 100μg to be loaded onto microbore columns as long as the sample is introduced in a weak solvent.

VI. DESALTING

Removal of salt from a protein solution may be necessary prior to chromatography. For example, injection of high concentrations of salt would result in irreproducibility in retention of early eluting peaks on an ion-exchange column. Concentrated salts may also be incompatible with some organic mobile phases in reversed-phase chromatography. Common desalting techniques are listed in Table 11.8.

TABLE 13.8 Desalting Techniques

- Dialysis
- Size exclusion chromatography
- Reversed-phase chromatography
- Ion-retardation
- Ultrafiltration

Dialysis is the traditional but time-consuming method for removing salts, chaotropic agents, and detergents from protein fractions. The speed and efficiency can be improved by using hollow fiber techniques (1). Size exclusion chromatography is a fast and effective alternative to dialysis for desalting, using small pore, cross-linked, dextran or polyacrylamide gels so that proteins are excluded while salts are retarded by permeation. Desalting gels are available as low pressure prepacked columns (29, 30) or packed centrifuge tubes with centrifugation as the driving force (33). Size exclusion HPLC columns can also be used for rapid desalting; however, operation with low ionic strength mobile phases may result in adsorptive interactions for some proteins.

Octyl and octadecyl reversed-phase sorbents are frequently used for desalting aqueous protein fractions because salts and chaotropic agents are unretained under conditions where proteins exhibit strong interactions. After sample application, salts and other small polar compounds can be washed from the sorbent with the weak solvent, followed by elution of protein with an aqueous/organic mixture. Desalting can be performed off-line using mini-columns or cartridges or on-line in an HPLC gradient method, after elution with a volume of weak solvent. Reversed-phase desalting is advantageous if it must be followed by a concentration step because the elution solvent is generally volatile.

Ion-retardation resins, as discussed previously for detergent removal, are also used for desalting (26). These resin copolymers consist of adjacent anionic and cationic sites that, in the absence of counterions, exist in a self-adsorbed state. On application of a protein-salt solution to the resin, ionic species in the sample adsorb to the resin in an ion-exchange process, disrupting the resin self-association. Salts of certain monovalent ions like sodium chloride, potassium chloride, and Tris-HCl are retained, whereas proteins can be readily eluted with water. Sodium acetate, sodium citrate, sodium phosphate, and ammonium sulfate do not compete as effectively with the self-adsorption, and thus they may not be completely removed from the proteins. After desalting, the resin can be regenerated by treatment with 1M hydrochloric acid, followed by 1M ammonium hydroxide containing 0.5M ammonium chloride, and, finally, by extensive washing with water.

The ultrafiltration devices used for sample concentration can also be effective for desalting (34). A sample that has been concentrated by ultrafiltration should be diluted 10-fold in buffer or water and subjected again to ultrafiltration; each ultrafiltration cycle removes about 90% of the salt, resulting in reduction by 99% in two or three cycles.

VII. FILTRATION

Filtration is an essential process in sample preparation. For complex samples, it may be implemented at more than one step in the procedure, but it is always one of the final steps before injection onto an HPLC column. Particulate matter will not only damage HPLC columns, it may also interfere with other purification steps. Usually the liquid from the filtration is collected, but sometimes the precipitate is the desired fraction to undergo further processing.

The primary considerations in filtration are the pore size and composition of the filters (35). The choices are based on the specific analyte and sample matrix. As mentioned previously, filters of 0.45µm should be used prior to injection. The composition of the membrane must be compatible with both the sample and the solvent. Mismatch with a solvent can result in disintegration of the filter or leaching of some of its components. Mismatching with a sample can result in adsorption; for example, proteins may bind to either very hydrophobic or very polar surfaces.

VIII. CONCLUSIONS

Many sample preparation methods are available for prefractionation of protein and peptide samples before HPLC analysis. Frequently the sample preparation methods of choice, especially when LC-MS is used, are precipitation and HPLC methods of collection and reinjection. It is not only important that methods adequately clean up the samples, but also that they be as simple as possible.

IX. REFERENCES

1.   A.M. Rizzi and I. Maurer-Fogy, "Sample Preparation" in *High Performance Liquid Chromatography: Principles and Methods in Biotechnology* (ed. E. Katz), John Wiley & Sons, Chichester (1996) 233.
2.   R.E. Majors, *LC-GC* 14 (1996) 754.
3.   C.T. Wehr, "Sample Preparation" in *HPLC of Biological Macromolecules* (ed. K.M. Gooding and F.E. Regnier), Marcel Dekker, New York (1990) 215.
4.   C.T. Wehr, "Sample Presentation and Column Hygiene" in *HPLC of Proteins, Peptides and Polynucleotides* (ed. M.T.W. Hearn), VCH Publishers, New York (1991) 37.
5.   "Guide to Sample Preparation" in *LC-GC* 14(10) (1996) supplement wall chart.
6.   R.E. Majors, *LC-GC* 14 (1996) 88.
7.   R.E. Majors, *LC-GC* 9 (1991) 78.
8.   D.R. Gere and E.M. Derrico, *LC-GC* 12 (1994) 352.
9.   D.R. Gere and E.M. Derrico, *LC-GC* 12 (1994) 432.
10. R.E. Majors, *LC-GC* 14 (1996) 936.
11. S.D. Power, M.A. Lochrie, and R.O. Poyton, *J. Chromatogr.* 266 (1983) 585.
12. R.A. Wolfe, J. Casey, P.C. Famillette, and S. Stein, *J. Chromatogr.* 296 (1984) 277.
13. L.E. Henderson, J.F. Hewetson, R.F. Hopkins III, R.C. Sowder, R. Neubauer, and H. Rabin, *J. Immunol.* 131 (1983) 810.
14. J. Ohara, S. Lahet, J. Inman, and W.E. Paul, *J. Immunol.* 135 (1985) 2518.
15. R.E. Majors and D.E. Raynie, *LC-GC* 15 (1997) 1106.
16. H.P.J. Bennett, *J. Chromatogr.* 359 (1986) 383.
17. J.D. Pearson, D.B. DeWald, H.A. Zurcher-Neely, R.L. Heinrikson, and R.A. Poorman in *Proceedings of the 6th International Conference on Methods in Protein Sequence Analysis, Seattle, WA, Aug. 17-21, 1986* (ed. K.A. Walsh), Humana Press, Clifton (1986) 295.
18. S. Englard and S. Seifter, *Meth. Enzymol.* 182 (ed. M.P. Deutscher), Academic Press, New York (1990) 285.
19. G.W. Welling, R. van der Zee, and S. Welling-Wester, "HPLC of Membrane Proteins" in *HPLC of Biological Macromolecules* (ed. K.M. Gooding and F.E. Regnier), Marcel Dekker, New York (1990) 373.
20. G.W. Welling, R. van der Zee, and S. Welling-Wester, *J. Chromatogr.* 418 (1987) 223.
21. G.E. Tiller, T.J. Mueller, M.E. Dockter, and W.G. Struve, *Anal. Biochem.* 141 (1984) 262.
22. A.J. Furth, H. Bolton, J. Potter, and J.D. Priddle, *Meth. Enzymol.* 104 (1984) 318.

23. L.M. Hjelmeland, *Meth. Enzymol.* 182 (ed. M.P. Deutscher), Academic Press, New York (1990) 277.
24. D. Wessel and U.I. Flugge, *Anal. Biochem.* 138 (1979) 153.
25. W. Konigsberg and L. Henderson, *Meth. Enzymol.* 91 (1983) 254.
26. Bio-Rad Technical Bulletin 1005, Bio-Rad Laboratories, Richmond, CA.
27. S.N. Vinogradov and O.H. Kapp, *Meth. Enzymol.* 91 (1983) 259.
28. Extractigel D Technical Bulletin, Pierce Chemical Co., Rockford, IL.
29. Bio-Rad Technical Bulletin 2068, Bio-Rad Laboratories, Richmond, CA.
30. Pharmacia Technical Bulletin, Pharmacia AB, Uppsala, Sweden.
31. P.C. Billings and A.R. Kennedy, *BioTechniques* 5 (1987) 210.
32. E.C. Nice, C.J. Lloyd and A.W. Burgess, *J. Chromatogr.* 296 (1984) 153.
33. R.I. Christopherson, *Meth. Enzymol.* 91 (1983) 278.
34. Amicon Technical Bulletin 522, Amicon Division, W.R. Grace & Co., Danvers, MA.
35. R.E. Majors, *LC-GC* 13 (1995) 364.

CHAPTER 14
QUANTITATION AND DATA HANDLING

Fred Klink, Scientific Training and Marketing

I. INTRODUCTION

The real purpose for doing an HPLC analysis is to extract quantitative or qualitative information; therefore, the final step of the chromatographic process is to derive the data required for the laboratory. Qualitative techniques, such as ultraviolet spectroscopy, tell what is present. Quantitative techniques, such as weighing, tell how much is there. The purpose of an HPLC analysis is to find out how much of which components are in a sample. The process of chromatographic data handling consists of two distinct steps: 1. integration, which gives a basic measure of the physical size of the peak; and 2. quantitation, which translates the physical size of the peak into a measure of how much of the compound of interest is present in the sample.

A. Peak Integration and Height Determination

The area and/or height of a chromatographic peak is proportional to the amount of the compound being detected. In a totally resolved peak like that in Fig. 14.1A, the height and area are well-defined; however, in the fused peaks seen in Fig. 14.1B, the overlapping can produce uncertainty, particularly for peak area.

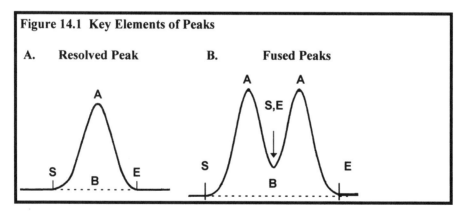

Figure 14.1 Key Elements of Peaks

A. Resolved Peak   B. Fused Peaks

In order to evaluate either type of chromatographic peak, four parameters must be determined:
Peak Apex (A) is the topmost point of the peak stated in terms of elapsed time from the injection. This is the retention time of the peak.
Peak Start (S) is the point at which a peak begins to rise from the detector signal baseline or, in the case of a fused peak, the point at which the previous peak ends and the current peak begins.
Peak End (E) is where the current peak signal ends and returns to detector baseline or, in the case of a fused peak, where the next peak begins. In the case of fused peaks, the same point, called a valley point, represents the start of one peak and the end point of the previous peak.
Baseline (B) is a straight line which defines the bottom edge of the peak. Height is measured vertically from this line up to the peak apex. The shape of a chromatographic peak, and thus its area, is determined by the peak apex, start, end, and baseline. Baseline is a very important parameter because its placement can have a profound effect on area or height determination.

B. Quantitation

Determining the amount (or quantity) of a compound from the peak height or area is called quantitation. The first step in performing quantitation with chromatography is to calibrate the chromatographic system. Calibration requires that standard solutions be made with known amounts of each compound to be quantitated. In the simplest instance of calibration, these standard solutions are injected and the area or height of each peak is determined. A calibration curve is then constructed from the data. A simple calibration curve for a single peak, using height, is shown in Fig. 14.2. Two standards, one containing 2mg of the compound and the other 4mg, produce peak heights of 1.5cm and 3.0cm, respectively. The calibration curve is created by drawing a "best fit" straight line through these two points.

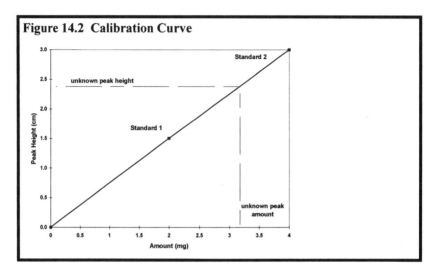

Figure 14.2 Calibration Curve

The amounts of unknown samples can be derived from the calibration curve. For example, in Fig. 14.2, an unknown sample which has a height of 2.4cm correlates to 3.2mg. This type of quantitation is known as the external standard method.

II. CLASSICAL DATA HANDLING TECHNIQUES

Computers have made chromatography data handling faster, less labor intensive, and better documented but they have not necessarily improved the accuracy of the results obtained. Simple tools such as rulers, pencils, and graph paper are all that is needed to perform accurate chromatographic data processing. Knowledge of these simple techniques gives a better understanding of how chromatography data is processed and also provides a fallback system to process data or to help diagnose what has gone wrong when the computer fails.

A. Peak Height

The simplest form of data handling uses a ruler to measure peak height on a chromatogram from a chart recorder. For the sake of good documentation, the baselines used to measure the peak height should be drawn directly on the chromatogram and the measured peak height noted. A metric ruler should be used to measure the peak heights rather than using the printed grid on the chart paper. Units of the metric system are easier to plot or calculate than those of the English system.

Calibration is performed by plotting the peak heights vs. the known amounts in the standard peaks (assuming external standard calibration is used) on scientific graph paper, as shown in Fig. 14.2. Each peak generally has a separate calibration curve. Sample peak heights should be measured with the same ruler used for the calibration runs and the unknown amounts can be determined from the calibration curves, as shown above. It is important to maintain the same attenuation and gain values on both the chart recorder and detector for all runs.

B. Peak Area

The cut and weigh technique is a manual method for measuring area which is based on the consistent weight per unit area of the chart paper. Each peak is cut out and weighed on a balance. This weight is used as a measurement of area. Consistent gain and attenuation settings must be used for all runs. A fairly fast chart speed avoids significant errors by ensuring that the peaks are large enough to handle, mark, and cut. A large, wide peak is easy to cut out accurately. All baselines and weights should be marked directly on the cut out and a photocopy of the intact chromatogram should be kept with the cutouts.

III. BASIC CONCEPTS OF ELECTRONIC DATA HANDLING

The human mind is the finest computer for processing visual or graphic imagery that has ever existed. It can instantly see the peak apex, start and end, as well as imagine a reasonable baseline for any chromatogram. Manual processing is not practical, however, when hundreds or thousands of samples go through a laboratory every day. In these cases, a computer can perform these actions rapidly by reducing the data to mathematical expressions.

A. The Computer's View

The four essential elements for peak integration (peak apex, peak start, peak end and baseline) must be defined mathematically for a computer. The key term in computer data handling is the slope, or how much a signal changes in the vertical direction relative to the change in the horizontal direction (dy/dx). In a chromatography peak, the slope changes continuously during the run, as seen in Fig. 14.3A. The fact that the slope can be positive or negative allows creation of a computer program that can integrate chromatograms (1). The computer program uses the first derivative of the chromatographic signal to determine slope; a plot of the first derivative, as shown in Fig. 14.3B, gives a continuous picture of the slope.

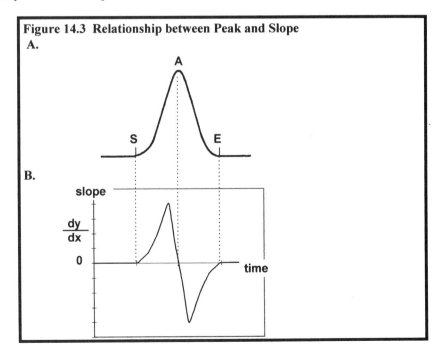

Figure 14.3 Relationship between Peak and Slope
A.
B.

The peak apex, start, and end, as well as the baseline, can be redefined using the concept of slope because they all represent points of zero slope, as seen in Fig. 14.3B. The baseline has zero slope because there is no change in y as x changes. The peak apex (A) is the point where the chromatographic trace goes from positive to negative slope. On the first derivative plot, this is the point where the line crosses zero in a negative (downward) direction. The peak start (S) is the last point of zero slope (baseline) before the chromatographic trace begins to exhibit increasingly positive (upward) slope. The peak end (E) is the first point of zero slope after the chromatographic trace has shown negative or downward slope. In the case of fused peaks, the valley (end of leading peak and start of following peak) is the point where the chromatographic trace goes from negative to positive slope in a positive (upward) direction.

B. Digital Chromatographic Data

A chart recorder representation of a chromatogram is analog data, a continuous variation of voltage with time. Computers must convert analog data to digital data or a series of discrete points defined mathematically. Plotting these points and "connecting the dots" reproduces the analog version of the chromatogram. An integrating analog-to-digital converter (ADC) continuously integrates the area under the detector signal (2), as shown in Fig. 14.4. Each of the trapezoidal shapes is a slice of the chromatographic signal and is stored as a data point which represents two important numbers: (1) the time since the beginning of the chromatographic run and (2) the accumulated area under the chromatographic curve since the previous data point.

Figure 14.4 Transformation of Analog Data

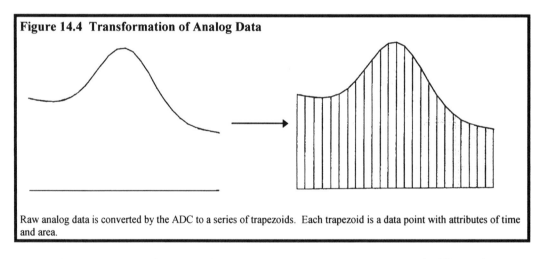

Raw analog data is converted by the ADC to a series of trapezoids. Each trapezoid is a data point with attributes of time and area.

The ADC has three important characteristics: data rate, range and least significant bit. The data rate is the maximum number of area points the ADC can generate in one second, stated in hertz (Hz), meaning "events per second". Typically, ADC rates in chromatography are not greater than 100 Hz. The data rate can affect processing accuracy, as discussed below.

An ADC has a limited linear range of input voltages. Voltages outside this range may produce nonlinear values and may even cause physical damage to the ADC electronics; therefore, the range of an ADC should be matched to the detector input range.

The least significant bit (LSB) is the smallest value an ADC can measure in voltage. For example, an ADC with a 1V range may have an LSB of 10^{-6} V or 1 μV which is the smallest detectable change in the incoming signal. This presents another reason for matching the ADC range as closely as possible to the detector input range. In this example, the 1V ADC can generate 10^6 points across the full 1V range, but if the detector has only a 1mV range, the resolution is limited to only 1,000 points or 0.1% of the available ADC range.

IV. PEAK INTEGRATION USING THE DATA SYSTEM

The manual techniques for determining peak areas and heights seem very simple but, in fact, they require very complex interaction of visual and tactile senses with intellectual understanding of the task. The steps required by both the human and the computer to perform the "simple" task of peak integration are shown in Table 14.1.

Table 14.1 Steps in Peak Integration

- Identify which elements of the chromatogram are peaks.
- Identify the noise and drift in the chromatogram.
- Determine retention times for the peaks.
- Determine which peaks represent compounds of interest and identify them.
- Determine an appropriate baseline for each peak of interest.
- Calculate the area and/or height for each peak.

A. Peak Width Factors

1. General

Most chromatography data systems acquire chromatographic points and determine peaks without any user intervention because settings are defaulted to "typical" values for chromatography. It is essential, however, for accurate data acquisition, that these parameters be understood so that they are set correctly for any particular chromatographic data.

Nearly all data systems have some means of inputting the width of the expected peaks at baseline or at half-height. This value is used by the data system to set the data acquisition or "bunching" rate for data points and to determine the chromatographic noise and drift.

2. Data Acquisition Rate

Peak width is used to set the data acquisition rate because the width is directly related to the number of data points required to accurately reproduce the peak in the analog-to-digital conversion process. Usually 10 - 20 points are required between peak start and peak end to accurately represent a peak (2). A peak which is 2 seconds wide at the base thus requires a data rate of 5 - 10 Hz. A peak which is 16s wide requires a data rate of only 1.25 Hz to acquire 20 points across the peak.

If the data rate is too low, the points will not accurately show the profile of the original chromatogram and the area for the peak will not be accurately calculated, due to information lost in the analog-to-digital conversion process. Wide variations in either peak shape or resolution between peaks can seriously impact the optimum data rate. Gaussian peaks are usually about twice as wide at the base as at half height. If a severely tailing peak is 2s wide at half-height but 8s wide at the base, a value of "8" for peak width will result in too few points being acquired for the peak. For this reason, many data systems use the peak width at half-height which biases data acquisition towards too many points rather than too few.

3. Bunching

Data systems acquire data at a constant rate and then bunch them to obtain the correct data rate and to increase sensitivity (2). Table 14.2 shows an example of a system in which the ADC acquires data at 40 Hz and data points are bunched to obtain 10 points per peak for peaks of various widths.

Table 14.2 Bunching Points to Achieve Specific Data Rates

Expected Peak Width at Base (s)	Number of Points Bunched	Effective Data Rate (Hz)	# of Points per Peak
1	4	10	10
2	8	5	10
4	16	2.5	10
8	32	1.25	10
16	64	0.625	10

ADC is operating at 40 Hz.

The bunching rate is only doubled when the peak width doubles. If the ADC is run at its maximum data rate for the entire chromatogram, too many data points may be collected. Bunching of data points through proper entry of the peak width value aids the data system in correctly finding only true chromatographic peaks. Additionally, bunching of data points has the effect of filtering out high frequency noise, thereby increasing the sensitivity of the system to peaks of the desired width. Incorrect point bunching may result in missed peaks and lower sensitivity. With present computers which have large hard drives and memory, storage of all the data points is not the problem it was formerly.

B. Noise and Drift

Chromatographic noise, peaks, and drift are only different in the perspective from which they are viewed or the peak widths expected in the analysis. Fig. 14.5 shows three chromatographic signals which are very short term (A), medium term (B), and long term (C) variations. The interpretation of which is noise, drift, or a chromatographic peak depends on the expected peak width. If peaks of 3s base width are expected, then A is noise, C is drift and B is the chromatogram. If peaks are expected to be one-half second wide, then A may be the chromatographic signal and B and C drift or other baseline anomalies. Finally, if 20s peak widths are expected, then C is the peak and A and B are noise.

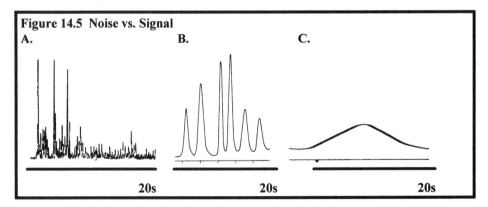

Figure 14.5 Noise vs. Signal
A.   B.   C.

20s   20s   20s

A simple definition for noise is signal variations of very high frequency relative to the expected peak width. Noise is generally of very low amplitude relative to that of the expected peak. Peak amplitudes must be at least 2 - 3 times greater than the noise. Drift is defined as signal variations of very low frequency relative to the expected peak width. The peak width value deals with noise and drift by telling the system to ignore signals whose widths fall outside the specified peak width range. Most data systems can reliably detect peaks that are 0.25 - 4 times the set peak width value. Narrower signals are considered noise and wider are regarded as drift.

C. Slope Sensitivity

In Fig. 14.3, it was seen that the data system does not look at the "raw" data but rather at a continuous plot of the slope (first derivative) of the chromatographic signal. The slope of the side of a peak is related to its width; therefore, peaks within a given range of peak widths also fall within a fixed range of slope values. Slope Sensitivity is the range of upper and lower slope values that correspond to approximately 0.25 - 4 times the set peak width value. When a slope within this range is detected, the signal is treated as a chromatographic peak, but slopes outside the range are ignored. Slope sensitivity varies between data systems. Peak width values in one manufacturer's data system may not correspond exactly to those in another and, therefore, an integration method cannot be easily transferred between different systems.

D. Detection Threshold

Most data systems have the option of automatically monitoring the baseline noise at the beginning of a chromatographic run by observing a fixed number of bunched points, usually 40 - 60, before the chromatographic injection. The peak width setting is again relevant because bunched points are monitored. A user-entered multiplier value called "signal-to-noise", "threshold" or "detection limit" is then applied to the noise. Only peaks whose apex amplitude exceeds the detection threshold value: (system-derived noise) x (user-entered multiplier) are reported. To evaluate only the major peaks in a chromatogram and eliminate small contaminant peaks, the threshold value should be set to a high number, such as 5 - 10 times noise. For trace analysis, the threshold value should be set as low as 2 - 3 times the noise. The optimum setting is that which permits consistent detection of all peaks of interest. Chromatographic peaks of amplitude less than two times the noise value cannot be reproducibly evaluated.

A noise value may be manually entered without the actual noise being monitored if the baseline noise before injection is considerably different from that encountered later in the chromatogram. The differences may result from changes in flowrates, solvent composition, detector settings, or valve switching. Manually entered noise values should be measured under specific elution conditions and periodically rechecked. The method for measuring and re-checking manually integrated noise should be part of the laboratory's Standard Operating Procedure (SOP).

In addition to the detection threshold setting, many systems also allow the user to specify minimum peak heights or areas. This is not the same as a detection threshold and it is not used to control peak detection, but rather it rejects peaks based on their area or height after integration is completed. The minimum area/height rejection reduces the list of reported peaks when there are numerous minor peaks, but it may also erroneously eliminate peaks of interest which are lower than this minimum area/height value.

E. Peak Detection and Retention Time

After slope sensitivity and threshold values have been established, it is possible to process individual peak data. The first step is to identify the peak apex, start, and end. Peak apex or retention time must be determined first so that the peak can be integrated and quantitated. Data processing may occur in real time or after the data has been stored electronically. In either case, the computer looks at the data points sequentially starting from zero time and proceeding to the end of the chromatogram. A buffer of about 60 data points back from the point currently being processed is kept at all times. As the computer moves to the next point, the last point in the buffer is dropped and the new point is added, keeping the total number of buffered points constant. It is important that the computer view a portion of the chromatogram greater than one data point to prevent it from assuming that a given point is the start, end, or apex of a peak. Computer evaluation proceeds as in Table 14.3.

TABLE 14.3 Computer Evaluation of Chromatographic Data

1. Recognize a point that is within the slope sensitivity range and greater than the threshold value.
2. Look back through the data point buffer and find the peak start point, which is the last point of zero slope.
3. Save the point of zero slope as a possible peak start.
4. Identify the peak apex (when the new point in the buffer is at negative slope and the one before it was positive slope).
5. Determine the exact retention time or the exact point of zero slope through interpolation with a least squares fit of the straight line through the points.
6. Find the peak end point which is the next point of zero slope in the buffer.
7. Draw the baseline and determine the area and height.

When the computer looks for the peak start and end, it is looking for the slope to be zero, just as it does for the peak apex. In these cases, interpolation is again used to determine the exact start or end of a resolved peak or the valley point between fused peaks. If a peak end point is at a lower absolute signal level then the peak start, then the peak end is not simply a return to or passing through zero slope but it is rather the point corresponding to the intersection of the lowest tangent line drawn from the peak start point, as illustrated in Fig. 14.6.

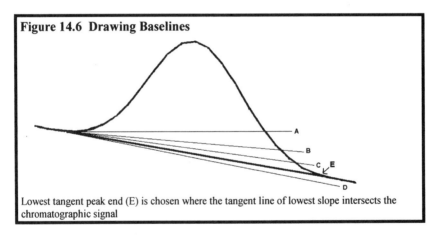

Figure 14.6 Drawing Baselines

Lowest tangent peak end (E) is chosen where the tangent line of lowest slope intersects the chromatographic signal

If, after a peak start point is detected, no peak apex is found within the window established by the peak width value, the peak start point is ignored and the data system continues to treat the incoming points as baseline. Such may occur in the case of a drifting baseline or a baseline offset. For a peak to be detected and reported, all three key points of peak start, end, and apex must be found.

F. Assigning Peak Baselines

1. General

Baselines must be established to determine the area or height for each peak of interest in the chromatogram. Assigning baselines is the most difficult aspect of data handling, requiring both experience in chromatography and knowledge of the specific analysis. Automatic baseline assignment uses pre-set rules for each data system. Most systems deal accurately with simple, well resolved peaks and can adjust for drifting baselines. However, it is often necessary to change the pre-set rules to force proper peak integration. Sometimes this is done by programming baseline events in a timed events table. It is often possible to use the graphics functions to interactively move and/or draw baselines after the data have been collected. It is best to program the data analysis method to draw the baselines correctly for most samples to minimize baseline modification and recalculation.

Fig. 14.7 illustrates common varieties of baselines. It is simple to assign a baseline to well-resolved peaks like c and d in Fig. 14.7A, which is why total baseline resolution is preferred in chromatography.

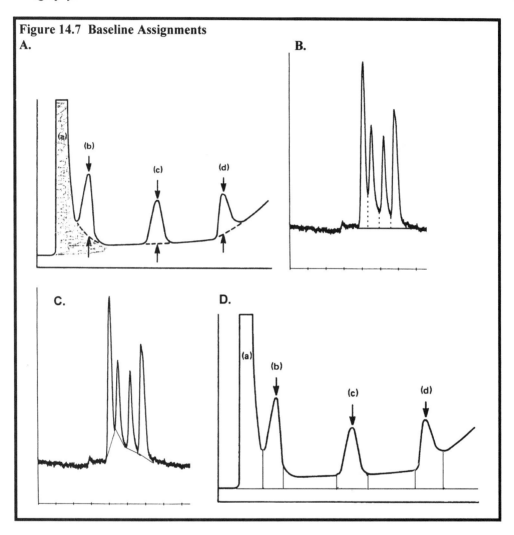

Figure 14.7 Baseline Assignments

Fig. 14.7B and C illustrate fused peaks baselines. Fig. 14.7B assumes that the cluster of unresolved peaks has a flat baseline, as verified by running a blank analysis. To integrate the peaks, a baseline is drawn from the first fused peak start to the last fused peak end and perpendicular lines are dropped from each peak valley point. This method does not precisely integrate the overlapped peaks, but it has been found to be a good approximation. It is also simpler to integrate using perpendicular drops than to mathematically deconvolute or pull apart each individual peak.

Fig. 14.7C shows the same group of fused peaks, but in this case the peaks are considered resolved peaks on a baseline which is rising and falling, as confirmed by a blank run. Baselines are drawn directly from valley to valley for each peak.

All data systems have the capability of programming baselines such as those in Fig. 14.7A, B and C. Integration as shown in Fig. 14.7A is usually automatic. Integration of peaks like those in Fig. 14.7B and C will be specified in the manual under programming baseline assignments.

2. Tangent Peaks

The small fused peak (b) shown in Fig. 14.7A may be considered a tangent or skimmed peak because it is much smaller than the large first peak, generally referred to as the mother peak. In this case, the area of the mother peak (shaded area) includes the area under the tangent peak baseline; only the small area under the tangent peak and above the tangent baseline is assigned to a tangent peak. Although generally it is correct to integrate very small peaks next to a large peak this way, the exact size where a peak should be considered a tangent must be determined from experience and experimentation. Peaks are usually considered tangent candidates when their height is less than 10% of the mother peak.

In most data systems, tangent peaks are defined as a proportion of the height of the mother peak. The manual will specify the tangent capabilities, the definition of tangent peaks which elute after the mother peak apex and sometimes those which occur before the mother peak apex. If tangent peaks are defined by a certain relative size, most systems will override other baseline assignments when they are found. For example, even though fused peaks have been programmed to integrate as shown in Fig. 14.7B, if tangent peaks are set in a data system at 10% of the mother peak height, then integration as shown in Fig. 14.7A (peaks a and b) will occur if the second and third peak heights drop below 10% of the height of the first peak.

Quantitative analysis on tangent peaks should be approached with caution, especially if the peak heights in the sample vary enough that some peaks are processed as tangents and others as fused or resolved. It is best to adjust the separation so that small peaks are well separated from adjacent large peaks. For the best quantitative results, a peak of interest should be processed with the same baseline assignment in every sample.

3. Horizontal Baseline

One further baseline assignment is the forced horizontal baseline illustrated in Fig. 14.7D. This baseline is rarely used in biotechnology applications but it does have specific uses in gas chromatography and in some size exclusion analyses. The horizontal baseline is usually programmed by specifying a starting and ending time. Most systems offer horizontal baselines in the forward direction only, drawn from an earlier time to a later time point. It is best to specify the starting time as immediately after the apex of the peak preceding the horizontal baseline's desired start point so that the horizontal baseline will start as soon as the next peak start is found. Horizontal baselines will often result in inaccurate integration because they are flat regardless of the shape of the chromatogram.

4. Manual Adjustment

It is usually possible to readjust the baseline at any point in the chromatogram after a baseline upset, such as a solvent front or valve switch. An instantaneous baseline point can be programmed to occur after the baseline has returned to its normal level following the baseline upset.

G. Peak Area and Height Determination

As discussed earlier, each data "point" actually represents two values as shown in Fig. 14.4: (1) the time since injection and (2) the accumulated area under the chromatographic curve since the previous data point. The initial determination of peak area is the sum of all slices between peak start and peak end down to the zero level of the ADC, as illustrated in Fig. 14.8. After the baseline is established, the data system subtracts the area below the baseline from the peak area so that the final reported area for each peak is that bounded by the chromatographic curve and the baseline (the more lightly shaded area).

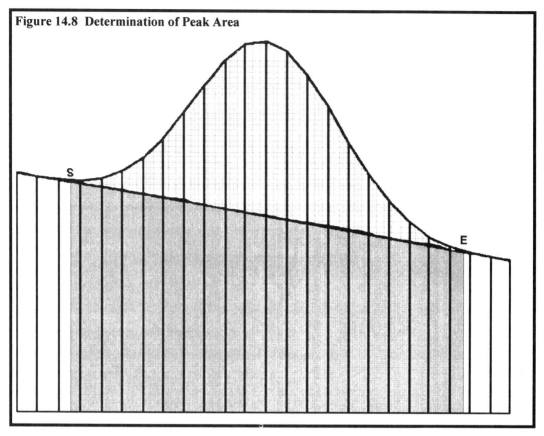

Figure 14.8 Determination of Peak Area

Height calculations are determined by the length of a line drawn vertically from the exact, interpolated peak apex to the chromatographic baseline. Most data systems report area and height in terms of the "least significant bit" (LSB), which is the smallest unit which the ADC can detect or the value "1" in the digital data. The LSB is generally a voltage, but may be specific to the detector being used, for example, Absorbance Units (AU) in the case of a UV detector. A height unit is simply an LSB. Area units are stated as the LSB multiplied by time in seconds.

H. Selection of Area or Height Measurement

Area and height of a chromatographic peak can each be affected by changes in the HPLC system. Because system changes which affect one measurement usually do not affect the other, it is important to understand when each should be used. Most HPLC detectors are concentration-dependent because they respond to the concentration of analyte in the flow cell, not to the absolute amount. Responses of this type of detectors are not sensitive to flowrate changes. In fact, if flow is stopped completely with analyte in the flow cell, the detector response will remain constant. The mass-dependent detector, on the other hand, has a response which depends on the absolute mass of analyte in the detector, and is highly sensitive to flowrate changes. Mass dependent detectors for HPLC include some mass spectrometers, evaporative light scattering, and electrochemical detectors (3). The following discussion assumes the use of concentration-dependent detectors.

In an HPLC system where retention time, area, and height variations are minimal, either peak area or peak height measurements will give acceptable results. When HPLC systems are not absolutely consistent, however, the correct choice of area or height is critical for both precision and accuracy of results. The choice of peak height or peak area measurement for quantitation is based on the sources of variation in the HPLC system, as shown in Table 14.4.

Table 14.4 Measurement Selection Based on Operating Parameters		
Primary Source of Run-to-Run Variation	**Typical Causes**	**Most Precise Measurement**
Flowrate	Imprecise pumping system Leaks Flow programming Mass-dependent detector	**Peak Height**
Mobile/stationary phase variation	Column changes compression contamination loss of stationary phase Volatile mobile phase Gradient analysis Temperature changes	**Peak Area**

Tailing or asymmetric peaks present a special problem, especially if they are not totally resolved. In cases where resolution is not complete, peak height may be most accurate (4). When resolution is good, peak area is commonly used because chromatographic variations tend to effect peak height. Flowrate problems, which effect peak area, should always be corrected immediately (3).

I. Background Subtraction

Many data systems allow the subtraction of a background signal from the raw chromatographic data. If the baseline change caused by gradients is severe, subtracting it out of the chromatogram may allow more accurate baseline assignment, as well as present a more visually acceptable chromatogram. Generally, subtracting background from a chromatogram without complete documentation would violate GMP requirements.

The most important requirement of background signal subtraction is that the signal being subtracted must be absolutely reproducible. Random drift, baseline upsets, and, in most cases, unwanted chromatographic peaks or noise spikes cannot be subtracted. Only a relatively gradual baseline change which is absolutely reproducible from run-to-run is a candidate for subtraction. For background subtraction, a blank run like that in Fig. 14.9a must be run without sample injection so that the data system can store it and later subtract it from chromatograms like that in Fig. 14.9b prior to data processing.

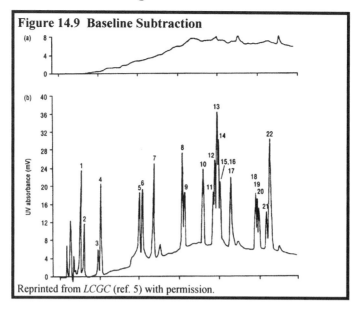

Figure 14.9 Baseline Subtraction

Reprinted from *LCGC* (ref. 5) with permission.

In a GLP environment, it is essential that the raw chromatographic data does not have the background profile subtracted from it; therefore, it must be possible to reprint the original raw chromatographic data.

J. Integration of Capillary Electrophoresis Data

The peak profiles and baselines observed in capillary electropherograms exhibit several differences compared to HPLC chromatograms, which should be taken into account when performing quantitative analysis with CE. In chromatography, all analytes migrate through the detector flow cell at the same rate, i.e. the flow velocity of the mobile phase. Therefore, quantitative estimates of analyte concentration can be obtained directly from peak area. In CE, analyte migration velocity past the detection window is the sum of electroosmotic flow velocity and electrophoretic mobility. Peak area will be a function of both analyte concentration and residency time in the detector window (which is inversely proportional to migration velocity). Variations in migration time can, therefore, introduce significant error if peak area is used for quantitation. These variations can be eliminated by normalizing peak area to migration time, and area/t_m values should always be used instead of raw area for calibrated analysis. Alternatively, peak height can be used for quantitation.

CE peaks may be qualitatively different from typical HPLC peaks. In fast separations, they can be quite narrow, requiring adjustment of detector time constants and integrator filtering to prevent peak distortion. Peaks may be asymmetric due to electrodispersion (often seen in the analysis of high-mobility species such as inorganic ions) and peak detection parameters may require optimization for correct baseline assignments.

Application of high voltage at the beginning of a CE run often causes spikes in the detector response which can interfere with correct peak integration. Inhibiting integration for the first few seconds after injection will prevent integration artifacts caused by spikes. In HPLC, baseline noise is often sampled automatically by the data system prior to injection, and the noise value used to determine the threshold. In CE, the noise signature before injection (before application of high voltage) can be quite different than that following injection (high voltage applied). To accommodate this, some data systems designed for CE applications provide for noise sampling after injection during a characteristic baseline segment of the analysis, or in a prior blank run.

V. QUANTITATION USING THE DATA SYSTEM

Two important aspects of chromatographic peak quantitation are the limit of detection (LOD) and the limit of quantitation (LOQ) which are given in terms of signal to noise ratios (S/N) (6). LOD at $S/N = 3$ defines the smallest peak that can confidently be judged to be a peak. LOQ at $S/N = 10$ defines the smallest peak which can be measured with acceptable precision (3). These criteria must be met to accurately identify (LOD) or quantitate (LOQ) a peak, sometimes necessitating larger injection volumes or concentrations.

The simple procedure for quantitation involving calibration with standard solutions to determine the concentration of unknown sample solutions is also followed by the data system:
1. Develop the chromatographic method and identify each peak of interest in terms of retention time.
2. Decide which quantitative calculation will be used.
3. Decide which calibration scheme will be used.

A. Identifying Peaks of Interest

1. Identification

The goal of HPLC method development is to elute the peaks of interest well separated from other peaks and contaminants within an acceptable time limit. If the data system parameters are set correctly, all of the peaks in the run will be detected (referred to as detected peaks). Once this goal is achieved, a "peak table" must be created for the data system to identify the peaks in the chromatogram. The set of peaks which are in both the peak table and the chromatogram will be called the identified peaks.

The entries in a peak table consist of the component names and corresponding retention times, which are required by the data system to identify peaks. Proper identification is essential; therefore, it is vital to understand the process so it can be controlled and modified.

Retention times vary from one run to the next due to random fluctuations in flowrate, temperature, and solvent composition; wear and tear on seals, columns, and mechanical devices; and human error. In many applications, retention time variation over the short term is less than 1% relative standard deviation (RSD). A peak with an average retention time of 14.875min with a RSD of ±1% may elute from 14.726 - 15.024min. Sometimes the average retention time value will drift over time even though the time variation remains at ±1%. The data system must be programmed to reliably find such a peak run after run.

All data systems will allow the user to specify a time "window" for each peak within which all detected peaks are candidates for identification. The window is stated as a percentage of the retention time of the peak of interest and/or as an absolute value in time units. The window should be set two to three times wider than the observed retention time variation to insure that the peak is found. Within each window, the data system will identify the peak of interest in one of two distinct ways: 1. the peak whose apex is closest to the center of the window or 2. the largest peak within the window. Generally, data systems will default to identifying the peak as the one closest to the center, which is usually the most reliable method. The second method of finding the largest peak in the window is normally reserved for internal standard and retention time reference peaks. It can also be used if there are numerous background peaks which are always smaller than the peak of interest. The largest peak method permits use of a larger retention time window, which may be valuable in cases of significant retention time variation.

2. Retention Time Variation

Retention time reproducibility should be checked periodically. If the retention times and peak areas of five injections are within stated limits, the system is operating satisfactorily. The relative standard deviation of the retention time should always be less than 2% and optimally, no more than 1%.

If there are variations in retention, the flowrate, column, or mobile phase may be changing. Variations in area can usually be attributed to the autosampler or injection.

The effect of column aging on retention time is generally monotonic, or in one direction. For example, as a column is used repeatedly, retention times may become smaller. Data systems can track this type of change by means of a retention time reference peak, which is always the largest peak in the retention time window. Only peaks like internal standards, which are always present in the chromatogram at fairly high and constant levels, should be used for reference. Compounds which are totally resolved from peaks of interest may also be added to a sample as references.

Data system manuals will specify the use of reference peaks. Generally, reference peak retention times are used to modify the retention time windows for all peaks in the peak table, using a formula which changes the center value of the retention time window in the direction of the retention time change. This is often the percent of change of the reference peak. For long chromatograms, two or three reference peaks spaced evenly throughout the run should be used. Reference peak updating is usually optional.

B. Quantitative Calculations

1. Area Percent

The simplest quantitative calculation simply adds up the areas for all of the identified peaks and reports what percentage of the total is represented by each. Mathematically, the result for a peak (Pk) is stated as:

$$\text{Area\%}_{Pk} = (\text{Area}_{Pk} / \Sigma\text{Area}) \times 100 \qquad \text{Eq. 14.1}$$

In most data systems, Area% results are based on all detected peaks, even if no peaks are identified. Area% is not routinely useful for quantitation, but it is often used during method development to maintain a rough idea of the relative amounts of each component present. Area% is sometimes used as a measure of protein purity using UV detection at 280nm or 254nm. Area% is also the common format of reporting the amount of hemoglobin A_{1c} present in hemolyzed blood analyzed by ion-exchange chromatography with detection at 415nm (7).

2. External Standard

The external standard quantitation method, as described earlier, uses the relationship between detector response and the amount of a known standard to determine the quantity in an unknown sample (2). Calibration requires one or more standard solutions containing known amounts of all peaks of interest. Each standard is run and a calibration curve of area vs. amount for each peak is determined. In the simplest case, the calibration curve is a straight line which has an intercept at (0,0), as illustrated in Fig. 14.2. After calibration, the response factor (RF) for each peak can be calculated as follows:

$$RF = x/y = \text{Amount} / \text{Area} \qquad \text{Eq. 14.2}$$

After the data system has calculated the area for each peak of interest in an unknown sample, the amount can be calculated by the response factor:

$$\text{Amount} = RF \times \text{Area} \qquad \text{Eq. 14.3}$$

Additional factors may be used to compensate for dilution or purity of the standard (See Section V.F.). An external standard is very simple to understand and to use; however, it has a dependence on injection volume. If the volumes of standard and sample are not the same, then the concentration will not directly relate to mass. Autosamplers generally achieve good precision using external standards. Another drawback of external standards is that they cannot compensate for errors in sample preparation.

3. Internal Standard

Internal standards are physically added to each sample and are thus subjected to the same sample processing as the other components. In this way, compensation is made for losses due to technique or steps in sample preparation. The quantitation process for internal standards also involves calibration with standards and calculation of response factors. The only apparent difference from the external standard method is in the preparation of the standards and samples; however, the standard curves are different in that they use relative values instead of absolute.

The first step in internal standard analysis is to choose a suitable standard compound(s) which meets the criteria in Table 14.5.

Table 14.5 Requirements for an Internal Standard

1. Chemically similar to the compounds of interest.
   a. similar detector response.
   b. similar recovery from sample preparation methods.
2. Chromatographically separated from all other peaks in the analysis with
      a similar retention time to the compounds of interest.
3. Cannot occur naturally in any of the samples.
4. Stable.
5. Commercially available in high purity or HPLC grade.

Standard solutions must contain known amounts of every peak of interest, as well as a known amount of an internal standard compound. The internal standard is added to every standard and sample at the same level. When serial dilution is used to make up standard solutions, the internal standard may be included in the diluent or added after dilution.

It is important to add the internal standard in an amount which gives approximately the same peak area/height as the compounds of interest. If there is wide variation in the areas/heights of the analytes in the sample, multiple internal standards should be implemented. If samples with higher concentrations are diluted, then a multiplier must be used to obtain the true concentrations for these samples.

For internal standards, the calibration curve points are the ratio of the amount or area of the peak of interest (Pk) to that of the internal standard peak (IS):

$$y = \text{Area}_{Pk} / \text{Area}_{IS} \qquad\qquad \text{Eq. 14.4}$$
$$x = \text{Amount}_{Pk} / \text{Amount}_{IS}$$

The key advantage of internal standards is that these ratios, and thus accurate quantitation, are independent of injection volume. Additionally, if the internal standard is added to the samples prior to sample preparation, any losses which affect the compounds of interest will similarly affect the internal standard so that the quantitative results will accurately reflect the amount originally in the sample. This is only true if the chemical behavior of the internal standard is similar to that of the sample, as confirmed experimentally. If they are not similar, the internal standard must be added after sample preparation and another type of standard, called a surrogate, must be used to track recovery from sample preparation. In this case, the internal standard only compensates for injection volume variation.

For a linear calibration with a (0,0) intercept, the calibration process determines a relative response factor (RRF) for each peak:

$$RRF_{Pk} = \frac{y}{x} = \frac{Area_{Pk} / Area_{IS}}{Amount_{Pk} / Amount_{IS}} = \frac{(Area_{Pk})(Amount_{IS})}{(Area_{IS})(Amount_{Pk})} \qquad Eq.\ 14.5$$

The RRF is simply the slope of the calibration line. For each analysis, the amount of internal standard in the sample (Amount$_{IS}$) is entered into the system which calculates the Amount$_{Pk}$ after the peak areas are determined:

$$Amount_{Pk} = \frac{(Amount_{IS})(Area_{Pk})}{(Area_{IS})(RRF_{Pk})} \qquad Eq.\ 14.6$$

If multiple internal standards are used for a sample, each standard applies to a subset of the identified peaks. Multiple internal standards should be considered under the following circumstances:

1. No single internal standard can be found which is chemically similar to all the compounds of interest in the analysis. In this case, compounds of interest are associated with a standard for each chemical class.
2. A very long chromatographic run is being used, which requires two or more standards that elute in the early, middle and/or late portions.
3. Compounds of interest are likely to be present at vastly different concentrations, necessitating standards at concentrations similar to each expected level.

C. Selecting the Calibration Curve

Up to this point, only linear calibrations which intercept the point (0,0), like that shown in Fig. 14.2, have been discussed. Other mathematical relationships are sometimes encountered which either intercept the y-axis at a point other than the origin or which have nonlinear portions of the curve. Although nonlinear equations can be dealt with mathematically, it is general practice to operate in the linear range of a curve by adjusting concentrations. The linearity of the calibration curve should always be confirmed experimentally for the range of concentrations being used.

Most HPLC detectors are quite linear within the normal ranges used. For example, UV detector response for the entire operational range of 0 to 2.0 absorbance units (AU) results in a second order curve. However, the typical operating range is below 1.0 AU and well within the linear portion of the response curve.

The mathematical minimum number of calibration standards is one plus the order of the curve; therefore, a linear curve requires a minimum of $(1 + 1) = 2$ calibration standards to define it. Higher order curves would require additional calibration standards. The mathematical minimum can never adequately ensure accurate quantitation.

D. Selecting the Calibration Scheme

A calibration scheme stipulates the number and composition of standards, the order in which they are run relative to the samples, and the final calculation of the results. The decisions which must be made in designing a calibration scheme are listed in Table 14.6.

Table 14.6 Factors in Calibration Scheme Design

- Number of compounds of interest in each standard
- Number of standard concentrations or amounts (levels)
- Number of replicate injections for each standard
- Frequency of re-injection of the standards for re-calibration
- Use of the re-calibration results
- Method for calculation of the unknown samples

---

1. Number of Standard Levels

The number of levels to be run is defined by the calibration curve type, the laboratory SOP and/or the regulatory agency SOP. At least one more standard than the order of the curve is always required. Beyond that, most SOPs will require at least one additional standard to add statistical reliability to the results.

2. Amounts in Each Standard Level

The amounts chosen for each standard level should be spaced equally throughout the range of expected concentrations of the unknown samples. The lowest standard concentration must be somewhat below the minimum expected sample level. Similarly, the highest standard concentration must be above the maximum expected. Samples whose concentrations fall outside the calibration range may not yield a linear response.

3. Replicate Injections

One way to add statistical reliability is to run replicate injections of each standard at each level and to use the average of the replicate results as the calibration curve point. A statistical test is often applied to each replicate result to ensure that it is within the expected range. If it is not, the data system may issue a warning to look for the cause, e.g. an incorrectly made standard, injection failure, or detector failure. Standards failing this statistical test are not included in the average result. Regulatory agencies or standard procedures within a laboratory will dictate the number of replicates to be run, but a minimum of two at each standard level is needed to ensure accurate calibration.

4. Frequency of Recalibration

It is necessary to recalibrate periodically to guarantee continued accuracy of quantitative results because HPLC systems will change over time. Detector lamps will lose intensity, columns will give different retention times, and flowrates or solvent composition may change. In many laboratories, calibration results are tracked statistically to determine when the HPLC system needs maintenance or repair.

The number of samples which can be run before recalibration is necessary is usually specified in the SOP. When the system is recalibrated, all calibration standards at every concentration level must be rerun because it is not scientifically valid to use only a subset of the standards when sample concentrations vary widely across the calibration range. Regardless of how many samples have been run, recalibration should be implemented whenever the HPLC system has been shut down and restarted or if a method is changed.

5. Use of Recalibration Results

It is very important that the calibration schemes required by the specific laboratory SOP not only be used for an analysis, but also supported by the data system and implemented properly. Three simple calibration schemes, replacement, averaging and bracketing, cover most chromatography laboratory requirements. In the examples shown in Table 14.7, two replicate standards are injected at each of two concentration levels along with eight samples (A-H). One calibration level is satisfactory for inspecting the sample when concentrations are confined to a narrow range within the calibration plot. For example, in formulation assays, only specific levels are expected.

Injection	Replacement		Averaging		Bracketing	
		1		**1**		**1**
1	Cal. level 1		Cal. level 1		Cal. level 1	
2	Cal. level 1	Calculate	Cal. level 1	Calculate Curve	Cal. level 1	Calculate Curve
3	Cal. level 2	Curve	Cal. level 2		Cal. level 2	
4	Cal. level 2		Cal. level 2		Cal. level 2	
5	Sample A		Sample A		Sample A	
6	Sample B	**2** Apply to	Sample B	**2** Apply to	Sample B	
7	Sample C	Samples	Sample C	Samples	Sample C	**3**
8	Sample D		Sample D		Sample D	Apply
9	Cal. level 1		Cal. level 1	Calculate Curve	Sample E	Averaged Curve to
10	Cal. level 1	**3** Calculate	Cal. level 1	**3** and	Sample F	Samples
11	Cal. level 2	Curve	Cal. level 2	Average with	Sample G	
12	Cal. level 2		Cal. level 2	Previous Calibration	Sample H	
13	Sample E		Sample E		Cal. level 1	Calculate Curve
14	Sample F	**4** Apply to	Sample F	**4** Apply to	Cal. level 1	**2** and
15	Sample G	Samples	Sample G	Samples	Cal. level 2	Average with
16	Sample H		Sample H		Cal. level 2	Previous Calibration

Table 14.7 Calibration Schemes

a. Replacement Calibration

The first two columns of Table 14.7 show the injection sequences for recalibrating after four samples are run. In the replacement scheme, the recalibration completely replaces the initial calibration. The first set of standards is applied to the first four samples. The second set of standards is applied to the second set of samples. In this scheme, results of the second calibration should be compared statistically to the first to ensure that they are reasonable. If this statistical test fails, the data system should warn the system operator. This scheme is best applied when the expected change in calibration results with time is monotonic or changing continuously in one direction. In this case, the logic for replacement of the previous calibration is that the most recent calibration is the most accurate for the point in time.

b. Averaging Calibration

In the second column of Table 14.7, the same injection sequence is applied as in the first, but the average of the second and first calibrations is used to calculate Samples E - H. This scheme is best applied when the variation from one calibration to the next is random so that averaging dampens out the variations. Before the averaging is performed, statistical tests should be applied to determine whether the new calibration is consistent with the previous results.

c. Bracketing Calibration

The third column of Table 14.7 shows a calibration scheme common to clinical analyses. In this scheme, called bracketing of standards, calibrations are run before and after the series of samples. Commonly, the two calibrations are averaged for calculation of the sample results. Because samples are quantitated by recalculation, this scheme requires that all the data files be stored until the second calibration is completed.

In a variation of this scheme, the sample results may be calculated from the first calibration set and the second calibration set used as a verification. If the second calibration is within some statistical limit of the first, no further calculations are done. If the second calibration is statistically different from the first, the samples may be recalculated using the second calibration or the entire sample set may be re-run.

d. Other Calibration Schemes

In another calibration scheme, comparison of each individual calibration standard with the previous results for that data point determines whether or not to include the new data point in the calculation of the curve. This sometimes results in only selected data points from a new calibration being included in the overall average calibration equation.

Another calibration scheme runs samples called check samples, test samples, or verification samples containing known amounts of the compounds of interest. These samples are calculated as if they were unknowns and the results are compared statistically to the actual amounts in the sample. The sample may be identical to one of the calibration standards or it may contain different amounts. Either method is an excellent way to test and document the validity of the calibration curve.

E. Units of the Results

Calibration of the chromatographic method and calculation of the results for unknown samples have been discussed thus far without mention of the units. Obviously, quantitative data must consist of both a number and its units. The three most common quantitative specifications used in chromatography are mass, concentration, and "parts-per". Mass is simply the absolute amount of the compound in grams. Concentration is mass per unit volume. Parts-per-million (ppm) means that one unit of the compound is present in every one million units of the sample, in any unit of measure. For example, 1 ppm can be $1pg/\mu g$ or $1\mu g/g$. In a solution, the density of water ($1g/ml$) can be used to restate ppm so that 2 ppm of a drug in solution could be translated to $2\mu g/ml$.

It is also important to know **how much** is **in what**. What does a result of 3.5mg for a drug assay mean? Were 3.5mg injected onto the column? Are there 3.5mg in the vial of prepared sample? Is 3.5mg the total amount in each tablet? Was 3.5mg recovered in a patient urine sample? Each of these could be the correct answer, but each is very different. It is important to document what the quantitative result means. The units and the meaning of the final result will be determined by the purpose of the analysis and must be stated in the SOP for the method.

F. Modifying Results Post-Calculation

In most methods, the final results will be stated in terms of the amount in an original sample of some type, e.g. weight of a drug in a single dosage unit or concentration of a substance in plasma. These units may not be convenient to use during the chromatographic calibration which is often given in terms of the concentration of standards in the calibration solutions. Many data systems provide multiplier or factor values which can be applied to the calculated results to obtain the final desired format. Use of multipliers can be illustrated with an example of a drug substance in a tablet whose final results are to be stated in "percent-of-claim" or what percentage of the amount of drug stated to be present in the tablet is actually there. This is a common quality control procedure in pharmaceutical analysis.

In this example, each tablet is formulated to contain 500mg of the drug. Sample preparation consists of dissolving the tablet in 100ml of water, filtering, and injecting 10μl onto the column. Standards of the drug made up in water at 4mg/ml, 5mg/ml and 6mg/ml are used to determine the calibration line. One of the injected samples returns the result 4.8mg/ml, which must be translated back into percent-of-claim. Because the original sample preparation consisted of dissolving the tablet in 100ml of water, the amount of drug in the original tablet is given by:

$$4.8mg/ml \times 100ml = 480mg$$

The percent-of-claim result is calculated by determining what percent of 500mg is represented by 480mg:

(480mg/500mg) x 100 = 96%

A general formula for transforming each quantitative result in mg/ml to percent-of-claim for this particular drug assay would be:

$$\frac{\left(\text{Result}_{mg/ml}\right)\left(100ml\right)}{500mg} \times 100\%$$

All data systems provide capabilities of modifying calculated results in this way. Some may have multiplier and divisor factors which can be put directly into the peak tables and applied to the calibration results. In other systems, a simple computer program or spreadsheet can modify results after the calculations. This latter type of processing is very flexible but it may be overly complex for a simple multiplier application like that shown above.

VI. REFERENCES

1. G.I. Ouchi, *LC-GC* 9 (1991) 628.
2. G.I. Ouchi, *LC-GC* 9 (1991) 474.
3. N. Dyson, *Chromatographic Integration Methods*, The Royal Society of Chemistry, 1990.
4. L.R. Snyder and J.J. Kirkland, *Introduction to Modern Liquid Chromatography*, John Wiley and Sons, New York, 1979, Chapter 13.
5. B. Jimenez, J.C. Molto, and G. Font, *LC-GC* 14 (1996) 968.
6. I. Krull and M. Swartz, *LC-GC* 15 (1997) 534.
7. U-H. Stenman, "Determination of Hemoglobin A_{1c}" in *HPLC of Biological Macromolecules* (ed. K.M. Gooding and F.E. Regnier), Marcel Dekker, New York, 1990.

CHAPTER 15
TROUBLESHOOTING

I. INTRODUCTION

Troubleshooting HPLC systems can appear to be very difficult because of the multiplicity of components; however, a good understanding of both their operation and the consequences of malfunctions makes the task easier. Many HPLC problems are first manifested by inconsistencies in the output signal; however, the detector itself is only infrequently the cause. More often the source is the sample, pump, column, gradient former or mobile phase. Experience is the handiest source of problem solving but there are excellent manuals which guide an inexperienced user, as well as offer suggestions for new problems. The scientists at LC Resources provide invaluable guidance through their books (1, 2), articles in *LC-GC* (3) and computer programs (4). Instrument manuals also contain troubleshooting sections which list symptoms and detailed repair procedures for specific components. Troubleshooting CE will not be discussed in this book; however, there is a guide to troubleshooting and a discussion of band spreading (5) in a recent issue of *LC-GC.*

This chapter will briefly discuss some common problems that are encountered in HPLC operation. A table of symptoms, their sources and solutions, as categorized by separation, detection and quantitation is at the end of the chapter. Only general column problems will be discussed because those which are mode-specific are detailed in the respective chapters. It is important to remember that all columns will die, systems will leak, seals and check valves will fail, and detector lamps will burn out. The frequency of those problems will depend substantially on performance of routine maintenance and compliance with manufacturer's guidelines.

Keeping a log for each HPLC system and for each column is highly recommended. The records expedite troubleshooting by describing problems and their solutions for future reference. All HPLC systems and columns should be evaluated initially, and periodically thereafter, to determine deviations from initial performance. For some high precision systems or analyses, calibration may be necessary daily or even more frequently. By testing after repairs or maintenance, correction of the problems can be confirmed, as well as new specifications defined, in the event of replacement of a major component.

Table 15.1 Recipe For Minimal Downtime

- Follow manufacturer's guidelines
    1. operation
    2. routine maintenance
- Keep a log for each HPLC
- Keep a log for each column
- Test system/column performance
    1. initially
    2. during troubleshooting
    3. after repairs

II. LEAKS

A. Fittings

It is very important that the whole HPLC system be leak-free to attain consistent flow and performance. Anytime a component, such as a column or a fitting, is replaced, it should be checked for leaks after operation is resumed. For nontoxic solvents, this is often accomplished by feel. Lab wipes can also be touched to the connection and observed for wetness. For small diameter columns and low flowrates, leak detection can be very difficult because of the low volumes generated. Leaks of volatile solvents may produce a cold fitting.

As mentioned in Chapter 3, a persistent leak is rarely stopped by further tightening, but usually requires replacement of a bad fitting or ferrule. A stock of the fittings and ferrules used in each instrument should be kept on-hand. If column fittings are leaking, they should first be tightened. If that is ineffective, and new fittings are available, it is sometimes possible to replace them by careful removal without disturbing the bed. Overtightening check valves during installation can result in a cracked valve seat or seal. It is important to only tighten the valve enough to effect a seal. As a general rule, a 6" long wrench should be used and 5 - 10lbs of force applied. Self-priming check valves from Analytical Scientific Instruments are not susceptible to damage by overtightening. Additionally, they will not lose prime, even when large air bubbles are present.

B. Injectors

Leaks from injectors can be caused by overpressurization, worn seals, or scratches on the surface of the rotor face. Precipitation of salt accelerates seal degradation and sometimes causes scratching of the rotor face. Frequent washing of the injector can minimize salt residue and its deleterious effects. Seals must be replaced periodically due to wear. The rotor face can only be repaired by fine polishing, usually by the manufacturer.

Table 15.2 Sources of Leaks

- tubing connections
- injector
- piston seal
- column fittings

III. PRESSURE

A. Elevated Pressure

High pressure can be caused by blockage in any system component past the pressure sensor. Isolation of the problem is most easily achieved by systematically loosening the connections, beginning with the last component of the system, which is usually the detector. When the source of the blockage is found, the pressure goes down.

Blockage of a frit or column usually produces a gradual pressure increase, whereas blockage of other components results in an instantaneous pressure surge - culminating in shutdown. If the elevated pressure is caused by a filter or column, replacement of the frit is the most common remedy. The same size and mesh of frit should be used. If the blockage is in a guard column and replacement of the front frit is ineffective, the guard column should be replaced or repacked. Any manipulations of a column's end fittings or bed can potentially harm its performance; therefore, such measures should only be undertaken as a last resort. In some cases, the front frit of an analytical column can be carefully replaced without damaging the column; removal of the back frit has higher probability of damaging the bed. Back frit blockage usually signals the presence of small particles in the column, probably caused by its degradation; therefore, its replacement is likely to be a stop gap measure to somewhat extend usage.

An analytical column which generates pressure substantially higher than its specifications must usually be replaced. For some columns, backflushing or reversal of flow may decrease pressure; however, if there is a small void or channeling in the column, this procedure is liable to totally disrupt the bed. The manufacturer's recommendations about backflushing should be followed. Sometimes it is possible to dig out and replace a portion of contaminated column packing at the inlet (1). This should be considered a temporary measure before column replacement. It is also an indication that a guard column should be used in the future.

Blockage of tubing, injectors, or detectors can often be freed by backflushing; however, there may be pressure restrictions for some detectors. Manufacturer's guidelines should be followed. If components after the column become plugged by particulate matter, the column should be inspected for leakage of support. It is best to wash out columns into a beaker rather than the detector to avoid contamination of the detector and to observe any abnormal effluent components, such as precipitates.

Table 15.3 Sources of High Pressure

- connecting tubing
- inline filter or frit
- injector
- guard column - front frit or packing
- analytical column - front frit, back frit, or packing
- detector

B. Low Pressure

A common cause of low system pressure is that the mobile phase has run dry or has a blocked inlet filter (this is usually due to the sinker filter in the column reservoir). Low pressure can also indicate a major leak or pump problem. These are all fairly easy to identify and repair. After fixing the problem, it is frequently necessary to reprime the pump to remove air, following the manufacturer's recommendations.

Table 15.4 Sources of Low Pressure

- empty solvent reservoir
- blocked inlet filter
- major leak
- check valve malfunction
- piston problem

C. Fluctuating Pressure

It is common for pressure to vary during a gradient, especially when the solvents have different viscosities. Such variation should be predictable and consistent from run to run. Single piston pumps may produce cycling pressure which can be modulated with a pulse dampener; this cycling should be a standard feature of the system. A sudden presence of pressure fluctuation may signify the failure of the dampener.

Pressure variations are often due to air bubbles in the check valves. These may be caused by a plugged inlet filter or the need for degassing the mobile phase, especially when using low pressure mixing. A defective or dirty check valve may also result in cycling pressure.

Pumps require routine maintenance which will minimize the need for major repairs. This maintenance is detailed by the manufacturer and includes lubrication, flushing out salts from the system and sometimes washing the back ends of pistons (6).

```
┌─────────────────────────────────────────────┐
│ Table 15.5  Sources of Fluctuating Pressure  │
│                                               │
│  •  air bubbles                               │
│  •  blocked inlet filter                      │
│  •  faulty check valve                        │
│  •  lack of pulse dampening                   │
└─────────────────────────────────────────────┘
```

IV. VARIATIONS IN DATA OUTPUT

A. General

Other than leaks and pressure variations, most HPLC problems are discovered by variations in the data output. These range from irreproducibility in quantitative or qualitative data to baseline inconsistencies. The source of the problem can be virtually any component of the system, a complexity which is the basis of most troubleshooting discussions. If an HPLC is used by multiple chromatographers, it is good practice to check the columns and settings before each use. In all circumstances, HPLC systems should be cleaned and ready for operation at the end of each day.

B. Peak Abnormalities

Peak abnormalities include any deviations in the chromatogram caused by components injected on the column. The discussion in this section deals with general peak and column problems. Issues relating to specific modes of chromatography are detailed in the respective chapters.

1. Extra Peaks (Table: Section 1.7)

Extra peaks are usually due to contamination of some component of the system, including the sample and mobile phase. Many samples and mobile phase solutions will degrade with time, yielding spurious peaks. The syringe and injector are frequent sources of impurities, especially if they are not cleaned after each use. Flushing with mobile phase does not always satisfactorily remove contaminants, and a stronger cleaner, such as dilute TFA, may be periodically necessary. The final rinse should be mobile phase to prevent deviations in retention. A dirty syringe can also contaminate a sample.

Ghost peaks from previous injections will occur in the chromatogram if part of the sample is nonspecifically adsorbed or only partially eluted during the analysis. Elimination of ghost peaks may require washing with a stronger mobile phase or, more radically, replacement of the column or mobile phase with one which does not exhibit these characteristics.

2. Broad or Tailing Peaks (Table: Section 1.4 and 1.8)

The cause of peak shape problems is most effectively diagnosed by running a standard (after eliminating simple remedies like cleaning the injector and syringe). The chromatogram of the standard will allow comparison with initial specifications for the column and system.

In an isocratic run, broad peaks which seem unrelated to the sample may result from peaks which did not elute during a previous run; either the elution conditions must be made stronger, a gradient must be used, or the column must be washed between runs. If retention times are correct but peaks are broad or tailing, some component of the analysis may be dirty or degraded. A syringe, injector, or column may cause these problems, especially when contaminated with the same sample run previously. A sample or mobile phase which has degraded may also cause peak skewing.

Split peaks are often due to a void or channeling in the column. Overloading generally causes band broadening and sometimes fronting, but overloading of biopolymers may also yield split peaks due to viscous fingering. Loss of efficiency with increased pressure can signal column bed failure or, at least, frit blockage.

3. Excessive Retention (Table: Section 1.2)

Longer retention than normal is caused by reduced flowrate, mobile phase strength or excessive interaction with the column. Reduced flow can result from system leaks, faulty check valves or seals, or blockage of a frit, as discussed previously. Inadequate mobile phase strength can be caused by degradation, improper preparation, or problems with the gradient former, mixer, or evaporation of organic modifiers. For isocratic elution, it is usually most reproducible to premix mobile phases rather than to allow the system to do so. If the stirring mechanism of a dynamic mixer is not operational, the gradient composition will be erratic, especially if the mobile phase components are not readily miscible. A faulty check valve can also cause a system to deliver the wrong mobile phase mixture.

Columns can produce excessive retention if they are degraded or contaminated. As base deactivated columns degrade, basic molecules will become retained on the free silanols. Columns can become contaminated by residual sample components or sometimes leakage of hydraulic fluid from a pump or pulse dampener.

4. Reduced Retention (Table: Section 1.1)

Some of the causes of reduced retention are related to those of excessive retention. The gradient formation, mixer, mobile phase, or column may be the source of the problem. During gradient operation, it is imperative that the column be reconditioned adequately, usually with at least ten column volumes of the initial mobile phase. Insufficient conditioning will result in low and/or erratic retention. A highly contaminated column may also yield low retention; such a column can often be rejuvenated by washing with a strong mobile phase before reconditioning. If washing and reconditioning do not restore the retention, it is possible that some of the stationary phase has leached; this is especially common for reversed phase columns. Frequently such columns must be replaced, although procedures for restoring the bonded phase by on-column silylation have been advanced (7).

5. Changed Selectivity

Selectivity is altered by a change in the column or in the mobile phase composition. When deviation of selectivity is observed, the column should be washed and adequately reconditioned. If selectivity is not restored, the units which control mobile phase composition (solutions, pumps, mixer, and gradient former) should be evaluated with a standard and a functional or new column to check proper performance of the system.

V. CONCLUSIONS

The importance of setting initial performance criteria cannot be overemphasized. Standards for isocratic and gradient operation should be run on each column used on a system. The chapters discussing each mode of HPLC suggest samples and conditions for these standards. For many troubleshooting procedures, injection of a standard can quickly isolate a problem, as well as confirm its repair. If component substitution is used for troubleshooting, only one unit should be exchanged at a time to definitively isolate the problem.

Recording details of HPLC problems and their solutions in a log produces an invaluable resource. Such a log is absolutely necessary for multi-user instruments, but is also extremely important for those with a single user. It is "common knowledge" that problems recur just after all details of their past occurrence have faded from the memory.

Table 15.6 Commandments for Effective Troubleshooting
•   Use standards to evaluate systems •   Use standards to evaluate columns •   Change only one variable (component) at a time •   Keep a log for each system •   Keep a log for each column

VI. REFERENCES

1. J.W. Dolan and L.R. Snyder, *Troubleshooting LC Systems*, Humana Press, New Jersey (1989).
2. J.W. Dolan and L.R. Snyder, *Troubleshooting HPLC Systems*, LC Resources, Lafayette, CA (1986).
3. J.W. Dolan, "LC Troubleshooting", *LC-GC*, monthly feature.
4. T.H. Jupille and B. Buglio, *The HPLC Doctor*, LC Resources, Lafayette, CA (1986).
5. C.A. Lucy, K.K.-C. Yeung, X. Peng, and D.D.Y. Chen, *LC-GC* 16 (1998) 26.
6. J.W. Dolan, *LC-GC* 15 (1997) 110.
7. C.T. Mant and R.S. Hodges, "On-line Derivatization of Silica Supports for Regeneration of Reversed-Phase Columns" in *High Performance Liquid Chromatography of Peptides and Proteins* (ed. C.T. Mant and R.S. Hodges), CRC Press, 1991, p. 57.

1. Problems With Separation

Problem	Situation or Clue	Mode	Possible Cause	Remedy
1.1 Decreased Retention Time	increased pressure	all	flowrate too high	confirm flowrate with graduated cylinder and adjust, if necessary
	increased pressure and decreased efficiency	all	loss of stationary phase	1. replace guard column 2. replace inlet frit and/or top off 3. replace column
	decreased pressure	not SEC or HIC	increasing temperature	control column temperature
	retention varies with injection volume	not SEC	injection solvent causing elution	use starting mobile phase as injection solvent
	large sample	all	column overload	use smaller sample
	gradient elution	not SEC	inadequate reconditioning	increase reconditioning time and volume
	gradient elution or mixed solvents	all	too much strong solvent	1. control solvent composition 2. prepare new mixture 3. check solvent proportioning
	many injections	all	contaminated column	1. flush with strong solvent 2. replace guard column 3. replace analytical column
	spiked samples yield different retention than standards alone	all	components in sample or sample matrix cause elution	1. prepare standard in sample matrix 2. remove matrix interference 3. use internal standard
	new buffer solution	IEC	1. buffer too concentrated 2. incorrect buffer 3. incorrect pH	prepare new buffer verify buffer composition verify pH calibration
	old buffer solution	IEC	buffer degraded	1. prepare new buffer 2. add stabilizing agent or bacteriostat 3. refrigerate between uses

Problem	Situation or Clue	Mode	Possible Cause	Remedy
	only ionic compounds affected	RPC, ion-pair	not enough ion-pair reagent	1. increase concentration of ion-pair agent 2. insure accurate preparation and appropriate storage
	retention of some (but not all) peaks changes with amount of sample (especially proteins)	all	sample structure varies with concentration	1. dilute or concentrate to minimize effect 2. change sample solvent
1.2 Increased Retention Time	pressure lower than normal and t_0 longer	all	1. flowrate too low 2. pump malfunction	check flowrate and adjust, if necessary repair check valve or seal
	wet or cold fitting and longer t_0	all	leak	tighten (or replace) fitting; slow leaks with volatile solvents may produce cold fitting.
	back pressure increasing	all	column contaminated	1. wash column 2. replace guard column 3. replace inlet frit 4. replace top mm of packing
	using volatile solvents	RPC	decrease or evaporation of volatile component	1. control solvent composition 2. prepare new mobile phase 3. check solvent proportioning system
	gradient elution or proportioned solvents	not SEC	one solvent not flowing properly: 1. empty reservoir 2. plugged inlet frit 3. air bubble in line 4. gradient former malfunction, including programming errors 5. inoperative mixer	fill reservoir and prime pump replace inlet frit purge air; check inlet filter fix gradient maker, including programming errors repair mixer
	increased pressure	all modes	plugged component or line	working from detector to pump, open each fitting to identify source of pressure; repair
	increased pressure	not SEC or HIC	decreasing column temperature	control column temperature

Problem	Situation or Clue	Mode	Possible Cause	Remedy
	new buffer solution	IEC	1. buffer too dilute 2. incorrect buffer 3. incorrect pH	prepare new buffer verify buffer composition check pH calibration
	old buffer solution (especially acetate and phosphate)	IEC	buffer degraded	1. prepare new buffer 2. add bacteriostat like azide 3. refrigerate between uses
	ionic compounds only	RPC, ion-pair	too much ion-pair reagent	1. decrease concentration of ion-pair reagent 2. insure accurate preparation and appropriate storage
	old sample	SEC	sample decomposition by bacteria or enzymes	1. ultrafilter 2. store at low temperature 3. change sample matrix to one which inhibits degradation
1.3 Decreased Resolution	trend observed over several runs	all	1. slow build-up of contaminants on column 2. solvent degradation	wash column control solvent composition
	retention times unchanged	all	column efficiency decreasing	1. verify plate count; if lower than initially, check for bed settling 2. replace guard column 3. replace column
	probable change in retention	all	1. loss of endcapping 2. loss of ligand chain 3. contamination with hydraulic fluid	replace column or add ion-pair agent replace column or resilylate (6) repair leaking unit and wash or replace column
	gradient elution or proportioned solvents	not SEC	solvent delivery malfunction	confirm operation of each pump, gradient and mixer
	retention shorter	all	flow too high	check flowrate; readjust flow control or electronics
	gradient elution	not SEC	1. inadequate reconditioning 2. gradient malfunction	recondition with at least 10 column volumes of initial solvent check mixer and pumps, then repair

Problem	Situation or Clue	Mode	Possible Cause	Remedy
	recently changed solvent	all	1. incorrect solvent composition 2. dirty container	carefully prepare new eluent wash and rinse thoroughly
	large sample load	all	column overload	dilute sample
	recently changed solvent systems, particularly from one mode to another	all	inadequate flushing of lines	1. use longer wash 2. use intermediate solvent, such as isopropanol, in wash cycle
1.4 Peak Tailing Tailing	used column	all	void at top of column	1. fill in void with suitable packing 2. replace column 3. avoid high flowrates and pressure
	new column	all	poorly packed column	confirm by comparison with initial plate count and/or good column; replace
	used column	all	column channeling or phase loss	replace
	old column or complex samples	all	frits partially plugged or rusted	replace frits
	old or unstable mobile phase	all	degraded mobile phase	prepare fresh
	complex samples	all	contaminated column producing mixed mechanisms	1. wash column 2. remove source of contamination 3. use guard column, replacing as necessary
	ionic solutes	RPC	interaction with free silanols	1. add 0.1% ion-pair agent, i.e. tetramethylamine 2. use endcapped column 3. control pH with TFA, formic or another acid
	large sample load	all	column overload	1. determine if tailing acceptable 2. if unacceptable, reduce sample size
	old or unstable sample	all	1. degraded sample 2. contaminated syringe or injector	1. prepare fresh sample 2. add stabilizing agent clean syringe and injector after each use

Problem	Situation or Clue	Mode	Possible Cause	Remedy
	unstable sample	all	sample degradation on column	1. decrease column temperature 2. convert sample into stable product 3. change mobile phase 4. change column packing
	ion-pair	RPC	slow reaction kinetics	use column with monomeric bonded phase
	polymer, proteins and combinations of biopolymers	SEC	separation of variants may be occurring	check column with monodisperse standards like proteins
	polymer, proteins and combinations of biopolymers	SEC	elution of secondary component(s) after main component	improve resolution with slower flowrate, longer column or different mobile phase
	elution not solely related to ionic concentration or pH	IEC	mixed mechanism (hydrophobic or adsorption)	1. add low concentration of organic modifier, such as isopropanol, to mobile phase 2. for adsorption, increase column temperature 3. for hydrophobicity, decrease column temperature
	lipids or detergents	all	micelle formation	1. use more polar solvent 2. increase salt content
	system plumbing recently changed	all	extra-column dead volume, especially due to fittings, ferrules or tubing	1. remove dead volume 2. use 0.010" tubing 3. use short lines 4. match fittings and ferrules
	changed detector or recorder	all	slow electronics	1. select shorter time constant on detector 2. select appropriate settings on data system or recorder 3. assure recorder and detector have matched impedance 4. consult service engineer

Problem	Situation or Clue	Mode	Possible Cause	Remedy
1.5 Fronting Peaks		all	poor column	replace column
1.6 Double Peaks	all peaks are double	all	column settling or forming void	1. check for void and fill 2. replace column
	some peaks	all	isomerization during sample preparation	1. determine if acceptable 2. change sample preparation procedures
	biopolymers 	SEC	viscous fingering	1. match sample solvent to mobile phase 2. dilute sample in less viscous matrix
1.7 Ghost Peaks	no injection with gradient elution	RPC	1. water impurities (*see* Chapter 13) 2. impurities in nonpolar solvent 3. degradation of mobile phase	1. confirm by pumping twice as much water for next run 2. use HPLC grade water 1. determine interference 2. use HPLC grade solvent prepare fresh
	extra peaks with sample	all	1. sample degradation or contamination 2. syringe or injector contamination	make fresh sample clean column well after each use
	isocratic run, peaks too broad 	all	residual peaks from a previous injection	1. run longer or with stronger solvent or gradient 2. wash column after each run

Problem	Situation or Clue	Mode	Possible Cause	Remedy
	gradient or proportioned solvents	not SEC	mixer not working	turn on or repair mixer
	gradient elution using pumps with large volumes of pressurized liquids	not SEC	concentration/viscosity oscillation	consult manufacturer
1.8 Poor Peak Shape	large sample	all	detector overloaded	1. decrease sample size 2. use cell with shorter path length 3. use different wavelength
		all	detector cell problem 1. leak between sample and reference cells	repair cell
			2. flow goes through reference after sample cell	correct plumbing
			3. photocell temperature variance 4. vacancy effect	consult manual or manufacturer use pure solvent (1)
	large sample	all	overloaded column	1. may be acceptable in preparative applications 2. decrease sample quantity
	large sample	all, especially IEC	other sample components displacing peak of interest	1. may be acceptable 2. decrease displacing ions or components

2. Problems With Signal

Problem	Situation or Clue	Mode	Possible Cause	Remedy
2.1 No peak	detector or recorder problem	all	1. no power	plug in or replace fuse
			2. wrong settings	adjust
			3. lamp malfunction	adjust or replace
			4. recorder disconnected from detector	connect
	sample problem	all	1. poor injection	improve accuracy or repair injector (*see* Section IIB)
			2. sample degraded	prepare fresh or stabilize
			3. sample not totally dissolved	sonicate, heat or change solvent
			4. sample too dilute	increase concentration
	pump malfunction	all	1. flow stopped	*see* Section IIIB
				1. gradient not operational
			2. no strong solvent	2. pump failure for strong solvent
2.2 Low Sensitivity	detector problem	all	1. absorbance settings wrong	adjust
			2. incorrect λ or lamp intensity	adjust
	sample problem	all	1. poor injection	1. improve accuracy
				2. repair injector (Sect. IIB)
			2. sample degraded	prepare fresh or stabilize
			3. sample not totally dissolved	sonicate, heat or change solvent
			4. sample too dilute	increase concentration
	decreased retention	all	flow too fast	adjust
2.3 Negative Peaks	RI detector	all	refractive index of solvent is between that of some sample components	1. may be acceptable
				2. use solvent with RI much different than sample

Problem	Situation or Clue	Mode	Possible Cause	Remedy
	all are negative	all	detector/recorder problem 1. polarity switch reversed 2. recorder leads reversed 3. sample going through reference cell	change polarity change leads reroute through sample cell
	near t_o, using UV detector	all	normal operation 1. RI of injection solvent different than that of mobile phase 2. temperature fluctuation associated with injection	use same solvent for sample and mobile phase control temperature of column and detector
2.4 Drift	instrument just turned on	all	instrument warming up	normal situation; allow more time
	old lamp	all	aging lamp	replace
	change in mobile phase	all	previous mobile phase not completely removed	use longer solvent change, possibly including purge with miscible solvent
	subambient temperature	all	condensation in optical path	1. purge detector with dry gas 2. use detector heater
	visible leak or condensation in optical path	all	slow leak in sample cell	repair
	broad, large peak	all	peak from previous run	1. use more effective column regeneration program 2. clean up sample before injection
	inert gas purge of mobile phase	all	1. the purge selectively removed one component of solvent 2. inert gas contains impurities that alter solvent	decrease flowrate of purge gas to lowest useful level use higher purity gas

Problem	Situation or Clue	Mode	Possible Cause	Remedy
	pressurizing cell changes direction of drift	all	1. slow dissolution of contaminant in cell 2. slow formation of bubbles	clean cell according to manual; flush with strong solvent 1. degas solvent 2. increase pressure on cell within specified limits
	using mixed solvent	all	selective evaporation of one component of solvent	1. may be acceptable 2. control evaporation by capping reservoir, etc.
	gradient	all	change in composition not transparent	1. change detection specifications 2. change mobile phase
	variable temperature	all	change in room or instrument temperature	1. control temperature of column and detector 2. move instrument to location with more uniform temperature
	conditioning column	all	1. inadequate conditioning 2. leaching bonded phase 3. leaching adsorbed material	condition longer ($\geq$ 10 column volumes) 1. use less destructive mobile phase; 2. replace column wash column
	no degassing	all	oxygen absorption	degas
	oscillations stop when pump stops	all	pump pulsations	1. repair pump check valve or seal 2. add/replace pulse damper
	do not stop when pump stops	all	detector lamp or other electronics	replace

2.5 Noise

a. high frequency

Problem	Situation or Clue	Mode	Possible Cause	Remedy
	other instruments on electrical circuit	all	another component causing noise	1. use separate circuit for all components of HPLC (recommended) 2. use isolation transformer for HPLC

Problem	Situation or Clue	Mode	Possible Cause	Remedy
	appears only on detectors mounted on board of instrument	all	1. poor grounding	plug all components into common grounded source
			2. excessive electrical leakage in part of HPLC	contact service representative
b. low frequency	recently changed solvent systems	all	old solvent still being flushed from the system	wash system with common miscible solvent like isopropanol
	changed solvents or numerous samples	all	dirty detector window	flush and clean: 1. as specified in manual 2. use 0.1 - 1% TFA until absorbance levels off at a minimum
	new column or one with recently replaced outlet frit	all	column packing leaking from column	1. confirm by removing from detector and collecting effluent in watch glass; look for column packing in liquid or in residue after evaporation 2. replace column
	sharp spikes using new or different solvent	all	bubbles in detector	1. increase flowrate after removing back pressure device 2. replace pressure restrictor 3. check for loose connector
	pulses correspond to pump stroke	all	pump pulsing	replace or repair pulse dampener
	accompanied by longer retention	all	leaks in system	tighten or replace fittings

3. Problems in Quantitative Analysis

Problem	Situation or Clue	Mode	Possible Cause	Remedy
3.1 Variable Peak Height	UV detectors	all	temperature-dependent absorbance (up to 1% per °C)	control column and detector temperatures
3.2 Excessive Variation of Peak Height	definite trend to decreasing peak height	all	1. column efficiency declining 2. retention increasing	clean column use stable column *see* Section 1.2
	all peaks	all	variation in sample volume	1. check syringe for plug or bubbles 2. wash injector between injections 3. injector leaking 4. needle seal needs tightening 5. poor syringe technique
	all peaks	all	poor or inconsistent recovery from sample work-up	use internal standard with similar chemistry to sample
	some to all peaks	all	sample degradation (time-dependent)	1. keep sample at lower temperature 2. store sample in dark 3. store sample under inert gas
	early peaks have largest variation	all	detector or recorder time constant too long	shorten time constant
3.3 Excessive Variation of Peak Area	all peaks affected proportionately	all	variable flowrate; slow flowrate increases area and fast flow decreases area and retention time	improve flow control
	all peaks	all	variation in sample volume	1. check syringe for plug or bubbles 2. wash injector between injections 3. injector leaking 4. needle seal needs tightening 5. poor syringe technique 6. injection volume too small for syringe or injector
	digital integrator	all	1. integrator not set up properly 2. area overflow	consult manual decrease signal or sample
3.4 Retention Time	increasing	all	*see* Section 1.2	

Problem	Situation or Clue	Mode	Possible Cause	Remedy
Variation				
	decreasing	all	*see* Section 1.1	
	not sufficiently reproducible	all	random errors in solvent preparation	1. use volumetric procedures including Grade A pipettes and volumetric flasks 2. protect solvents from evaporation, contamination, etc.
	with column usage	all	1. column contamination 2. bonded phase leaching	wash column with strong solvent replace column
3.5 Trends in Data		all	usually caused by problems related to retention time, resolution, etc.	refer to appropriate sections in Section 1
3.6 Linearity	new component(s) of assay	all	response not linear	1. use absorbance, not %Transmittance, with optical spectrophotometers and photometers 2. prepare standards with similar concentration to sample so that required linearity is minimal 3. all detectors do not give linear response; consult manual 4. recorder/detector mismatch; consult manual or service engineer
	poor linearity between runs	all	sample carryover	thoroughly wash injector and syringe, especially between runs with different concentrations
	nonlinear peak height measurements	all	column overloaded	1. use peak area 2. decrease sample size 3. use working curve 4. use larger column

Problem	Situation or Clue	Mode	Possible Cause	Remedy
3.7 **Impossible Results**	no quantitative correlation	all	1. standards are not the same as sample	use correct standards
			2. calibration or sample solutions are degrading	stabilize
			3. interfering compound adding to sample response	separate peaks using different operating conditions
			4. using peak with $k = 0$ to quantitate	use peak with $k > 0$
			5. math error	correct
	calibration curves do not have intercept at zero	all	1. interference with another compound	separate from interfering peak
			2. standards not representative	replace with more appropriate standards
			3. standards not stable	stabilize
			4. error in measurement of peak height	measure again
			5. response not sufficiently large to neglect noise	1. increase concentration
				2. use column with smaller ID
	peaks seen when no sample injected (see Section 1.7)	all	1. contaminated solvents	change or purify solvents
			2. contaminated injector	flush injector; disassemble, if necessary
3.8 **Poor Accuracy**	accuracy is function of sample concentration	all	1. detector response is nonlinear	reduce concentration or path length to get in linear region
			2. determination is made near noise level of the detector	increase amount or concentration to improve signal/noise ratio
			3. sample structure varies as a function of concentration	change concentration or solvent to maintain integrity of sample
3.9 **Poor Precision**	retention times not reproducible	RPC, ion-pair	reagent or column degradation	1. use fresh reagents
				2. keep flow through column when not in use
	small samples	all	sample volume too small for reproducible measurement	increase size of sample and internal standard (25-100μl is recommended)
	long retention	all	column too long, diluting sample unnecessarily	1. shorten column
				2. use coupled columns
				3. use narrow ID column

Problem	Situation or Clue	Mode	Possible Cause	Remedy
	small samples	all	insufficient detector response	1. use detector with better detection limits 2. derivatize to improve detection limits 3. increase sample concentration or injection volume 4. change separation mode so that small peaks elute at beginning of chromatogram
	new assay	all	1. desired accuracy may be beyond state-of-the art	nothing
			2. insufficient data	run more samples and standards
			3. variable recoveries	use internal standards
			4. gradient elution	isocratic or coupled columns (often lowers C.V. by 50%) (7)
			5. malfunctioning equipment	1. check by running stable test sample 2. repair, as necessary
3.10 Uncertain Peak Identification	unexpected peaks	all	ghost or artifactual peaks	see Section 1.7
	variable retention of complex mixture	all	1. insufficient precision of retention time to discriminate	1. improve precision with better control of flowrate and/or column temperature 2. many samples will show consistent elution order even when times are variable
			2. complex chromatogram	1. rely on relative response factors using marker peaks 2. use response ratio technique 3. use more selective detector for compounds of interest

CHAPTER 16
METHOD DEVELOPMENT AND VALIDATION

I. INTRODUCTION

The wide use of HPLC in the pharmaceutical and biotechnology industries has dictated the establishment of sound methods. In the past few years, specific attention has been focused on HPLC by the USP, ICH (International Conference on Harmonization), and other agencies so that all methods follow similar criteria for excellence (1, 2). There has been some effort to evaluate capillary electrophoresis methods similarly (3).

II. METHOD DEVELOPMENT

Although method development must occur before validation, it is most efficiently carried out with the requirements of validation in mind. Development of a new method is necessary under many circumstances, including new analytes, poor existing methods, and emergence of new technology. The goals of the method should be determined at the beginning of the process so that they are incorporated into the method. These may be qualitative or quantitative measurements, cost, speed, or automation.

Most frequently, new methods are adapted from those published in the literature or from similar methods used in the laboratory. Suitable instruments must be chosen and selectivity demonstrated using standard solutions. During optimization of a method, each of the analytical parameters must be evaluated to ensure that the method has the potential for validation. Generally, the sequence for evaluating those parameters is robustness, linearity, precision, and limits of detection (2). Efficient and effective method development is a complex subject which has been discussed in depth in a book by Snyder, Glajch, and Kirkland (4).

III. VALIDATION

If the criteria for validation are understood, then method development is a logical process which encompasses each of the requirements for validation. Validation consists of at least four parts: software validation, hardware validation, method validation, and system suitability (1, 2).

A. Software and Hardware Validation

Software and hardware validation are simplified if the vendors conform to ISO 9000 standards and have written or software driven procedures for in-house maintenance and qualification. Assuming that the products have been qualified by the vendor, it is the further responsibility of the owner to do functional qualification before and after use (5). Before use, instruments and software must be subjected to installation qualification, operational qualification, and performance qualification (6). Installation qualification (IQ) checks whether the instrument works the way the manufacturer specified. Operational qualification (OQ) evaluates whether the instrument will work for the specific purposes of the laboratory by comparing the operation with the user-designed specifications. Performance qualification (PQ) periodically tests whether the instrument continues to comply with the requirements. After use, instruments must routinely undergo maintenance qualification, operational qualification, and performance qualification under a timetable specified in the laboratory SOP. It is always necessary to requalify an instrument after it has undergone maintenance procedures.

B. Method Validation

Method validation provides an assurance of reliability during normal use of the method (2). It is the process of providing documented evidence that the method does what it is intended to do (7).

During method validation, data are gathered and specifications are set on seven aspects of the method, as listed in Table 16.1. These terms are commonly referred to as analytical performance parameters or figures of merit. Although each of these terms is routinely used in laboratory parlance, their official definitions and implementation must be understood before compliance can be attained. The specific words and terms are still under discussion by ICH and may continue to evolve and change with time.

Table 16.1 USP Steps of Method Validation

- Accuracy
- Precision
- Limit of Detection
- Limit of Quantitation
- Specificity
- Linearity and Range
- Ruggedness
- Robustness

---

1. Accuracy

Accuracy is the measure of exactness of an analytical method. It describes the closeness of agreement between the accepted (or true) value and the value found (2). Accuracy is measured as the percent of analyte recovered using spiked samples in a blind assay. The ICH guideline recommends collecting data over a minimum of three concentration levels covering the specified range. It should be reported as the percent recovery of the known or the difference between the mean and true values with confidence intervals. It is sometimes difficult to assess accuracy because it requires carrying out the test on a sample with known composition, which in turn requires an alternate validated method (8). If there is no alternate method, the purity of the standard must be assessed by the presence of detectable impurities. For example, the Area% at 210nm is sometimes used.

2. Precision

Precision is the measure of the degree of repeatability of an analytical method under normal operation, expressed as the percent relative standard deviation (%RSD) for a statistically significant number of samples (1, 2). Precision is often performed at three stages which are called repeatability, intermediate precision, and reproducibility. Repeatability evaluates the results under the same conditions over a short time interval, generally with at least nine determinations over the specified range of the procedure. Intermediate precision encompasses within-lab variations due to different periods, analysts, or equipment. Reproducibility evaluates results from collaborative studies among laboratories. Results from all three measures of precision should include standard deviations, %RSD (coefficients of variation), and confidence levels.

3. Limit of Detection

As discussed in Chapter 4, the limit of detection (LOD) is the lowest concentration of an analyte which can be detected in a sample. It is usually expressed as a concentration at a specified signal to noise ratio (often 2 to 1 or 3 to 1). Another method of calculation uses the standard deviation (SD) of the response at the detection limit and the slope (S) of the calibration curve: LOD = $3.3(SD/S)$ (1, 2). The method for determination of LOD should be documented and supported using 5 - 6 samples.

4. Limit of Quantitation

The limit of quantitation (LOQ) is the lowest concentration of an analyte in a sample that can be determined with acceptable precision and accuracy under the stated operating conditions. It is expressed as a concentration in conjunction with the associated precision and accuracy. A 10 to 1 signal to noise ratio is typically used as the conditions of measurement, as shown in Fig. 4.1; LOQ = $10(SD/S)$. Because LOD and LOQ are related to signal to noise ratio, and thus, height of the peak, they are dependent on column efficiency. They will be optimal when a column yields narrow peaks and may change with column aging or replacement (1). In effect, the LOQ is 3.3 - 5 times higher than the LOD.

5. Specificity

Specificity is the ability to measure accurately and specifically the analyte of interest in the presence of other components. It is thus a measure of the degree of interference from other ingredients, impurities and degradation products (2). Specificity is measured by the resolution from other components. It can also be evaluated by verifying peak purity with photodiode array detection, mass spectrometry, chromatography with other modes, or other techniques, such as CE. ICH recognizes two kinds of specificity. For identification, it is the ability to discriminate between compounds with closely related structures. For assay and impurity tests, specificity is demonstrated by the resolution of the two closest eluting compounds, usually the main component and an impurity. If possible, it should be demonstrated that the results are unaffected by spiking with other sample components, such as excipients or impurities. Some of the USP and ICH descriptions of specificity are similar to the term "selectivity" in chromatography but "specificity" is the denoted term.

6. Linearity and Range

Linearity is the ability of a method to elicit test results that are directly proportional to analyte concentration within a given range (1, 2). Linearity is generally based on peak area rather than peak height so that variations in peak width and assymetry are less critical (8). Linearity is the variance of the slope of the regression line, and range is the interval between the upper and lower levels of analyte that have been demonstrated to be determined with precision, accuracy and linearity using the method. Visual inspection of the data is often a good indication of linearity. Linear regression coefficients greater than 0.995 are usually acceptable. Five concentration levels should be used within specified ranges of the target concentration: 80 - 120% for assay tests, 70 - 130% for content uniformity testing, and ±20% for dissolution testing.

7. Ruggedness

Ruggedness is the degree of reproducibility of the results, expressed as %RSD, obtained under a variety of conditions, to include distinct laboratories, analysts, instruments, reagents, and periods (1, 2).

8. Robustness

Robustness is the capacity of a method to remain unaffected by small deliberate variations in method parameters, such as pH, ionic strength, organic concentration, or temperature (1, 2). Robustness should be considered during method development and any significant variations should be noted in the documentation. There is currently a movement towards allowing leeway for certain operational specifications, such as pH, column length, or particle diameter (9). The refinement would allow a range for each parameter wherein a method can be adjusted without requiring revalidation. The proposal by William Furman will be published in *Pharmacopeial Forum* in the near future.

9. Requirements for an Assay

It is not necessary to evaluate all of the parameters of method validation for every type of assay. For example, limits of detection and quantitation are unnecessary in assays which quantitate only major components and active ingredients. The specifications for USP are shown in Table 16.2. The ICH requirements are generally the same, with the addition of an identification category which only mandates specificity (1, 2).

Table 16.2 USP Required Parameters				
Analytical Performance Parameter	Assay Category 1 Quantitation of Major Components & Active Ingredients	Assay Category 2 Impurities & Degradation Products		Assay Category 3 Determination of Performance Characteristics
		Quantitative	Limit Tests	
Accuracy	Yes	Yes	*	*
Precision	Yes	Yes	No	Yes
Specificity	Yes	Yes	Yes	*
Limit of Detection	No	No	Yes	*
Limit of Quantitation	No	Yes	No	*
Linearity	Yes	Yes	No	*
Range	Yes	Yes	*	*
Ruggedness	Yes	Yes	Yes	Yes

* may be required, depending on the specific test.
Reprinted from *LCGC* (ref. 2) with permission.

For content determinations, standard reference materials (SRM) are usually available from USP, NIST, or other organizations, but that is not the case for many other assays. Assignment of purity is ideally determined by comparison with a method using a standard of assigned purity. In the absence of such a standard, then it is assessed on the basis of 100% detected impurities.

Determination of impurities is optimally carried out by comparison with reference standards for the impurities. If this is not possible, then approximations must be made. When impurities are found at levels less than 0.5%, it is best to use a diluted sample (often 200 fold) to measure the major peak. Undiluted, it is likely to be above the linear range (8).

C. System Suitability

System suitability tests are used to verify that the resolution and reproducibility of the whole system are adequate for the analysis to be performed (1, 2). Sometimes sensitivity is also specified. System suitability requires testing the system as a whole, including hardware, software, analytical operations, and samples. Parameters such as plate count, tailing factors, resolution, and reproducibility (%RSD of retention time and peak area for six repetitions) must be determined and compared with the specifications. Some general recommendations for acceptable methods are that the peak of interest have a capacity factor (k) greater than 2 and be well resolved ($Rs > 2$) from both the void volume and other peaks (2, 7). The peak efficiency should be at least 2000 with a tailing factor less than 2. The RSD of retention time and peak area should be less than 1%. Software which will document these factors and also help troubleshoot a method is available from certain HPLC system vendors.

One test of system suitability requires injection of an analyte spiked with an impurity which has a close retention time (8). Retention time, resolution, plate count, and assymetry are compared to the specifications for the method. Another test of system suitability uses 3 - 4 analyte solutions spiked with an impurity, each injected once (8). In this case, resolution and detector response (peak area/analyte concentration) are used to determine compliance.

D. Method Transfer

After the requirements for method validation are met, it can be transferred to other facilities. Documentation accompanying the method should state the intended purpose and general acceptance criteria (1, 2). It should also include a detailed written procedure with a method validation report and system suitability criteria so that the end users can verify the method performance before incorporating it into their SOPs.

E. Revalidation

At times it is necessary to revalidate a method because of changes that have occurred. Alterations of factors like raw materials, manufacturing, sample preparation, etc. invoke a reactive response for revalidation. On the other hand, proactive revalidation may be desirable when new technology which can improve the results or the costs becomes available. Usually revalidation primarily examines the specific variances due to the new factor rather than beginning the method validation process over again.

IV. STANDARD OPERATING PROCEDURE

All of these components of method validation and development must be included in an SOP. A working copy of an SOP regarding validation procedures for HPLC methods in a contract laboratory follows:

1.0	**PURPOSE**

To describe a standard operating procedure to be followed for the validation of analytical methods. This procedure defines which analytical methods require validation, the level of validation required, the general procedures to be followed, and the acceptance criteria for each step. The detailed procedures were written with HPLC in mind; similar parameters need to be investigated for other types of techniques and apparatus. Exceptions to this procedure may occur in specific cases where variations are required because of product specific characteristics. A method is validated for the type(s) of samples to be analyzed, thus changes in drug product formulations may require additional validation steps for a method.

2.0 REFERENCES

2.1   *USP 23*, pages 1982-1984

2.2   *USP 23*, pages 1768-1779

2.3   Darwin R. Williams, "An Overview of Test Method Validation", *Biopharm*, November 1987, p.34-37

2.4   *Technical Review Guide: Validation of Chromatographic Methods*. Linda L. Ng, HFD-150, Version 3, December 1993.

3.0 RESPONSIBILITIES

3.1   It is the responsibility of the analyst(s) performing the method validation to follow this procedure.

3.2   It is the responsibility of the Laboratory Director or designated alternate to review and approve all method validation reports.

4.0 DEFINITIONS

4.1   Levels of Validation: The type of method validation required is determined by the stage of development of the drug product and the attribute being tested. The three categories and associated performance parameters for each drug product stage are included in Appendix I.

4.2   Accuracy: The measure of agreement between an experimental result and the true or accepted value. Accuracy is a measure of the exactness of an analytical method and may be expressed as percent recovery of the assay by known, added amounts of analyte.

4.3   Precision: The degree of agreement among individual test results when the procedure is applied repeatedly to multiple samplings of a homogenous sample. Precision is usually expressed as the standard deviation or relative standard deviation (coefficient of variation) and is a measure of the degree of reproducibility under normal operating circumstances.

4.4   Linearity: The measure of the degree to which the analytical curve approaches a first order linear relationship over the range of interest. Linearity is usually expressed in terms of the variance around the slope of the regression line.

4.5   Sensitivity (Limit of Detection): The lowest concentration of analyte in a sample that can be detected, but not necessarily quantitated, under the stated experimental conditions. The limit of detection is usually expressed as the concentration of analyte in the sample.

4.6   Limit of Quantitation: The lowest concentration of analyte in a sample that can be determined with acceptable precision and accuracy under the stated experimental conditions.

4.7   Specificity: The degree to which the analytical method can accurately and specifically measure the analyte in the presence of components that may be expected to be present in the sample matrix. Specificity may be expressed as the degree of bias of the test results obtained by analysis of samples containing added impurities, degradation products, related chemical compounds, or placebo ingredients when compared to test results from samples without added substances. Specificity is a measure of the degree of interference (or absence thereof) in the analysis of complex sample mixtures.

4.8   System Suitability: A regularly performed test of instrument function to ascertain the adequacy of the system to provide valid test results when the analytical procedure is followed. System suitability can be determined by precision, resolution, peak tailing, capacity factor, and efficiency. Exact criteria can only be determined as the assay is developed. Some criteria are listed in Section 5, but individual methods may vary.

4.9   Ruggedness: The degree of reproducibility of test results obtained by the analysis of the same samples under a variety of normal test conditions, such as different laboratories, different analysts, different instruments, different lots of reagents, different days, etc.

4.10   Peak Resolution, R_s: A quantitative measure of the relative separation between two closely eluting peaks.

(1)
$$R_s = \frac{2(t_2 - t_1)}{w_1 + w_2}$$

where: t_1 and t_2 are the retention times (in seconds) of peaks 1 and 2, and w_1 and w_2 are the corresponding widths (in seconds) of the bases of the peaks obtained by extrapolating the relatively straight sides of the peak to the baseline. (See Figure 1, Appendix II)

4.11   Peak Tailing Factor, T: A quantitative measure of the asymmetry of a peak.

(2)
$$T = \frac{w_{0.05}}{2f}$$

where: $w_{0.05}$ and f are as defined in Figure 2, Appendix II.

4.12   Theoretical Plate Count (N): A quantitative measure of chromatographic column efficiency.

(3)
$$N = 5.54\left(\frac{t_R}{w_{0.5}}\right)^2$$

where: t_R is the retention time, and $w_{0.5}$ is the width (in the same units used for t_R) of the peak at half height.

4.13   Capacity Factor (k): A measurement comparing the time of elution of the peak of interest and the time of elution of the dead volume (indicated by elution of unretained components).

$$(4) \qquad\qquad k = \left(\frac{t_R}{t_o}\right) - 1$$

where t_R is the retention time of the peak of interest and t_o is the retention time of an unretained substance measured at the time of the first baseline disturbance from the solvent injected.

4.14   Excipient: A matrix or formulation containing the same ratio of ingredients as the material under investigation, excluding the analyte of interest.

5.0   PROCEDURE

The following procedures will be performed by the assigned analyst as appropriate for the method being validated. These procedures were written with HPLC in mind; similar parameters need to be investigated for other types of techniques and apparatus. The criteria set forth here are based on reasonable expectations, but still require well-functioning equipment and expert laboratory technique. Criteria that differ significantly from these should be explained in the validation report. Before each validation test is performed, system suitability parameters (see section 5.10) should be measured, even if final criteria have not been set.

5.1   Intra-assay Precision (Repeatability)

      5.1.1   Procedure: Make no less than (NLT) 5 injections of the same sample of analyte. Calculate the average, standard deviation, and relative standard deviation (RSD) for peak area or height and retention time.

      5.1.2   Criteria: RSDs of 2% or less indicate acceptable intra-assay precision. If the RSD is accepted at >2%, it should be based on NLT 6 injections. For some assays, smaller RSDs may be desirable. Larger RSDs may be acceptable for small peaks (e.g. impurities or degradation products).

5.2   Accuracy (HPLC Column Recovery)

      5.2.1   Procedure: Make one blank injection (excipient or diluent) and NLT 5 injections of the analyte at a concentration sufficient to produce a signal well within the linear range of the detector (e.G. $S/N > 50$). If performing a gradient analysis, calculate the gradient concentration where the analyte elutes as follows:

$$(5) \qquad\qquad C = \left(\frac{P_2 - P_1}{t_G}\right) x \ t_R$$

where: P_1 and P_2 are the initial and final gradient percentages (e.g. 0 and 100%), t_G is the time (in minutes) it takes to reach P_2, and t_R is the analyte retention time (in minutes). Program the pump to deliver this concentration. Replace the HPLC column with a length of Teflon tubing (e.g. 600 cm of 1/16", 0.3 mm ID) and install the column prior to the sampling device in order to maintain consistent system pressure. Make NLT 3 blank injections and NLT 5 injections of the analyte.

      5.2.2   Calculations: calculate the column recovery by dividing the average total peak area (the sum of all non-artifact peaks) obtained with the HPLC column by the average obtained without the column. If the blank area is greater than 2% of the analyte area, subtract the average area of the blank injections from the area of the analyte obtained without the column.

      5.2.3   Criteria: Recoveries of 95-105% indicate that the HPLC column is not irreversibly adsorbing a significant amount of the analyte.

5.3   Linearity

    5.3.1   Procedure: Prepare standards in the range from 50% of the lowest to 150% of the highest normally used concentration, keeping in mind the solubility limits of the analyte. If a quantitation limit is to be set (see Section 5.6 below) the lowest concentration should be near the expected limit, and this sample should be injected in triplicate. A minimum of 6 standards (not including a blank) should be used, with multiple measurements (NLT 2) made on each standard.

        For impurities, the linearity test should be performed in the presence of the main component.

    5.3.2   Calculations: Using the method of least squares, perform a linear regression analysis using peak area or height versus analyte concentration. Calculate the linear least squares regression equation ($y=mx+b$), correlation coefficient (r^2), and standard error of the y-intercept (b). Do not include zero as a standard, since the response of a standard containing zero analyte may not be zero, but only below the detection limit of the assay.

    5.3.3   Criteria: r^2 should be ≥ 0.999; the y-intercept should be within ± 3 times the standard error of the intercept of zero or an intercept outside this range should be explained. The r^2 value may be lower for impurities.

5.4   Accuracy (Spike recovery)

    5.4.1   To be performed for drug products when appropriate.

        5.4.1.1   Procedure: To a placebo formulation (excipient), accurately add the drug substance at 80%, 100%, and 120% of label claim. (This must be in the linear range of the assay and solubility range of the drug.) For each concentration, prepare samples for analysis in triplicate and analyze according to the method.

        5.4.1.2   Calculations: Calculate the mean and RSD for each concentration.

        5.4.1.3   Criteria: The RSD should be $\leq 2\%$; the mean should be within 2% of the expected value.

    5.4.2   To be performed for impurities when appropriate.

        5.4.2.1   Procedure: To a drug substance solution free of impurities (if available), accurately add the impurity(ies) at the quantitation limit or 50% of the specification (whichever is higher) and at a higher concentration (typically $\geq 0.1\%$ by weight). For each concentration, prepare a sample for analysis and make NLT 3 injections of each.

        5.4.2.2   Calculations: Calculate the mean and RSD for each concentration.

        5.4.2.3   Criteria: The RSD should be $\leq 5\%$; the mean should be within 10% of the expected value.

5.5   Sensitivity (Limit of Detection)

This is required for limit tests which merely substantiate that the analyte concentration is above or below a certain level.

    5.5.1   Procedure: The limit of detection will be calculated using the peak height of the standard with the lowest concentration of analyte analyzed in the determination of linearity.

    5.5.2   Calculation:

(6)
$$LOD = \frac{c}{(h_1 / h_2) / 3}$$

where: h_1 is the height of the analyte peak, h_2 is the height of the noise (must be on same scale or correction for scale must be made), and c is the amount of analyte injected. The limit of detection (LOD) will be reported as analyte weight/injection (e.g. ng/injection).

5.6   Limit of Quantitation
        This limit is normally applied to related substances (impurities or degradation products) in the drug substance or drug product.
        5.6.1   Procedure: The limit of quantitation will be calculated using three injections of the standard with the lowest concentration of analyte analyzed in the determination of linearity or of spiked recovery. If a reference standard for a related substance is available, it should be used. If not, the reference standard of the drug substance may be used.
        5.6.2   Calculation: Calculate the average, standard deviation, and relative standard deviation (RSD) for peak area or height and retention time for the three analyses of the standard specified above.
        5.6.3   Criteria: If the RSDs for peak area or height and for retention time are NMT 5%, then the concentration of this standard will be reported as the limit of quantitation.
5.7   Specificity
        5.7.1   Procedure: A sample and standard of approximately the same concentration will be compared for specificity by taking on-line spectral scans throughout the peak of interest. The appropriate software will be used and the peak purities of both will be compared.
        5.7.2   Criteria: Both the sample and the standard should be qualitatively pure. If quantitation of the purity of the peaks is possible, the results should be included in the validation report.
        5.7.3   Depending on the analyte, other tests for specificity can be added at the discretion of the analyst/lab supervisor. These may include degradation experiments, oxidation, enzymatic digestion, contact with metals and cross validation experiments using other techniques such as capillary electrophoresis.
5.8   Sample and Standard Stability during Analysis
        The stability of sample and standard solutions during the time of analysis must be demonstrated.
        5.8.1   Procedure: For the longest expected analysis run, analyze (duplicate injections) the same solution of a standard and a typical sample at the beginning and the end.
        5.8.2   Calculation: Calculate the average, standard deviation, and relative standard deviation (RSD) for peak area or height and area % for the two injections at the beginning of the run, and the two injections at the end, for the standard and for the sample.
        5.8.3   Criteria: The area or height RSDs should meet the same criteria as set for area or height intra-assay precision. The area % RSD should be ≤ 1%. The method should state the maximum length of time a standard or sample solution may be held in the autosampler before analysis.
5.9   Sample and Standard Stability during Storage

If standard solutions or samples prepared for analysis are stored, they must be demonstrated to be stable during the time of storage.

> 5.9.1   Procedure: Prepare a standard and a typical sample, perform the analysis, and store both solutions. After the desired storage time, prepare a fresh standard and use it to analyze the stored standard and sample.
>
> 5.9.2   Calculation: Calculate the concentration and area % purity of the sample and standard before and after storage.
>
> 5.9.3   Criteria: The results should not vary by more than the intra-assay RSD determined in Section 5.1 above. The method should state the maximum length of time a standard or sample solution may be stored and the conditions of storage.

5.10   System Suitability Specifications

System Suitability includes precision (see Section 5.1 above), peak resolution, tailing factor, theoretical plate count, and capacity factor.

> 5.10.1   Procedure: System suitability will be determined using a representative injection of the analyte. A separate sample in which a closely eluting related substance is added to a sample of the drug may be required for measuring resolution.
>
> 5.10.2   Calculation: Using equations 1, 2, 3, and 4, calculate the peak resolution (R_s), the peak tailing factor (T), the theoretical plate count (N), and the capacity factor (k).
>
> 5.10.3   Criteria:
>
>> 5.10.3.1 R_s: >2 between the peak of interest and the closest potential interfering peak is desirable. If a standard for the closest expected peak is not available, another compound which elutes close to the peak of interest may be used to prepare a solution for the resolution check. Some analyses may not include two peaks so a resolution specification is not appropriate.
>>
>> 5.10.3.2   T: ≤ 2. Higher values are acceptable, but an effort should be made to verify that excess tailing does not compromise the analytical result.
>>
>> 5.10.3.3 N: >500 for very simple separations involving a single component well resolved from an excipient; >2000 for more complex separations.
>>
>> 5.10.3.4 k: The specified value for a method should be between 2 and 10. This insures that the peak of interest is well resolved from the void volume without wasting time. The measured value should be within 30% of the specified value.

5.11   Ruggedness (Intermediate Precision)

> 5.11.1   Procedure: Ruggedness will be determined by analysis of the same set of samples by two different analysts (where possible, using a different instrument). The type of sample set and number of samples shall be determined by the analyst.
>
> 5.11.2   Criteria: The results should not vary by more than the intra-assay RSD determined in Section 5.1 above.

6.0   DOCUMENTATION, REVIEW, APPROVAL, AND REVISION

A validation report based on these experimental results will be written by the analyst and must be reviewed and approved by the Laboratory Director or other appropriate laboratory manager. The report will be written according to SOP-005 and checked according to SOP-020.

Appendix I

Requirements for Assay Validation				
Parameter	Category I	Category II		Category III
		Quantitative	Limit Tests	
Accuracy	yes	yes	*	*
Precision	yes	yes	no	yes
Linearity	yes	yes	no	*
Sensitivity	no	no	yes	*
Limit of Quantitation	no	yes	no	*
Selectivity	yes	yes	yes	*
System Suitability	*	*	*	*
Ruggedness	yes	yes	yes	yes

*To be determined by analyst/lab supervisor

Category I: Analytical methods for quantitation of major components of bulk drug substances or active ingredients (including preservatives) in finished pharmaceutical products.
Category II: Analytical methods for determination of impurities in bulk drug substances or degradation compounds in finished pharmaceutical products. These methods include quantitative assay and limit tests.
Category III: Analytical methods for determination of performance characteristics (e.g. dissolution, drug release).

Appendix II

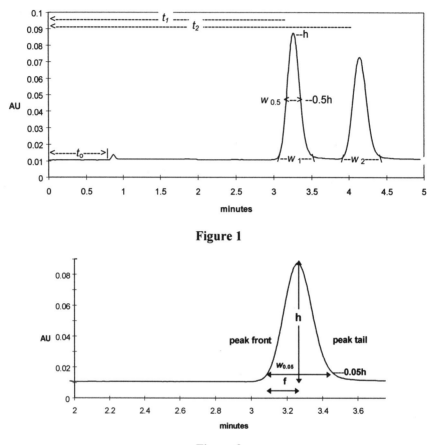

Figure 1

Figure 2

V. REFERENCES

1. M.E. Swartz and I.S. Krull, *Analytical Method Development and Validation*, Marcel Dekker, New York, 1997.
2. I. Krull and M. Swartz, *LC-GC* 15 (1997) 534.
3. H. Watzig and C. Dette, *J. Chromatogr.* 636 (1993) 31.
4. L.R. Snyder, J.L. Glajch, and J.J. Kirkland, *Practical HPLC Method Development (2nd Edition)*, John Wiley & Sons, New York, 1997.
5. I. Krull and M. Swartz, *LC-GC* 15 (1997) 842.
6. C. Burgess and R.D. McDowall, *LC-GC* 15 (1997) 130.
7. Center for Drug Evaluation and Research, *Reviewer Guidance: Validation of Chromatographic Methods*, U.S. Government Printing Office #615-023-11302/02757, Washington D.C., 1994.
8. W.J. Lough and I.W. Wainer, "Method Development and Quantitation" in *High Performance Liquid Chromatography, Fundamental Principles and Practice* (ed. W.J. Lough and I.W. Wainer), Blackie Academic & Professional, London (1995) 143.
9. L. Bechtel, R. Thackrey, and J. Palmer, "Updating and improving RP HPLC Methods for Recombinant Process Monitoring within Regulatory Guidelines", *ISPPP '97*, Rockville, MD, October 1997.

APPENDIX A

RESOURCES

Selected General HPLC References

1. L.R. Snyder and J.J. Kirkland, *Introduction to Modern Liquid Chromatography*, Wiley Interscience, New York, 1979. (basic reference about all facets of HPLC)
2. D.J. Runser, *Maintaining and Troubleshooting HPLC Systems, A User's Guide*, Wiley Interscience, New York, 1981.
3. C. Poole and S.A. Shuette, *Contemporary Practice of Liquid Chromatography*, Elsevier Science Publishers, Amsterdam, 1985.
4. L.R. Snyder, J. Glajch, and J.J. Kirkland, *Practical HPLC Method Development*, Wiley Interscience, New York, 1988.
5. J.W. Dolan and L.R. Snyder, *Troubleshooting LC Systems*, Humana Press, Totowa, NJ, 1990. (detailed troubleshooting guide)
6. B.A. Bidlingmeyer, *Practical HPLC Methodology and Applications*, John Wiley and Sons, New York, 1992.
7. *Chromatography, 5th Edition* (ed. E. Heftmann), *J. Chromatogr. Library,* Elsevier Science Publishers, Amsterdam, 1992.
   - Vol. 51A: *fundamentals and techniques* (in-depth discussion of theory and chromatography techniques)
   - Vol. 51B: *applications* (discussions of chromatography by solute class)
8. *High Performance Liquid Chromatography, Fundamental Principles and Practice*, (ed. W.J. Lough and I.W. Wainer), Blackie Academic and Professional, London, 1995. (textbook for HPLC course, includes several applications chapters)
9. L.R. Snyder, J. Glajch, and J.J. Kirkland, *Practical HPLC Method Development (2nd Edition),* John Wiley & Sons, New York, 1997.

Selected Books About HPLC of Biomolecules

1. "Enzyme purification and related techniques" in *Meth. Enzymol.* 104 (ed. W.B. Jakoby), Academic Press, 1984.
2. *Handbook of HPLC for the Separation of Amino Acids, Peptides and Proteins, Vol. I and II* (ed. W.S. Hancock), CRC Press, 1984.
3. *Methods of Protein Microcharacterization, A Practical Handbook* (ed. J. Shively), Humana Press, 1986.
4. *Proteins, Structure and Function* (ed. J.L. L'Italien), Plenum Press, 1987.
5. *Protein Purification: Principles, High Resolution Methods and Applications* (ed. J.-C. Janson and L. Ryden), VCH Publishers, New York, 1989.
6. *HPLC of Biological Macromolecules: Methods and Applications* (ed. K.M. Gooding and F.E. Regnier), Marcel Dekker, New York, 1990. (in-depth discussion of relevant techniques and specific biomolecule classes)
7. *High Performance Liquid Chromatography in Biotechnology* (ed. W.S. Hancock), John Wiley & Sons, New York, 1990.
8. *Analytical Biotechnology: Capillary Electrophoresis and Chromatography* (ed. Cs. Horvath and J.G. Nikelly), ACS, Washington, D.C., 1990. (biotechnology aspects of CE and HPLC)
9. *High Performance Liquid Chromatography of Peptides and Proteins* (ed. C.T. Mant and R.S. Hodges), CRC Press, Boca Raton, 1991. (short chapters about many relevant topics)

10. *HPLC of Proteins, Peptides and Polynucleotides* (ed. M.T.W. Hearn), VCH, New York, 1991. (in-depth discussion about relevant topics, including CE)
11. *High Performance Liquid Chromatography: Principles and Methods in Biotechnology* (ed. E. Katz), John Wiley & Sons, Chichester, 1996. (in-depth articles about relevant facets of HPLC)

Selected Journals and Magazines for HPLC

1. *Journal of Chromatography*
    a. regular issues
    b. proceedings of the *International Symposium on Protein, Peptide and Polynucleotide Analysis (ISPPP);* published about eight months after the autumn meeting.
    c. proceedings of the *International Symposium on Column Liquid Chromatography* (*HPLC '97, etc.*); published about eight months after the summer meeting.
    d. "Separation of Biopolymers and Supramolecular Structures" in *J. Chromatogr.* 418 (ed. M.T.W. Hearn and Z. Deyl), Elsevier Science Publishers, 1987.
2. *LC-GC Magazine*
3. *Journal of Liquid Chromatography*
4. *Journal of Microcolumn Separations*
5. *Chromatographia*

Selected Books About CE of Biomolecules

1. *Analytical Biotechnology: Capillary Electrophoresis and Chromatography* (ed. Cs. Horvath and J.G. Nikelly), ACS, Washington, D.C., 1990. (biotechnology aspects of CE and HPLC)
2. *Capillary Electrophoresis, Theory and Practice* (ed. P.D. Grossman and J. C. Colburn), Academic Press, San Diego, 1992. (covers theory and different techniques)
3. S.F.Y. Li, *Capillary Electrophoresis, Principles, Practice, and Applications, J. Chromatogr. Library*, Vol. 52, Elsevier Science Publisher, Amsterdam, 1992.
4. R. Weinberger, *Practical Capillary Electrophoresis*, Academic Press, Boston, 1992.
5. J. Vindevogel, *Introduction to Micellar Electrokinetic Chromatography*, Hüthig, 1992.
6. *Handbook of Capillary Electrophoresis: A Practical Approach* (ed. James P. Landers), CRC Press, Boca Raton, 1993.
7. *Capillary Electrophoresis, Theory and Practice* (ed. P. Camilleri), CRC Press, Boca Raton, 1993.
8. N. A. Guzman, *Capillary Electrophoresis Technology*, Marcel Dekker, New York, 1993.
9. *Capillary Electrophoresis Handbook: Principles, Operation and Applications* (ed. K.D. Altria), Humana Press, Totowa, NJ, 1995.

Selected Journals and Magazines for CE

1. *Electrophoresis*
2. *Journal of Capillary Electrophoresis*
3. *Journal of Chromatography*
4. *LC-GC Magazine*

Selected Internet Web Sites

	Source	Special Features
Magazines		
iscpubs.com	ISC	information on many suppliers
lcgcmag.com	LC-GC	reviews of books
Databases		
krscience.dialog.com	Knight Ridder	database, fee-based
stneasy.cas.org	STN International	database, fee-based
nlm.nih.gov	NIH	Medline, free search
Consultants/Vendors		
lcresources.com	LC Resources	consulting & services
sciquest.com	SciQuest	links to > 10,000 vendors
merck.de/chromatography/iupac/chmom.htm	E. Merck	IUPAC terms for chromatography
chemcenter.org		vendors

APPENDIX B

AMINO ACID DESIGNATIONS

	Single Letter Code	Three Letter Code	Molecular Weight	α-COOH	pK$_a$ Value* α-NH$_3^+$	Side Chain
alanine	A	ala	89.09	2.3	9.7	
arginine	R	arg	174.2	2.1	9.0	12.5
aspartic acid	D	asp	133.1	1.8	9.6	3.7
asparagine	N	asn	132.12	2.0	8.8	
cysteine	C	cys	121.16	2.0	10.3	8.2
glutamic acid	E	glu	147.13	2.2	9.7	4.3
glutamine	Q	gln	146.15	2.2	9.1	
glycine	G	gly	75.07	2.3	9.6	
histidine	H	his	155.16	1.8	9.2	6.0
isoleucine	I	ile	131.17	2.4	9.7	
leucine	L	leu	131.17	2.4	9.6	
lysine	K	lys	146.19	2.2	9	10.5
methionine	M	met	149.21	2.3	9.2	
phenylalanine	F	phe	165.19	1.8	9.1	
proline	P	pro	115.13	2.0	10.6	
serine	S	ser	105.09	2.1	9.2	
threonine	T	thr	119.12	2.6	10.4	
tryptophan	W	trp	204.22	2.4	9.4	
tyrosine	Y	tyr	181.19	2.2	9.1	10.1
valine	V	val	117.15	2.3	9.6	

*A. White, P. Handler, and E. L. Smith, "Principles of Biochemistry," Fifth Edition, McGraw-Hill, Inc., New York, 1973, p. 104.

ABBREVIATIONS

The abbreviations used in this glossary are generally taken from those defined by IUPAC, *CRC Handbook of Chemistry and Physics,* or *Chromatography, 5th edition* (ed. E. Heftmann), vol. 51A of the *Journal of Chromatography Library*. In cases where there are multiple accepted meanings (e.g., V_o), only one definition is used in this text.

A. Normal Type

a	front width of peak at 10% height
AAEE	N-acryloylaminoethoxyethanol
AAP	N-acryloylaminopropanol
ACN	acetonitrile
AEX	anion-exchange
AMP	adenosine 5'-monophosphate
API	atmospheric pressure ionization

AU	absorbance units
b	back width of peak at 10% height
BCA	bicinchoninic acid
BES	N,N'-bis(2-hydroxyethyl)-2-aminoethanesulfonic acid
Bis-Tris	bis(2-hydroxyethyl)iminotris-(hydroxymethyl)methane
bp	base pairs
BPTI	bovine trypsin inhibitor
BSA	bovine serum albumin
C4	butyl ligand for RPC
C8	octyl ligand for RPC
C18	octadecyl ligand for RPC
CAPS	3-(cyclohexylamino)-1-propanesulfonic acid
CE	capillary electrophoresis
CEC	capillary electrochromatography
CEC-HILIC	cation-exchange chromatography-hydrophilic interaction chromatography
CEX	cation-exchange
CGE	capillary gel electrophoresis
CHAPS	3-[(3-cholamidopropyl)dimethylammonio]-1-propane sulfonate
CHES	2-(N-cyclohexylamino)ethanesulfonic acid
CI	chemical ionization
CIEF	capillary isoelectric focusing
CLOD	concentration limits of detection
CM	carboxymethyl
CMC	critical micelle concentration
CMEC	capillary micellar electrochromatography (see MEKC)
CNBr	cyanogen bromide
Con A	concanavilin A
CPM	counts per minute
CV	coefficient of variation
CZE	capillary zone electrophoresis
Da	Dalton, measure of molecular weight
DEAE	diethylaminoethanol
DHPLC	denaturing HPLC
DMSO	dimethylsulfoxide
DPM	disintegrations per minute
dsDNA	double stranded DNA
DTT	dithiothreitol
EC	electrochemical
EDTA	ethylenediaminetetraacetic acid
EI	electron ionization
ELSD	evaporative light scattering detection
EOF	electroosmotic flow, electroendoosmotic flow
ESI	electrospray interface
f	front width of peak at 5% height
FAB	fast atom bombardment
FDA	Food and Drug Administration
FT-ICR	Fourier transform ion cyclotron resonance
GPC	gel permeation chromatography

GuHCl	guanidinium hydrochloride
h	height of a peak
Hb	hemoglobin
HEC	hydroxyethylcellulose
HEPES	N-(2-hydroxyethyl)piperazine-N'-(2-ethanesulfonic acid)
HFBA	hexafluorobutyric acid
HIC	hydrophobic interaction chromatography
HILIC	hydrophilic interaction chromatography
HMC	hydroxymethylcellulose
HPLC	high performance liquid chromatography
ICH	International Conference on Harmonization
ID	internal diameter
IDA	iminodiacetic acid
IEC	ion-exchange chromatography
IMAC	immobilized metal ion affinity chromatography
IMP	ion-moderated partitioning
IPA	isopropanol
IQ	instrument qualification
ITP	isotachophoresis
KDN	3-deoxy-d-glycero-d-galacto-2-nonulosonic acid
LALLS	low angle laser light scattering
LC-MS	liquid chromatograph-mass spectrometer
LEC	ligand exchange chromatography
LIF	laser induced fluorescence
LOD	limit of detection
LOQ	limit of quantitation
LPA	linear polyacrylamide
LSB	least significant bit
LSS	linear solvent strength model
mAb	monoclonal antibody
MALDI	matrix-assisted laser desorption ionization
MALLS	multi-angle laser light scattering
MC	methylcellulose
MCAC	metal chelate affinity chromatography
MDQ	minimum detectable quantity
MEKC	micellar electrokinetic chromatography
MeOH	methanol
MES	2-(N-morpholino)ethanesulfonic acid
MIC	metal interaction chromatography
MLOD	mass limit of detection
MOPS	3-(N-morpholino)propanesulfonic acid
MS	mass spectrometry
MW	molecular weight
NDA	naphthalenedialdehyde-cyanide
Neu5Ac	N-acetylneuraminic acid
Neu5Gc	N-glycolylneuraminic acid
NLT	no less than
OD	outer diameter

ODS	octadecylsilyl
OPA	o-phthalaldehyde
OQ	operational qualification
$P^-_{M/S}$	ion-pair ion in mobile or stationary phase
PAD	pulsed amperometric detection
PAGE	polyacrylamide gel electrophoresis
PC	peak capacity
PCR	post-column reaction
PCR	polymerase chain reaction
PDA	photodiode array detector
PEEK	polyetheretherketone
PEG	polyethylene glycol
PEI	polyethyleneimine
PEO	polyethylene oxide
pI	isoelectric point
PIPES	piperazine-N,N'-bis(2-ethanesulfonic acid)
ppb	parts per billion
PQ	performance qualification
PSDVB	polystyrenedivinylbenzene
Q	quaternary amine
R	functional group
RI	refractive index
RIU	refractive index units
RPC	reversed phase chromatography
RRF	relative response factor
RRT	relative retention time
RSD	relative standard deviation
S	sulfonyl
$S^+_{M/S}$	positively-charged solute in the mobile or stationary phase
SCX	strong cation-exchange
SDS	sodium dodecylsulfate
SDVB	sulfonated divinylbenzene
SEC	size exclusion chromatography
SFE	supercritical flow extraction
SOP	standard operation procedure
SPE	solid phase extraction
SRM	standard reference materials
SST	stainless steel
T	tailing factor
TBA	tributylammonium
TEA	triethylamine
TEAA	triethylammonium acetate
TEAP	triethylammonium phosphate
TED	tricarboxymethylenediamine
TEMED	N,N,N',N'-tetramethylethylenediamine
TFA	trifluoroacetic acid
TMA	trimethylammonium

TOF	time-of-flight
Tris	tris(hydroxymethyl)aminomethane
USP	United States Pharmacopeia
UV	ultraviolet absorption
VIS	visible absorption
X	reactive group

B. Italics

A	band broadening constant: eddy diffusion and laminar flow
A	absorbance
A	peak area
A_2	second virial coefficient
A_s	Asymmetry of peak
AU	absorbance units
B	band broadening constant: axial diffusion
b	cell path length
C	band broadening constant: mass transfer
c	concentration (mol/l)
C_1	total salt concentration
d_c	column diameter
d_f	thickness of stationary phase
D_M	diffusion coefficient of solute in mobile phase
d_p	particle diameter
D_s	diffusion in stationary phase
E	electric field strength
F	flowrate
h	reduced plate height
H	plate height
H_{min}	best column efficiency (at optimum linear velocity)
I	sample light intensity
I_o	reference light intensity
I_θ	light scattering intensity
I_S	intensity of emitted light
K	partition coefficient
k	capacity factor
k'	capacity factor (see k)
$k*$	average value of k during gradient elution
K_D	solute distribution coefficient
k_w	retention with water as the mobile phase
L	length of column
L	length of capillary from injector to detector, effective length
L_{tot}	total length of capillary
m	mass
m/z	mass to charge ratio
M	molecular weight
N	theoretical plates, separation efficiency
n	refractive index
N_A	Avogadro's number
n_{aq}	moles of aqueous phase
n_{mic}	moles of micelle
P	pressure
ΔP	pressure drop

pI	isoelectric point
$P(\theta)$	particle scattering function
pK	ionization constant
Q_{inj}	amount injected
r	radius of gyration; Stoke's radius
r	correlation coefficient
r_c	radius of capillary or column
R_θ	Rayleigh Factor
R_s	resolution
S	constant, RPC gradients
S/N	signal to noise ratio
S_i	sensitivity or calibration factor
t_d	delay volume
t_{eof}, t_o, t_n	migration time of neutral solute
t_G	time of gradient
t_{inj}	injection time in seconds
t_m	migration time
t_{mic}	migration time of micelle
t_o	deadtime of unretained solute, corresponding to total mobile phase volume
t_R	retention time
u	linear flow velocity
u_{eo}	electroosmotic velocity
u_{ep}	electrophoretic velocity
u_{opt}	optimum linear velocity in terms of efficiency
v	reduced velocity
V	volume
V	applied voltage
V_i	internal volume (within the pores); interstitial volume
V_M	total mobile phase volume in a column; deadvolume
V_o	void volume; excluded volume (outside the matrix)
V_R	retention volume
V_S	volume of the stationary phase
V_T	total volume in a column
w	peak width
$w_{0.5}$	width at half height
Z	charge
z	charge on ion or salt (in units of electronic charge)

C. Symbols

α	separation factor, selectivity
Å	Ångstrom, 10nm
ϕ	phase ratio
ϕ	volume fraction of organic solvent
ε	molar absorptivity, molar extinction coefficient
ε_r	dielectric constant of the mobile phase
Φ	volume fraction of organic solvent
η	mobile phase viscosity
λ	wavelength
λ_{em}	emission wavelength
λ_{ex}	excitation wavelength
μ	apparent mobility
μ_{eo}	electroosmotic mobility
μ_{ep}	electrophoretic mobility
θ	constant related to pressure
θ	angle of incidence or refraction
σ	standard deviation
ζ	zeta potential

TRADEMARKS

Company	Trademarks	Company	Trademarks
Amersham Pharmacia:	Sephadex	Macherey Nagel:	Nucleogen
	Sephacryl		Nucleosil
	Sepharose		Nucleogel
	Superose	MICRA Scientific:	SynChropak
	Superdex		SynChroprep
	Resource	Packard Instruments:	Radiomatic
	Mono Q	Perseptive Biosystems:	POROS
	Mono S	Pierce Chemical:	Extractigel
	HiTrap	PolyLC:	PolyEthyl
Beckman:	Spherogel		PolyPropyl
	System Gold		PolyCAT
	Ultrasphere		polySulfoethyl Aspartamide
Bio-Rad Laboratories:	UNO	Sedex:	ELSD
	Bio-Gel	Separations Group:	Vydac
	Bio-Sil	Supelco:	LC-HINT
	Aminex	ThermoSeparations:	Spectra Vision
Dionex:	CarboPak	Toso Haas:	TSK
E. Merck:	LiChrospher		Toyopearl
	Fractogel	Varian Associates:	LC Star
Eldex:	MicroPro		Micropak
Glycotech	Hy-Tach	Waters Corporation:	Delta-Pak
Hewlett-Packard:	ZORBAX		NovaPak
ISCO:	mLC-500	Wyatt Technology:	Optilab OSP
LC Packings:	Fusica		miniDAWN

INDEX

Second Printing: April 2003

If you have comments or suggestions for future editions, please contact:
Bob Cunico   rlcunico@baybiolab.com

ORDERING INFORMATION

Basic HPLC and CE of Biomolecules can be ordered from Bay Bioanalytical
Laboratory.
FAX: 510-724-8053
Phone: 510-724-8052
Mail: Bay Bioanalytical Laboratory, Inc.
              551A Linus Pauling Drive
              Hercules, CA 94547
E-mail: book@baybiolab.com or **Internet**: www.baybiolab.com.

_____

Name

_____

Organization

_____

Address

_____

_____

Phone Number

Price is $39.95 plus $7.00 shipping and handling for US orders. Please inquire
about non-US shipping. CA residents add 8.5% sales tax.

o Purchase Order #_____
o Check
o Credit Card o VISA o Mastercard

_____   _____

Credit Card #   Signature

_____

Expiration Date

Return Policy: This book may be returned for a full refund (less shipping and
handling) for any reason.